The Self-Evidencing Agent

The Self-Evidencing Agent

Mind, Existence, and Predictive Processing

Jakob Hohwy

The MIT Press
Cambridge, Massachusetts
London, England

The MIT Press
Massachusetts Institute of Technology
77 Massachusetts Avenue, Cambridge, MA 02139
mitpress.mit.edu

The MIT Press would like to thank the anonymous peer reviewers who provided comments on drafts of this book. The generous work of academic experts is essential for establishing the authority and quality of our publications. We acknowledge with gratitude the contributions of these otherwise uncredited readers.

This book was set in Stone Serif and Stone Sans by Westchester Publishing Services. Printed and bound in the United States of America.

Library of Congress Cataloging-in-Publication Data

Names: Hohwy, Jakob, author.
Title: The self-evidencing agent : mind, existence, and predictive processing / Jakob Hohwy.
Description: Cambridge, Massachusetts : The MIT Press, [2026] | Includes bibliographical references and index.
Identifiers: LCCN 2025004525 (print) | LCCN 2025004526 (ebook) | ISBN 9780262553896 (paperback) | ISBN 9780262385619 (pdf) | ISBN 9780262385626 (epub)
Subjects: LCSH: Psychology—Methodology. | Prediction (Psychology) | Cognitive science. | Neurosciences.
Classification: LCC BF38.5 .H59 2026 (print) | LCC BF38.5 (ebook)
LC record available at https://lccn.loc.gov/2025004525
LC ebook record available at https://lccn.loc.gov/2025004526

EU Authorised Representative: Easy Access System Europe, Mustamäe tee 50, 10621 Tallinn, Estonia | Email: gpsr.requests@easproject.com

Contents

Preface

One of the greatest privileges of my academic life is being able to work with and learn from amazing, inspirational PhD students and early career researchers, first in the Cognition & Philosophy Lab and then in the Monash Centre for Consciousness and Contemplative Studies (M3CS). Each of them has influenced the journey leading to this book. Throughout the book, I proudly highlight the research of my students and postdocs to inform and illustrate the various themes that I discuss—and to chart my intellectual debts.

I am grateful to all the past and present PhD students with whom I have had the pleasure of working. The inaugural lab: Bryan Paton and Colin Palmer. The CogPhil Lab first generation: Julian Matthews, Noam Gordon, Steve Gadsby, Ryoji Sato, Kelsey Perrykkad, and Andrew Corcoran. The CogPhil Lab second generation: Andy Mckilliam, Ricardo Perera, Niccolo Negro, Simon van Baal, Manu Kirberg, Iwan Williams, and Anikó Kusztor. Gen-M3CS: Kevin Berryman, Beth Fisher, Tom Darling, Adam Manoogian, Lars Sandved-Smith, Shawn Prest, Mengting Zhang, Joe Melling, and Vismay Agrawal. Students at M3CS: Ema Demšar and Jasmine Walter.

I am also grateful to have worked with and learned from a truly amazing group of scholars. Visiting researchers to the lab and M3CS: Eliane Descrijver, Tess Guthrie, Alessio Bucci, Manja Engel, Dan Williams, Lorijn Zaadnoordijk, Francesco Marchi, Laura Convertino, Simon Hviid del Pin, Zuzanna Skóra, Thomas Parr, Noor Sajid, Leonardo Barbosa, Paweł Gładziejewski, Filip Novický, Antoine Lutz, Christopher Whyte, Mette Høgh, Ruben Laukkonen, Kai-Yuan Cheng, Regina Fabry, Karin Kukkonnen, and Jakob Stenseke. Past postdocs: Mateusz Woźniak, Rafik Hadfi, and Roger Koenig-Robert. Collaborators and colleagues at Monash: Nao Tsuchiya, Antonio Verdejo-García, Claire Foldi, Tom Drummond, Neil Bailey, Linda Barclay, and Cammi Murrup-Stewart. Colleagues around the world: Olivia Carter, Anil Seth,

Alex Kiefer, Karl Friston, Becky Lawson, Pat Johnston, Joel Pearson, Rosalyn Moran, Andreas Roepstorff, Thomas Metzinger, Felix Blankenberg, Thomas Andrillon, Sid Kouider, Giovanni Pezzulo, Ryan Smith, Sara Lazar, and many more.

Since 2021, I have had the fortune of working at M3CS, funded generously through the Three Springs Foundation. The team of students, postdocs, educators, professional staff, and colleagues is just fabulous, and every day is a learning journey. My stellar philosophy colleagues at M3CS: Tim Bayne, Jennifer Windt and Monima Chadha. The M3CS postdocs and researchers: Mark Miller, Toby Woods, Jonathan Robinson, Milan Andrejevic, Andrew Corcoran, and Kelsey Perrykkad. The contemplative A-team: Craig Hassed, Richard Chambers, Jess Weijers, Megan Flamer, Sherelle Connaughton, Gidi Ifergan, and Luke Smith.

My intellectual debt to Karl Friston is obvious throughout the entire book. I am grateful for the opportunity to engage with Karl's research and work with him on several projects. I was first introduced to Karl's work in a reading group at Aarhus University set up by Andreas Roepstorff around 2004, where the first readings were Karl's "Beyond Phrenology" and "Learning and Inference in the Brain." The exciting philosophical potential of his approach immediately became clear to me. I am enthusiastic about taking predictive processing and the free energy principle—cast as self-evidencing—into philosophical debates and diverse areas of conceptual inquiry.

I am also grateful to Christopher Whyte, Bryan Paton, Andrew Corcoran, Alex Kiefer, Kelsey Perrykkad, Karl Friston, Kevin Berryman, Beth Fisher, Tom Darling, and Linda Barclay for their thoughtful and helpful comments on various drafts of chapters of the book; and the MIT Press's five reviewers offered excellent and knowledgeable feedback.

List of Figures

1 Introduction

Self-evidencing occurs when a model of the world exists because it is used to explain what is observed such that the explained observations become the evidence for the model. This book argues that there is good reason to cast the existence of biological agents, such as humans, in terms of self-evidencing—that is, we are self-evidencing agents.

If we begin from self-evidencing, then we can use it as a first-principles method to understand what we are, how we experience the world and ourselves, and how we engage in diverse ways with the world. Self-evidencing can help situate attention, action, and perception. It also furnishes compelling approaches to decision-making and rationality, to the self, and to preferences, volition, and values. It can build bridges to consciousness, and even to free will and to meaning and wisdom.

I want in this book to use self-evidencing to make sense of what I see as the whole, diverse spectrum of everyday experiences and our attempts to live well. We are embodied creatures, we are conscious, we make decisions, and we enjoy some degree of autonomy. We sometimes connect strongly and fluidly to ourselves, to others and the world around us. Other times, our bodies betray us, we feel disconnected from who we are ourselves or from others and the world, and we feel out of control or controlled by circumstances. We can often perceive the world accurately, but what we see is also filtered through what we want and believe. We can be fully connected throughout the flow of an absorbing task, caught up in the moment, but we also can be profoundly detached from the external world, lost in thought, mind-wandering, or dreaming. Although our individual human experiences have many features in common with those of other people, there are extraordinary differences and diversity in every person's lifelong trajectory through everyday experience, in our diverse sense of self and body, in the kinds of things we do to get what we believe we want, and so on. Life is full of ever-vacillating tensions and conciliations between internal and external, between self and other, between connection and detachment, and between body and mind, the conscious and the nonconscious, and freedom and determination. I am wary of philosophies that emphasize either the disconnected and internal or the connected and embodied side of everyday experiences. A good account of everyday experience should not paper over the full and diverse spectrum of everyday experience.

Self-evidencing, as I shall develop the concept, allows the full gamut of everyday experiences—warts and all. The simplicity of one process, self-evidencing, allows this lived cacophony to emerge from our causal connectedness to the world, but it also puts the burden on the agent's ability to actively maintain their internal model of themselves and the world around them. The grounding perspective that self-evidencing offers is one of inescapable causal embedding, but with profound epistemic solitude. Although the mind is highly active, the activity happens only within the system itself, blanketed behind our sensory organs and having to do everything through a kind of model-constrained trial and error process, using variations in the mind's model of the world to elicit the best possible feedback from parts of our bodies and our environment. Given self-evidencing, this solitude is the inescapable reality of existence. This can seem a dire or pessimistic philosophy, providing a disembodied and skepticism-prone account of the mind. I think it is in fact an honest picture of the way that each of us manages to go through life. However, this view is at the same time suffused, more subtly and realistically, with a modest optimism, enshrined in a belief that the world does enable our self-evidencing and existence.

From self-evidencing flows several insights concerning things that matter to us. Self-evidencing demonstrates the tractability of our pursuit of truth and reason, in spite of the intractability of exact computational inference. It allows vast individual differences in everyday experience and opens the door to our capacity for making wiser decisions and creating meaning in life, in spite of our diversity and occasional propensity for irrationality. It helps explain why we have ambivalent attitudes toward risk, how we can explore and shape our own preferences and even our selves, how emotion and salience connect to who we are, and what it takes to bring value and meaning into our conscious lives.

In the last decade or so, the predictive processing framework has arisen for cognitive science, theoretical neurobiology, machine learning, and philosophy of mind. Predictive processing proposes that the brain engages in a kind of hypothesis testing via precision-weighted predictive coding, which leads to perception. This was a denouement of ideas going back to Ibn al-Haytham (Alhazen), Immanuel Kant, and Hermann von Helmholtz presenting perception as a process of judgment or unconscious inference. Predictive processing has turned into a vibrant research paradigm, with many developments across computational theory, philosophy, experimental psychology and psychophysics, neuroimaging, and domains as diverse as physiotherapy and aesthetics. One of the most important developments in the predictive processing framework is the focus on active inference, which is now seen as more central than standard predictive coding approaches. In active inference, action policies for planning and decision-making are inferred based on prior beliefs about which policy will best reduce uncertainty about future outcomes. There is a wealth of exciting work on active inference emerging, and it will play a pivotal role in this book.

At the inception of the predictive processing framework, it was already clear that there were more ambitious principles at play, sitting behind the idea of the brain as an organ for minimizing prediction errors. This took the shape of the free energy principle, articulated and developed by Karl Friston. I then suggested that "self-evidencing," a concept borrowed from the philosophy of science, would be useful for speaking about these underlying principles. This book provides a full-fledged unfolding of the concept of self-evidencing. The book philosophically explores what self-evidencing might tell us about the mind, going beyond the kinds of questions and topics mostly discussed for predictive processing and delving more broadly into everyday experiences and things that I believe matter greatly for our understanding of ourselves and how we try to live well.

Prediction error minimization is intimately connected to self-evidencing, but predictive processing always conveyed a fairly idealized, uniform process of gradient descent on an error landscape, simply knocking down prediction error whenever it occurs based on a cortical hierarchy that recursively repeats the same predictive coding mechanism at each level. In contrast, self-evidencing is a notion that affords a more nuanced and open-ended set of processes, where agents can engage in diverse ways as they maintain their internal model of their world and themselves in various contexts and through varying levels of uncertainty. This makes a new, broader set of tools available for approaching a wide range of topics while still operating within the ontologically austere landscape of self-evidencing, provided by using the free energy principle as a method for understanding certain self-organizing systems—namely, ourselves. In this way, self-evidencing is an especially well suited concept for advancing from the original notions of predictive processing, straddling technical and philosophical domains. Technically, to self-evidence is simply to maximize the marginal likelihood or Bayesian model evidence by minimizing free energy. Philosophically, it paves the way for new epistemic and cognitive perspectives that are relevant to many debates.[1]

Methodologically, one can take what is sometimes (and nonpejoratively) labeled the "low-road" approach to predictive processing. Here, different phenomena are considered one after another, and we notice that often predictive processes such as predictive coding appear to be in play, either as perceptual inference or as active inference. On the winding "low road," the more phenomena we can explain with predictive processing, the more interesting the framework seems. This is a fruitful and valuable way of doing science. It is consistent with treating predictive processing as an empirical hypothesis, which could, for example, express that certain organisms have evolved to use predictive processes to solve certain kinds of problems and thereby increase fitness.

A more theoretically ambitious framing is in terms of self-evidencing, grounded on the free energy principle. Here, predictive processing and active inference are motivated directly from a fundamental understanding of the existence of self-organizing systems. It is this framing—or, as it is sometimes called, "the high road"—that is in play in this book. On the less winding but more exhaustingly uphill high road, self-evidencing is developed a priori, on the basis of conceptual analysis and mathematics. Here, the free energy principle is understood as similar to principles in physics, such as the principle of least action. It does not convey an empirical law of nature, which can be confirmed to various degrees through observation. The principle may or may not apply to certain systems when they are described in certain ways. In this sense, self-evidencing is a method or system for understanding existing things, such as humans. Empirical confirmation comes in when principles are applied to furnish process theories that describe particular systems, such as the anatomy and structure of the brains and bodies of humans, of other animals, or of machines. Because self-evidencing is a priori, it is intended as a way to describe any self-organizing thing that exists; therefore, it would not make sense to say, for example, that self-evidencing is something that some organisms and not others have evolved to maximize fitness.[2]

It would be nice if there turns out to be copious empirical evidence for the process theories applying to human anatomy and cognition that conform to self-evidencing and the free energy principle. I think that the signs here are promising, although the empirical jury is still out. Although I occasionally will appeal to some of the empirical evidence acquired through process theories, the main thrust of this book will be the theoretical case built at the level of principles. I am mainly—and somewhat unapologetically—going by the conceptual, high-road case for self-evidencing. The motivation for this is twofold: I think that the a priori arguments for conceiving existence in terms of self-evidencing are compelling, and comprehensively unfolding self-evidencing provides a worthwhile method for making sense of the diversity of everyday experience, while elucidating fundamental aspects of existence ranging from rationality to value to consciousness and beyond.

This plan for the book means that I will discuss some classic philosophical topics, such as rationality, volition and free will, self, value, consciousness, wisdom, and meaning in life. To make this agenda tractable, I will make some simplifying moves for each topic. For example, I opt for compatibilism about free will, I allow that realism for the self is just for the self to be as real as other humdrum things, I am content with a naturalistic account of consciousness rather than a full-scale assault on the metaphysical mind-body problem, and I consider wisdom and meaning through their recent operationalizations in psychology rather than their full unfolding through millennia of philosophy. When the explanatory targets are set up in these ways, self-evidencing can be deployed fruitfully, pointing to resolutions of long-standing debates or helping to change our conceptions of what those debates are about. I hope that building such bridges between self-evidencing and some of these long-standing debates can foster, and perhaps reorient and unify, conversations within philosophy and across interdisciplinary domains.

There is plenty of criticism of predictive processing, some philosophical or theoretical and some empirical. Criticism is important and welcome for a thriving research program like predictive processing. I will not engage systematically with all the criticism here, which is a task that I believe is better suited for more technical journal articles. But throughout this book, I will reference points of controversy and debate, suggesting ways of responding. Unsurprisingly, I am not swayed by the criticisms. When self-evidencing is seen in its full high-road scope, as a first-principles method for philosophical investigation, it has ample resources to prevent the various criticisms from getting traction.[3]

Self-evidencing is in a peculiar dialectical situation because it is often touted quite imperiously (by me and several others) as the one principle to which all theories must conform. But that ambition is in fact consistent with a welcoming, collaborative dialectic encompassing insights from other theories. Many theories from numerous scientific fields can be seen to conform to self-evidencing without this conformity detracting from their empirical status or their theoretical insights. This opens the opportunity for synthesis and unification, where those theories extend our conception of self-evidencing, and self-evidencing can help transform and unify debates without consigning all other theories to the trash heap.[4]

Self-evidencing then emerges as independently informative, but also as consistent with some existing approaches and liberal with respect to various explanatory projects that researchers from various disciplines may embark upon. That is, self-evidencing as a method wants to have its cake and eat it too—to be both exclusive and inclusive. Self-evidencing accomplishes this by unification, demonstrating that scientific approaches to various problems (such as rationality, attention, self, and volition, among others) need not appeal to fundamentally distinct and potentially inconsistent theoretical constructs, somehow cohabiting within our cognitive systems. Importantly, self-evidencing under the free energy principle is a valuable method for describing existing biological agents such as humans because a self-evidencing agent can be understood as self-supervised—that is, able to perceive, attend, decide, and act under their own steam and without relying on labeled training data or other external supervision. To me, this is significant because it means that any approach that conforms with self-evidencing has a chance of latching on to our fundamental existential condition, rather than presupposing that the computational challenges that agents confront have already been solved by processes external to the agent. In contrast, approaches that cannot be brought into conformity with self-evidencing can never be truly self-standing, as they will always ultimately depend on more fundamental guarantees that the agents in question can be self-supervised. The consequence is that whereas self-evidencing is quite ecumenical, it does pose a challenge to the comprehensiveness and uniqueness of approaches that do not conform to it.

The book pursues self-evidencing into domains that touch on biology, physics, and neuroscience, and some of the key moves in the argument, which I seek to explain and frame philosophically, are recognizable from other fields, such as information theory, machine learning, dynamical systems theory, theoretical biology, and statistical physics. I also bring self-evidencing into debates about ethics, morality, and contemplative studies. The book primarily reflects my outlook as a cognitive philosopher, and I am exploring self-evidencing as a method for philosophical inquiry. It takes the story of self-evidencing to a certain conceptual stage, which will hopefully be a useful point of contact for experts in other fields such as biology and contemplative studies. To situate self-evidencing in the overall discourse relating to predictive processing, I reference numerous published studies and discussions across multiple debates and disciplines. The literature on predictive processing is burgeoning and I cannot capture it all, but I hope that readers will be able to navigate from my reference points to the wider debates. I have collected all the references, as well as more technical remarks, in endnotes.[5]

Overall, this book is written as a first-principles argument for self-evidencing, and as advocating for self-evidencing as a fruitful method for approaching a broad, interdisciplinary array of debates about what we, as self-supervised human agents, experience and do. This project is in keeping with the excitement that I and many others around the world feel about predictive processing, active inference, and the free energy principle. That excitement is rapidly translating into amazing theoretical and computational advances, together with an increasing body of impressive empirical work on the process theories. Although the explanatory aims are ambitious for the book, I conceive the case that I build here with relative humility—that is, the case is only as strong as each step of the a priori argument. I see the arguments developed in this book as both providing a philosophical backdrop for much of the ongoing theoretical and empirical research and pointing forward, charting a conceptual map for self-evidencing and its connection to debates that are central to our self-understanding.

* * *

Here is the plan for the book. Chapter 2 will bring out the core argument that self-evidencing is key to an analysis of existence. The chapter seeks to convey the peculiar status of the idea of self-evidencing for theorizing about existence and mind, given its connection to the free energy principle. The first section provides a toolbox that will be useful for the entire book, with concise descriptions and examples of the notion of the existence of things and of self-evidencing and its applications. A key insight in this chapter is that we get an adequate method for investigating agents like us only if it can be used to capture how our cognitive and perceptual processes are self-supervised (i.e., the underlying process is tractable by the agents themselves, without the help of externally provided guidance, such as labeled training data). The free energy principle builds on variational methods for approximate inference, which I heuristically conceive as a kind of model-constrained trial and error process. The beauty of self-evidencing, with the free energy principle, is that the agent's self-supervised trial and error processes are assessed against a simple, tractable objective function—namely, free energy (where free energy can be thought of as the fit between the observations expected under an assumed model of latent causes in the world and the observations that they make, where the objective function will also be made to capture future observations and actions). The cost of that simplicity is that all the onus is on the agents' work to maintain their internal model of themselves and the world. That is where the complexity lies and where diversity among individuals emerge.

Chapter 3 takes the core conception of self-evidencing, understood in the light of the free energy principle, and asks what a self-evidencing organism might look like. The starting point here is how self-evidencing is about maximizing model evidence and how this evokes an idea from the philosophy of science about inference to the best explanation. The chapter builds a case that inference to the best explanation leads to just the processes entailed by the free energy principle—balancing complexity and accuracy and risk and ambiguity. This helps us see what is distinctive about self-evidencing. It further develops a key theme in the free energy principle—namely, that self-evidencing offers a synthesis or conciliation of descriptive and normative perspectives on cognitive processing. Finally, the chapter puts forward-looking processing at center stage, highlighting allostasis and introducing active inference.

Chapter 4 revisits themes that have been central to predictive processing, such as perception, attention, and the role of action in perception. In my previous book, *The Predictive Mind*, I told a mechanistic story beginning with predictive coding for perception, and then casting attention as precision-weighted predictive coding and adding active inference to account for ways in which action changes attention and perception. However, in light of self-evidencing, these discussions are reorganized and reconceived. Precision is now the basic mechanistic notion, understood in the setting of active inference as precision control. From precision control, attention, action, and perception flows. The general outlook is still that existing organisms must minimize uncertainty, but now this is forward-looking, focusing on expected uncertainty, before the inferences then unfold in real time. This reordering has deep implications for understanding what we do and what mental life is. It presents self-evidencing agents as epistemically solitary and yet causally embedded in the world.

Chapters 2, 3, and 4 thus set out the concept of the self-evidencing agent. The following chapters then explore self-evidencing through several themes and areas of debate, showing how in each case, self-evidencing can unify and illuminate our understanding.

"Self-evidencing" might appear perniciously circular. How can such a seemingly circular construct lead to accurate beliefs, perceptions, and representations? How can it be conducive to getting us what we desire if it is an epistemically flawed, circular construct? Chapter 5 brings out how self-evidencing facilitates accurate representation, noting that this kind of accuracy is consistent with a pragmatic, self-serving aspect of self-evidencing, forming an attractive package. This leads to useful insights into other kinds of apparently false inference, as well as more generally skeptical challenges to knowledge. Existence as self-evidencing carves out a comfortable and productive middle position between never being wrong and always being wrong. This chapter also considers the accuracy of the self-evidencing framework itself, considering its a priori status and role in an empirical research program.

After that, chapters 6, 7, and 8 use the fundamental notion of self-evidencing to develop novel perspectives on key concepts of rationality, self, and volition and value. Chapter 6 considers decision-making and rationality from the self-evidencing perspective. Self-evidencing provides, I will argue, an attractive new idea of rationality, modeled on inference to the best explanation. This is inference to the best decision, which provides a coherent and accommodating blend of descriptive accounts of tractable rationality and normative accounts of intractable, ideal rationality. Here, decision-making is based on a particular kind of model-constrained trial and error process rather than on predetermined algorithms or on bare heuristics. I liken rationality to a kind of abductive hill climbing, allowing much diversity among individuals.

Chapter 7 looks at what happens to the self in self-evidencing, distinguishing between the self-model and the broader internal model of the world. There is an important point to convey first—namely, that the "self" in "self-evidencing" refers in the first instance to the circular process of a model explaining some observation that becomes evidence for the model *itself*. The earlier chapters of the book argued that the overall internal model can be equated with the agent because if the agent exists, it must be a self-evidencing model. Chapter 7 argues that what we call the "self" is a subset of the internal nodes of this broad model, representing those causes of our observations that originate within us. This makes the self as real as any other thing that we deem real. The chapter brings out how we can close the action-perception loop to provide evidence for the specifics of the model of our own self, consolidate and clarify it, and perhaps change it—hence shaping ourselves. At this point, a picture of specifically human self-evidencing begins to emerge, in which we are, as I shall label it, agents of volatility. Such agents occupy and exploit an eco-niche characterized statistically by volatility or relatively abrupt state transitions. The notion of agents of volatility then informs parts of the following chapters.

Self-evidencing presents agents as wholly enmeshed in the causal order of the world around them, and operating through a mechanism of internal causal precision control. In chapter 8, I ask what happens to volition, emotion, and value in such a thoroughgoing causal picture. Much work on decision-making assumes that we act for value or utility, occasionally with some added room for epistemic value. I discuss how self-evidencing entails that resolving and reducing uncertainty is the fundamental outcome of action, from which utility and epistemic value naturally emerge, making sense of many of our patterns of behavior. I discuss how the sense of volition, consistent with compatibilist conceptions of free will, plays into this picture, particularly for agents of volatility such as humans. I then discuss issues of emotions and sensation, building on several existing discussions. The chapter finishes by delving into areas of ethics and value. The question there is whether self-evidencing, with its unwavering focus on the self, might be inherently egotistical. I argue that value and ethics cannot arise inherently from self-evidencing, but self-evidencing nevertheless has a place for them, emphasizing the role of democratic discourse. I also speculate that some of the self-evidencing tools can help navigate an interesting route in some contemporary debates about utilitarianism.

In chapters 9 and 10, several of the pieces from previous chapters are brought together to elucidate substantive debates on consciousness, as well as on meaning and wisdom. In chapter 9, I discuss how the active inference element of self-evidencing can be brought to bear in consciousness science. This chapter both prepares the ground such that self-evidencing as a method can apply meaningfully to consciousness science and lays out how the various elements from previous chapters together provide what I think is a fruitful account of consciousness. Thus, I first advocate for a particular methodological approach to the thorny issue of consciousness, which can both reveal underlying mechanisms for conscious phenomenology and let us glimpse why consciousness has its particular subjective feel. However, I make clear that such an account is not a solution to the hard, metaphysical problem of consciousness (and this is a good thing). I then build a case that active inference is necessary for changes in conscious content, providing an illuminating account of the underlying mechanisms of consciousness. My case proceeds by showing how the counterfactual processing of active inference, in agents of volatility in particular, provides the detachment that marks out our conscious experience. I then outline how active inference can explain conscious content in everyday perception and how this plays out as the policies selected in active inferences are enacted. I finish by highlighting elements of active inference that speak to subjectivity and unity of consciousness. Overall, self-evidencing begins to unify self, volition, and consciousness within agents of volatility.

The austere, a priori analysis of existence underpinning self-evidencing may seem a long way from human well-being and flourishing. Chapter 10 therefore considers some quintessential philosophical questions about wisdom and meaning, filtered through contemporary psychological debates. Can self-evidencing be used to recast ancient aspirations to be not just clever or rational but also wise? Can it be used to explain the nebulous idea that we can experience, and perhaps enhance, meaning in our lives? I will argue that these otherwise somewhat exalted notions of meaning and wisdom can be harnessed by self-evidencing and shown to be achievable in diverse ways for us all. The chapter ends with a more practical perspective, picking up on internal precision control as a cognitive mechanism for mindfulness.

The concluding remarks in chapter 11 collect the threads of all the other chapters, summarizing why self-evidencing makes sense of mind and cognition and the diversity of everyday experiences. We are cast into existence and keep ourselves afloat by (whether calmly or frantically) adjusting the beliefs of our internal model. We do this secluded behind the veil of our senses, and we can only rely on a trial and error process based on our own beliefs about the world and ourselves and assessed only against overall uncertainty. We manage our way through, sometimes going well and other times achieving more error than success. Although self-evidencing is inherently solitary, it optimistically allows knowledge, rationality, self, emotion, value, and consciousness. Although the self-evidencing agent can fail miserably, such agents can also live well, continuously learning how to act more wisely and how to create more meaning in life.

2 From Existence to Free Energy and Self-Evidencing

The story that puts self-evidencing at the core of our understanding of the mind and existence can perplexingly seem both trivial and far from trivial. It would be trivial because the reasoning is largely analytical, in the same ballpark as "a triangle has three sides"; and it would be nontrivial because this simple analytical meditation aspires to deliver a method for understanding the mind and the way that we experience ourselves, each other, and the world. I hope to avoid perplexity by moving stepwise from conceptual analysis of the notion of existence to the notion of self-evidencing organisms. Along the way, I introduce several key notions, such as models, self-supervision, free energy, approximate variational inference, and model evidence. The aim is to connect the notion of self-evidencing to Karl Friston's free energy principle, focusing on its philosophically salient elements. The stepwise order of explanation is important to show how self-evidencing emerges a priori, as a first-principles system.

I begin this chapter by describing initial, heuristic concepts of the two key notions of things and self-evidencing, which will help anchor the more technical arguments and discussions. I then offer an analysis in statistical terms of the concept of existence and describe the boundaries of an existing thing and the states that characterize its existence. The notion of states is connected to notions of sensory input and internal models, and then to the idea of maintaining the model through action. This in turn leads to the challenge of describing a system as genuinely self-supervised and able to function on its own, without external supervision—this is what brings in the free energy principle. Then self-evidencing is shown to capture all these notions under one concept. Next, chapter 3 will extract some of the philosophically distinctive aspects of the self-evidencing agent, expanding on how action comes into the picture.

Preparing the Ground: Things and Self-Evidencing

The starting point for this chapter is *things*. For a thing to be a thing—for it to exist—is just for it to be distinguishable from other things for some period of time. This is a liberal conception of things in which things are not limited to (in the philosopher J. L. Austin's classic phrase) "moderate-sized specimens of dry goods" such as chairs, tables, pictures, books, flowers, pens, and cigarettes. The notion of "things" includes squishy things, like jellyfish and sponges, and things without contiguous boundaries, such as universities, organizations, and neural populations. It also includes big things like planets and small things like cells or molecules, as well as things within things, such as cells within organs. There are technical debates about the "thingness" of some cases, some of which we will return to later in the book, such as those that Austin also lists—namely, "people, people's voices, rivers, mountains, flames, rainbows, shadows, pictures on the screen at the cinema, pictures in books or hung on walls, vapours, gases." There are important differences among things, and this book, in particular, will be preoccupied with things that have agency, such as people, not with things that do not have agency, such as rocks. To engage with the argument in this chapter, we set aside the problem cases for now and adopt a liberal notion of "thing."[1]

It will be useful to have a heuristic picture of what a thing is (see figure 2.1), where it has a *boundary* that distinguishes it from other things and some *internal states* that are characteristic of that thing. The boundary states may be affected causally from things outside of the thing in question, and in some things, some boundary states may cause effects in other things. Internal states may be affected causally by boundary states or causally affect boundary states. An example could be a cell, which has a cell membrane and nuclei, mitochondria, ribosomes, and other elements; or it could be a university, where the boundary is given by complex physical, organizational, financial, and legislative states and the internal states are different decision-making bodies, like boards and committees, students and staff, and labs, classrooms, examination processes, and others. This conception of things in terms of boundary states and internal states will be specified with the technical term of a *Markov blanket* and discussed in more detail in chapter 4. The key analytical point is that if no boundary can be identified in principle, then there is no thing.[2]

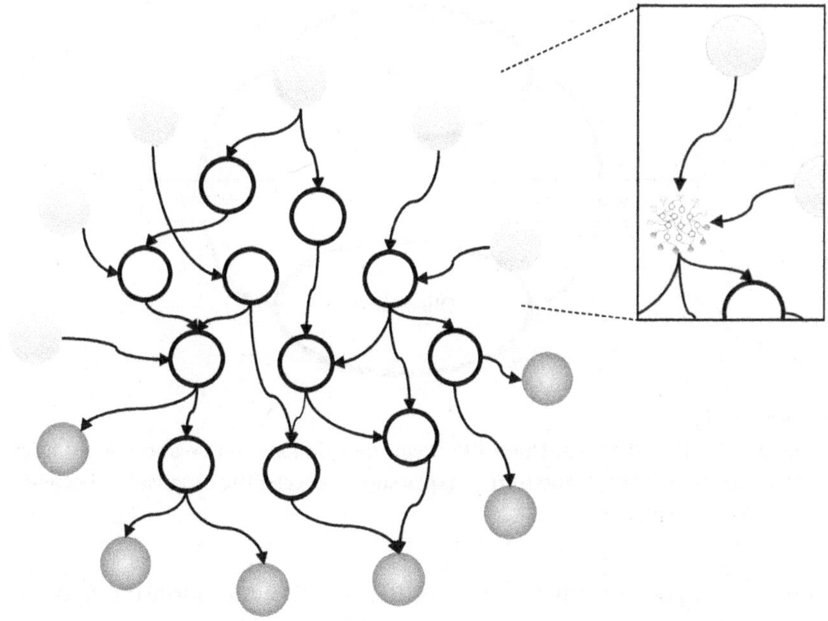

Figure 2.1
Depiction of a thing. Light gray nodes are input nodes of the boundary, dark gray nodes are output nodes of the boundary, and black circles are internal nodes. Arrows indicate causal relations. Things can be nested within each other, and there can be arrows going into input nodes and arrows going out from output nodes, as in the magnified section. This drawing also depicts a Markov blanket, which will be discussed in more detail in chapter 4.

The first objective, then, is to find a good method for investigating things so we can describe what characterizes one thing from another. This means that we begin with pointing at a thing and asking, "What makes it a thing?" and "What makes it the thing that it is?" In particular, the project is to point at things like us, human individuals (and other animals at times), and ask those questions. The argument is that the free energy principle and self-evidencing provide a good method for answering such questions.

The thoroughgoing theme of the entire book is the notion of *self-evidencing* and how it applies to things. Self-evidencing as a concept emerged in the philosophy of science, where it was first described by Carl Hempel in the 1960s. I will begin with a simple example. You see that the milk in the office fridge has begun to go bad surprisingly quickly, and to explain this

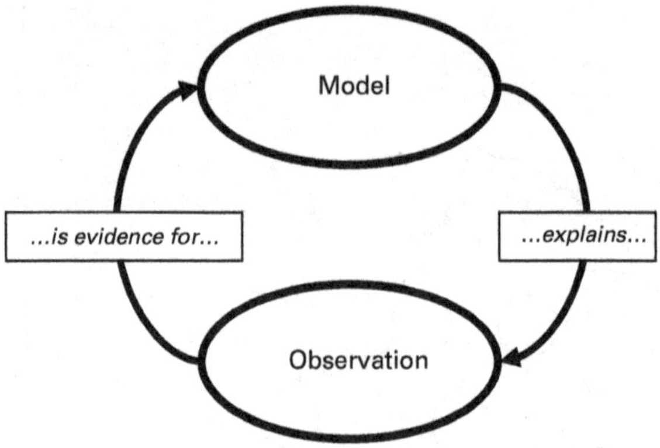

Figure 2.2
Diagram of self-evidencing. Upon the occurrence of some observation, a model is constructed to explain it, and in that explanatory process, the observation becomes evidence for the model.

observation, you construct a *model* of the situation and form the hypothesis that the power must be turning off at night for some unknown reason, perhaps having to do with maintenance works. This explains the observation, and the sour milk then becomes evidence for your model and its power shut-down hypothesis—the better the observation is explained, the more the observation becomes evidence for the model itself. At the same time, that model's hypothesis would not exist if not for the observation.

This strangely bootstrapping process is self-evidencing; it describes a model whose hypothesis exists to explain observations that, in turn, provide evidence for the model itself. In a simple sense, the upshot of self-evidencing is inference—coming to a reasonable belief about what caused observations (such as power outage causing the bad milk observations). Figure 2.2 depicts self-evidencing, which we will return to in several variations.[3]

Self-evidencing has several interesting epistemic properties, which will occupy us later in this chapter and the rest of the book. For now, though, notice that the evidence for the model increases for every observation that the model is able to explain. This means that if you use the model to generate a hypothesis about a new observation and you perform a successful test of this hypothesis, then the model will garner more evidence for itself. For example, you might hypothesize that if the power is actually turned off

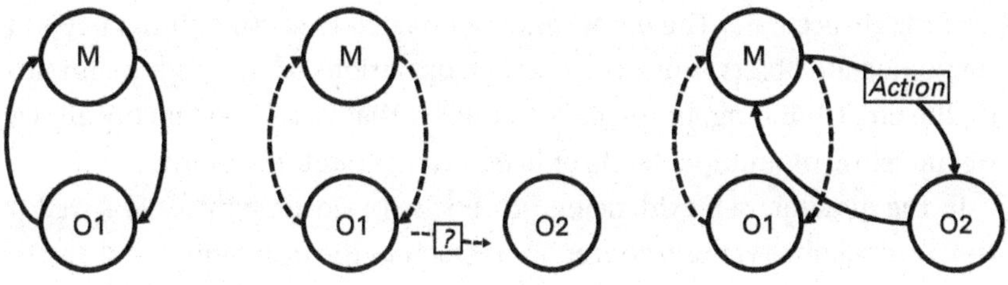

Figure 2.3
Increasing model evidence in self-evidencing through selective sampling of observations. Left: Initial self-evidencing, where model M explains observation O1, which becomes evidence for M. Middle: Forming the hypothesis (indicated by "?") that if M is accurate, then O2 should be observed. Right: Performing an action whereby O2 is in fact observed, increasing model evidence. Dotted lines indicate past or future self-evidencing.

overnight, then things should defrost quickly overnight, but not during the day; so you put a bag of frozen peas in the fridge and observe them indeed defrosting quickly overnight, but not during the day, which increases the *model evidence*. This active self-evidencing process of increasing model evidence is depicted in figure 2.3.

The depiction in figure 2.3 also anticipates the central concept of active inference, the free energy principle's account of planning and decision-making, which will play a large part in the rest of the book. Self-evidencing can occur both for observations that the model happens to be exposed to and for observations that are selectively sampled based on the model's generation of expected observations. As we will see, this relates to the free energy principle, where the objective functions guiding the updating of models pertain to the actual uncertainty from surprising observations (sour milk) and the expected reduction of uncertainty from performing an action (defrosted peas). This relates to the simple idea that what drives self-evidencing is the increase of model evidence or, equivalently, the reduction of uncertainty about observations given a model.

Our example also implies that self-evidencing is model-based since the initial observation of the bad milk can be surprising and worth explaining only if it collides with the prior expectations of a model. Together with the active sampling of observations, this illustrates the dynamic character of self-evidencing, where a model is maintained over time as it encounters or seeks out new evidence, increases its model evidence, and becomes increasingly accurate. The model may be updated to capture things beyond the immediate observations (such as the operations of the team maintaining the office building, in the milk example); that is, self-evidencing allows the modeling of multiple levels of *hidden causes* of observations.

In the dynamic self-evidencing process, self-evidencing goes less well if the expected observations do not occur (such as the peas failing to defrost), where the model cannot explain that new observation after all and model evidence does not increase. Now the model must be revised to come up with some other explanation of the bad milk and frozen pea observations, or one must reevaluate how clever the frozen pea experiment actually was. In a changing environment, where things can transition from one state to another (milk going bad, peas defrosting), self-evidencing becomes a cascading process, looping through actions and observations. Even though self-evidencing is just about maximizing model evidence, it comes with many degrees of freedom for how and what to explain, as well as how to evaluate the evidence that comes in and how to update the model in the light of surprising new observations.

Next, we combine our concept of existing things with the concept of self-evidencing. The kinds of things that we will be pointing to are active things, humans in particular, rather than passive things, like rocks. In this chapter and the next, therefore, we will move from a focus on things to a focus on active organisms, and then to human agents. Agents are the kinds of things that can form models to explain observations and use those models to selectively sample observations consistent with those models. In fact, we shall equate models with agents and equate the observations with our sensory input, acquired either serendipitously or in active sampling for information or for utility, as depicted in figure 2.4. Then we can say that to understand agents, our method should be to consider them as models that self-evidence.

Two more concepts will be useful to have on the table from the beginning: the generative model and the generative process. In the depictions of self-evidencing, the arrow from the model to the observations is meant to be explanatory since the model explains the observation. A good starting point for explicating what this means is to say that the model must be able to generate the observations; the model assumes such-and-such hidden, latent causes in the environment and specifies how, together, they give rise to the observations. It would be hard to see how a model could be an explanation

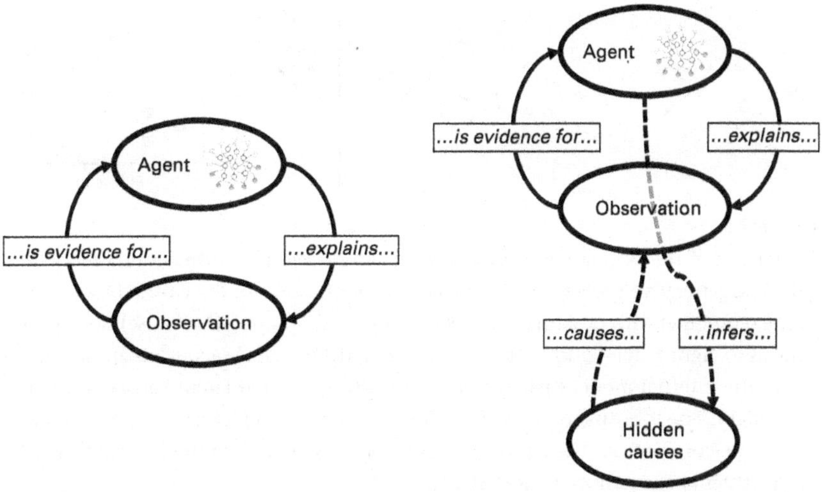

Figure 2.4
Left: The agent understood as the model in a self-evidencing system. Right: Embedding the self-evidencing agent in a causal structure, where there are hidden causes of sensory observations and the model infers hidden causes (dotted lines).

unless it can at least generate the observations; hence, we call it a *generative model*. Technically, the generative model is a probabilistic specification of the joint occurrences of latent or unobservable causes and observable, sensory consequences, which can be expressed in terms of a likelihood (the probability of the observations, given the causes) and a prior (the probability of the causes). When we get to chapter 3, we will see that this simple condition on explanation can be harnessed in a classic abductive explanatory setting of inference to the best explanation. On the other side of the model is the external world, which the model makes inferences about, having access only to the observations. This is the real process where the true external causes and states go together to generate the observations for the model; this is called the *generative process*. Perfect self-evidencing would have a generative model that recapitulates the generative process (we will discuss this matching of model and process in chapter 5). Figure 2.5 depicts the generative model and process, and it notes that observers (such as scientists) can observe them both, creating their own generative model that treats them as a combined generative process (with the quirk being that one of the latent causes of scientists' observations is itself a generative model).

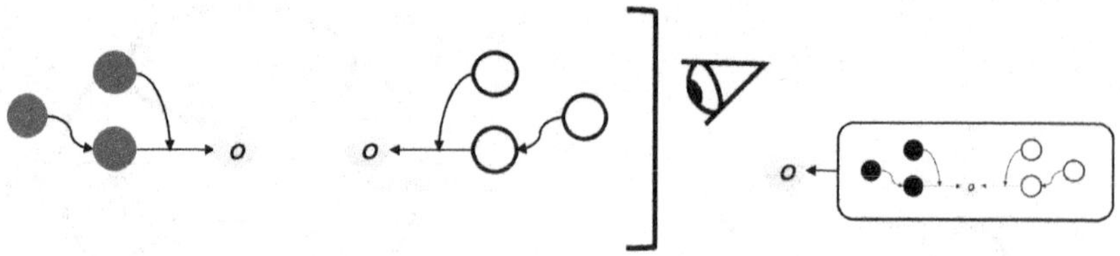

Figure 2.5
Depictions of the generative process, the generative model, and scientific modeling. Left: The generative process, where the actual latent causes in the world (black circles, here exemplified with proximal and distal causes and a modulating cause) cause observations (o, light gray). Middle: The generative model where observations are generated under the internal model's parameters representing inferred latent causes (white circles). Right: The scientific process of creating a generative model for scientists' observations (eye icon, o) from the entire generative process, including the generative model of the system or organism, treated as a latent state.

Throughout this book, we will see how this simple move of considering agents as actively self-evidencing, generative models provides a method to elucidate perception, attention, action, and decision-making, and from there on to rationality, self, value and volition, consciousness, meaning, and wisdom.

As we work our way through this chapter, these basic heuristics for the core constructs of things and self-evidencing will serve as reference points for the stepwise argument. Having prepared the ground, the journey to the free energy principle and self-evidencing can begin. Here, the argument is that if an agent exists as a thing, we must be able to describe it as a self-evidencing agent.

What Does It Mean to Exist?

What does it mean for a thing to exist rather than not exist? As foreshadowed in the previous section, the question is not, yet, about any particular thing. It is in the first instance calling for a conceptual analysis of the concept of *existence* of a thing. To begin this analysis, consider first what happens when a thing ceases to exist. For example, a rock is crushed, a tree burns, or a cell dies. The boundary of the thing ruptures or erodes, and what was the thing begins to disperse (dissipate, decay, die . . .). That is, the thing cannot be distinguished from other things—it is not a thing any longer. If we think of all possible states as a vast state space, inexistence means that the probability that the thing is in a limited set of these states is zero. If there is no limited set of states that is characteristic of the thing, then it is not a thing. Conversely, to say that a thing exists is to say that the probability that the thing is in some characteristic set of states is quite high. A rock is there now, not dispersed by wind, water, and sun. If a tree exists, we will only find it going through its life cycle rooted in the soil. A biological cell will be in its typical states, interacting with things happening in its environment, so long as its cell membrane can hold it together. Thus, if we consider all the possible states, and for each of them ask if the thing in question is likely to occupy it, the answer will be zero probability for the vast majority of states, and only a few states will have a high probability of the thing being in them. It is the same for every thing (even those with less obvious boundaries such as universities, as mentioned in the previous section).[4]

The analysis here is meant to be uncontroversial: existence implies that the thing in question can in principle be individuated or separated from everything else. We can begin to describe this as a set of nodes in a statistical/causal net, where each node captures various states. As we look at this collection of nodes that distinguishes the thing from everything else, some of the nodes must form a boundary between the thing and everything else; some nodes must be such that the next node that one looks at is not characteristic of the thing, and in that sense is "outside" the thing. Hence, there must be boundary states. The remaining, nonboundary states will be internal to the boundary, which we call "internal states." Figure 2.1, earlier in this chapter, depicts this structure (and also captures the technical notion of a Markov blanket, which we will return to in chapter 4). It is tempting to translate this into spatial terms since many things have spatial boundaries (such as cell membranes, for which intercellular water molecules would be external), but this is not always obvious (such as a university, for which governmental legislative states would be external). The description of boundaries in terms of causal nets coheres with our definition of a thing in terms of a state space, where thingness depends on the thing occupying some states and not others. To exist, things must be "attracted" to a subset of states—if they exist, things have an attracting set in state space. The limit case is when nothing exists, the probability of anything being in any state is zero—a cosmic soup of everything, and thereby nothing.[5]

The ensuing probabilistic or statistical analysis of existence, in terms of the probability of occupying some set of states, is extremely liberal. A small fluctuation in probability can suffice for existence in this sense, even if ever so transient from our perspective. And the boundaries of existing things are nested and overlapping; for example, cells making up trees, or organs a body. This is an advantage because we want to be able to consider, and disagree about, the existence of many things besides rocks, trees, and cells, such as coral reefs, universities, viruses, and slime-molds.

Once cast in statistical terms, this otherwise trivial analysis of existence immediately becomes loaded with epistemic aspects. If a thing exists, it will be unsurprising to find it in its characteristic set of states and surprising to find it in other states. "Surprise" here is statistical, in the sense of what state a thing is in, given its characteristic states. A probability distribution (or density function) hovers, as it were, over each thing, indicating the states that it tends to occupy and informing us what would surprise us about them—or indeed, what should surprise the thing itself.

In fact, we can flip things around and simply begin to characterize things in terms of the states that they are likely to occupy. We ask, "What kind of thing is that?" and the answer is, "It is the kind of thing that have these as its *characteristic states*, making it unsurprising to find it occupying those states, but for which all these other states would be surprising."

By saying that this train of reasoning from existence to characteristic states is analytical, I also mean to say that it is difficult to argue against it. Someone arguing against it would need to deny that existence implies being able to distinguish the existing thing from other things, and denying that would suggest that they do not really operate with the concept of existence. It would be something like encountering someone saying that a triangle does not have three sides, where we would reasonably doubt that this person is dealing with the concept of a triangle at all. It may appear odd and somehow naturalistically myopic to understand existence in terms of statistical states, but the steps there are straightforward. Once we add more considerations, drawing on discussions of Friston's free energy principle, a more substantial story emerges.[6]

States, Sensory Input, and Models

Consider first what a state is. It is tempting to think of states in spatial and temporal terms. Polar bears tend to occupy icy states around the Arctic, have done so for more than 100,000 years, and will do so likely until the Arctic becomes too hot. For biological organisms, one might characterize such states in terms of eco-niches. But, to begin with, we want to shrink the perspective and consider just the states of the boundary that we have agreed must demarcate the existence of a thing. Then we can talk about the states that we expect a thing to be in as the states that statistically cooccur with its boundary—that is, the thing's characteristic states. For different kinds of animals, this might include certain temperature ranges, setting aside states out in the world.

Different states have the potential to cause changes in the states on the inside of the boundary in different ways. Some of these states are consistent with the ongoing integrity of the boundary, and thereby the thing itself; in humans, this might be certain light and sound spectra, as well as some nutrients; other states are more pernicious to us, such as extreme temperatures. There is much variability in the characteristic states of things, even among biological organisms. Bacterial spores, for example, have an impressive range of states that are consistent with the integrity of their boundaries, going far beyond the ranges of characteristic states for humans. Spores are not very permeable, being able to remain dormant while withstanding extreme heat, ultraviolet irradiation, lack of water, and the action of antibiotics; they wake up only in response to a particular kind of nutrient. Humans, in contrast, wake up and display varied behaviors in a wide range of states. That is, bacterial spores and humans exist—are distinguishable from their environments—in very different ways.[7]

For biological organisms, this proximal notion of a state—what happens at the thing's boundary—captures its typical sensations or sensory input, broadly speaking and across all its senses. Hence, for biological organisms, there is a conceptual move from talking about existence in terms of an attracting set in state space to talking about characteristic sensory input. The conceptual move is to consider some nodes of the boundary as sensory nodes, simply because they are causally affected by external nodes (recall figures 2.1 and 2.4). Empirically, this alludes to notions of sensory epithelia, tissue covering organisms or organs that transduce external signals for internal processing (including input stemming from the body itself to the central nervous system, such as through our proprioception and the vestibular system, or the sensors for the preferred nutrient for the bacterial spore). A nonbiological thing such as a snowflake is exposed to causal effects at its boundary when the molecules in the air around it evince greater motion and it melts; this would be considered uncharacteristic input to it, although we

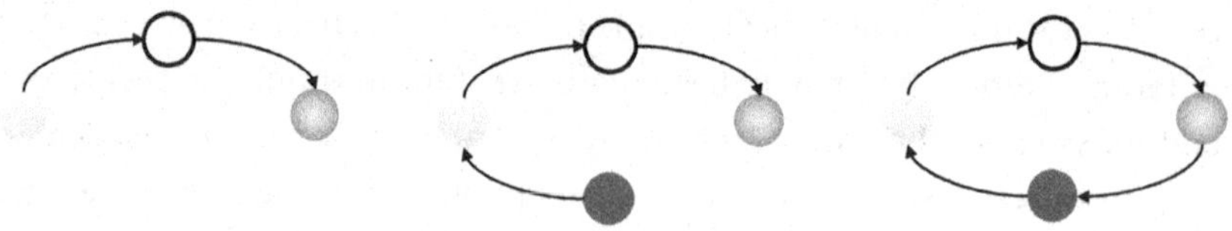

Figure 2.6
Left: A thing with proximal, (input or sensory) states in light gray, internal states in black circles, and output (active) states in darker gray. Middle: Adding a distal (hidden or latent) state in solid black, which causes changes in the proximal states. Right: Closing the action-perception loop, such that an active thing's actions can cause changes in the distal, environmental states, which can loop back and help determine the proximal sensory states.

do not in ordinary parlance regard this as "sensory" input to the snowflake. In contrast, humans have thermoreceptors and nociceptors that allow some causal impact to filter inward, but without disintegrating our boundaries (mostly, and up to a point).

Shrinking the focus to the states of the boundary of things is convenient because the opposite (casting the net broader, to more distal states) quickly becomes unwieldy. For example, there is no end to the constellations of states that could give rise to the temperature range characteristic of a biological organism (i.e., all kinds of meteorological, geographical, climate, heating/cooling, building, clothing states, and their interactions). It would be intractable to try to figure out which of these multifarious states are characteristic of the organism. In contrast, the temperature states that can be observed at its boundary would be easier to determine.

Now we can distinguish between proximal, *sensory states* of the boundary's input states and distal causes of this input. Distal states can be conceived as the causes of sensory input, which are often called "latent" or "hidden causes" because they are hidden from direct view of the sensory states (as in figure 2.4). We are going to be much preoccupied later in this book with this distinction, as well as the idea that it is difficult to determine which hidden causes are characteristic of an organism. Figure 2.6 provides the simplest depiction of a thing, adding hidden causes and then depicting the *action-perception loop*, where the organism can act on the world and thereby in turn help determine its own sensory input. It is easy to imagine how there could be additional layers to the hidden causes, such as climate change having an impact on seasonal patterns that cause an organism's thermal input.

Since the existence of a thing is analyzed in terms of its characteristic states, this set of states can be considered a statistical description of that thing. A statistical description of a thing is in a rudimentary sense a model of that thing. If we know the characteristic states (i.e., the probability distributions of states associated with a thing), then we can begin to model that thing; then we can say that it would be surprising to find the thing in certain states, given its model. For biological organisms, we have also focused on giving the characteristic states in proximal terms of the organism's sensory input. This means that the model of the thing is at least partly specified by the organism's characteristic sensory states. Then, when we are talking about states that would be surprising, we are talking about surprising sensory states. In information theory, this kind of surprise is known as *self-information*—namely, the implausibility of a particular sensory event, given the thing or model in question. This is still a relatively trivial idea—for example, making sense of the statistical surprise of finding an existing biological organism, such as a human, in an extremely hot sensory state, such as a state caused by the hidden state of lava flow, and then using that set of surprising/unsurprising sensory states to specify the organism in question.

Bringing sensory input into the story, as conceptually related to the fundamental conception of the existence of a thing, amplifies the epistemic properties of the story so far. Existence, for biological organisms in particular, is a matter of being in states with relatively unsurprising sensory input, where the level of surprise is determined by the statistical model capturing the organism's own characteristic states (hence the same input may be surprising to one organism and not to another). Epistemic concepts of believing and knowing, having evidence for beliefs, seeking information, and learning about hidden causes are then within our reach, while still operating on the slender basis of our analysis of what it means for a thing to exist. This begins to put into place some of the epistemic concepts for the formulation of self-evidencing that we are working toward (depicted initially in figures 2.2 and 2.3).

Our statistical formulation of existence is consistent with alternative methods for describing different kinds of things. For example, for some fields of inquiry, such as ecology, geology, and history, it may be more expedient for scientific progress to conceive states in terms of selected distal causal events (e.g., to understand the events around the Krakatoa eruption in 1883, one would usefully seek out temporally and spatially distal facts about plate tectonics, the local and global populations, ecosystems, atmospheric patterns at the time, and other contributing factors). Such distal formulations can in principle be transposed into proximal terms of flows of sensory input at the boundary of individual organisms (or inorganic things), insofar as they have any proximal effects, so proximal and distal descriptions do not preclude each other.

There is a technical sense in which giving a model of a thing in proximal terms of its sensory states preempts the modeling of more distal states. If we are interested in how some internal states of a thing occur, it will be useful to know everything about how the sensory states unfold because, as we have set up things, they are the causes of the behavior of the internal states. But if we already know how the sensory states unfold, then we cannot get more information about the internal states from adding distal, hidden causes to our model (as we will see later, it is also important to know the states on the other side of the boundary, the active states affecting the world). That is, the probability of the internal state being in a certain way is determined by the sensory state, so knowing that state renders the internal state informationally independent of the distal states (technically, this is conditional independence, also known as the *Markov property*, where the probability of the internal state being a certain way given the sensory and distal states is the same as the probability just given the sensory state, depicted in figure 2.7 left). This is useful, as previously mentioned, because figuring out the multifarious and interacting distal causes can be too difficult. But such a myopic perspective on what drives changes in the internal states is limiting if we want to know why the sensory states are as they are, and such knowledge can be critical for our ability to predict what will happen to the internal states over time, which can be too hard if we can only operate with the possibly noisy and confusing raw time series of sensory states (see figure 2.7 right). To illustrate, which image is seen on a computer monitor is determined by the pixels on the screen and adding information about which object is filmed will not change that; however, information about which object is filmed will be more useful for predicting how the image will change than just keeping track of each pixel.

Whether one adopts a proximal or distal approach is typically determined by pragmatic concerns about what one is interested in describing and what kind of explanation would be useful (we will return to the theme of pragmatics in chapter 5). The proximal approach is useful for modeling the existence of a thing in the first place, and without having to average over all the distal

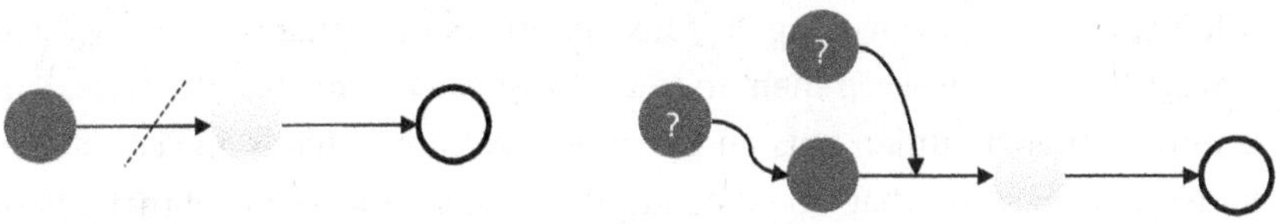

Figure 2.7
Depictions of the effects of proximal and distal states on internal states. Left: Given the proximal, sensory states (gray node), the distal states (black nodes) are rendered uninformative with respect to the internal states (black circle nodes); the dotted line indicates this conditional independence. Right: It is difficult to model interacting and more deeply hidden distal states (indicated with '?'), but they may be important for understanding and predicting how internal states will unfold.

states that could be affecting the thing. As we will see, the proximally defined model is also useful for understanding what existence is like from the organism's own perspective.

Surprise and Self-Supervision

Rocks and snowflakes exist—many rocks for a long time and many snowflakes for only a short while. Neither kind of thing moves around by its own agency—each one is what it is until its boundaries disintegrate. Rocks are formed under certain geological conditions, mostly persist in a wide range of conditions, and slowly weather away. Snowflakes appear and persist in specific atmospheric and temperature states, and they cease when the temperature increases. Biological organisms differ in a fundamental respect from rocks and snowflakes since they are endowed with capacity for action. This means that they can try to act in ways that maintain themselves in their characteristic states. For example, single-cell organisms like *Escherichia coli* can swim up a nutrient gradient or tumble in search of such a gradient. Humans engage in a plethora of actions, such as feeding themselves, drinking, and moving about so as not to be too cold or warm; we also play sports, participate in art or science, care for things or build them, create crafts or write code, help and compete with other people, and engage in other activities. In contrast, rocks do not seek shelter from the elements, and snowflakes do not drift away from heat. We will talk much more about action throughout this book because the focus is on self-evidencing as an active process for biological organisms (as depicted in figure 2.3). At this stage of the argument building up to self-evidencing, it is just an observation that some things act in ways that seem to keep them in their statistically characteristic states. For example, when a human puts on a coat on a cold day, that is an action that eludes freezing states that would be surprising, given the model of their characteristic warm state.

The question is: How can one describe the ability of biological organisms to act and stay within their characteristic states? This question must have an answer, since organisms do not manage to exist by serendipitously ending up in various states; if organisms like humans were just waiting passively to receive unsurprising sensory input, then we would quickly be surprised, succumbing to the dissipative forces of the world around us. Answering this question is going to be the driver of our conception of self-evidencing. As we answer it, we will also see the connection to the free energy principle.

An obvious first attempt at reaching an answer is to suggest that organisms can directly assess whether their sensory input is surprising, allowing them to act accordingly. But this attempted answer will not work. The model conveys information about which states the organism is expected to be in, across all possible states that it could be in. An organism, as the finite creature it is, cannot possibly directly assess this, as there are too many possible states to try to entertain. And it is not something that an organism could hope to learn by averaging over all states as it goes through life; the organism would perish if it investigated states that were, in fact, highly uncharacteristic for it. In artificial intelligence and machine learning, this averaging over states is called *marginalization*, and the ensuing probability of sensory data is known as the *marginal likelihood*. This marginal likelihood is the probability of observing those sensory data under the model of the characteristic states. In statistics, this quantity is known as *model evidence*; that is, as we have seen, the sensory evidence for a given model, which is what self-evidencing provides for a given model.

This is the problem: somehow the organism must be able to act to avoid surprising states, which implies that it must know in advance which states are surprising; but the organism cannot feasibly directly assess the surprise (or marginal likelihood) of the state that it is in. From an outside observer's perspective, perhaps we can amass evidence and form reasonable beliefs about the marginal likelihood of a given state for types of organisms such as bacterial spores or humans. But the organism cannot observe itself from the outside and ahead of time in order to determine if a given state is surprising. It must be able to solve the problem from its own perspective, under its own steam. That is, the organism has to be self-supervised, or in sole charge of supervising its own assessment of surprise and deciding what to do.

Self-supervision is a critical constraint on any fundamental account of a biological organism capable of acting purposefully in the world. If an account of decision-making, perception, attention, or other mental or cognitive process fundamentally relies on an assumption that the organism already can assess which states are surprising and which are not, then the account is not fundamental and cannot help us understand the basic conditions for existence of the organism. For example, if an account relies on labeled training data (provided by an external supervisor) equipping the organism with the means to sift through and determine the surprise of sensory input, then the account is assuming rather than explaining that the organism has performed the seemingly intractable task of marginalizing over all possible states. The problem with such supervised accounts is that, given that they have solved the organism's fundamental predicament by fiat, they might not be reflective of something that the organism could actually do.

As I see it, there is little attraction in pursuing accounts of action, perception, learning, and so on that are not portraying the organism as self-supervised, or that at least conform to a more fundamental account of the organism as self-supervised. There would be little point in building an elaborate account of the workings of an organism only to find that, at the end of the day, the account postulates processes that the organism is not capable of undertaking. Therefore, the best strategy is to solve the self-supervision problem of how the organism can assess its surprise and then build accounts of action, perception, learning and so on. Then we would be assured that our overall understanding of the organism reflects what the organism itself can do. As we will see as this chapter progresses, this is what self-evidencing accomplishes.

Free Energy and Variational Inference

The problem that we have encountered is that to be able to act to avoid surprising, uncharacteristic states, the organism needs to marginalize over all possible states, which is not tractable for it. Therefore, the organism seems unable to solve the problem under its own supervision.

A solution could begin by considering what an organism does in fact have access to—namely, its sensory input and the states internal to the boundary that distinguish it from other things. The task for the organism is to determine if some sensory input is surprising, and in principle it could do this by assessing the input against an internal model of its characteristic states, encoded in the internal states. But, as we have seen, the model of the characteristic states is not directly available to be encoded in the internal states. The solution is to let the organism simply assume a model—essentially make a guess at what its characteristic states might be—and use that assumed model to guide its assessment of the surprise of the sensory input. We are thus saying that the internal states of the organism should embody an assumed model of its characteristic states.

Assuming an internal model of its characteristic states and using that to assess the surprise of the input would be feasible for the organism, in the sense that the organism in fact has access to these quantities. But the problem now is that there is no guarantee that this assumed model is anywhere near accurate, and if it is not accurate, it may be misleading in what it tells about the surprise of the sensory input. A human agent will not exist long if, for example, it makes the inaccurate assumption that extreme heat is characteristic of itself.

To solve this problem and show how the organism's assumed model of its characteristic states can come to accurately guide its assessment of surprise, I will heuristically bring out some of the formal components of the *free energy principle*. First, we identify the quantity that we are searching for, but which the organism does not know—namely, the surprise of the sensory input given the model ("How surprised should I be about this observation?"); this is the negative log probability of the input given the model. Next, we introduce another quantity that the organism also does not know—namely, the probability of the actual model given the sensory input ("Does this input in fact provide evidence for my model?"). This is the quantity that an ideal Bayesian reasoner would arrive at (i.e., using Bayes' rule to calculate the posterior probability of the model given the observation). Then we introduce the quantity that the organism does have access to and can change—namely, its assumed model.[8]

Now we have three things in play: (1) the assumed model, (2) the exact inference of the probability of the model given the observation, and (3) the actual surprise of the observation given the model. Only the first of these is known by the organism. It turns out that there is a critical relationship among these quantities, where what is called *free energy* is the divergence between (1) and (2), plus (3):

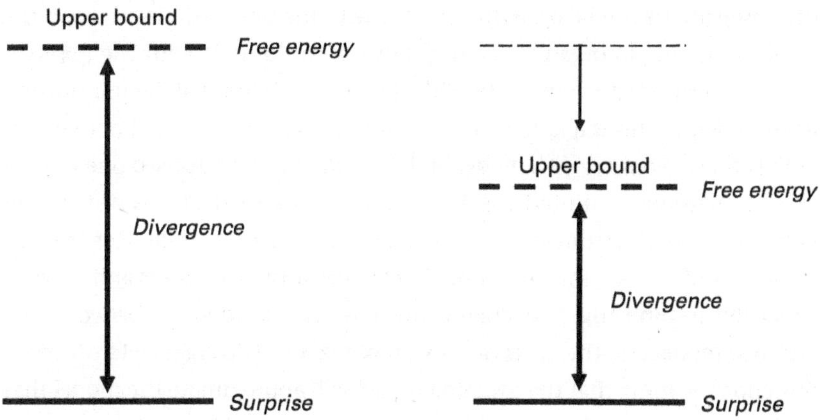

Figure 2.8
Left: Illustration of the free energy upper bound on surprise, as described in equation (2.1). Right: Given that free energy is an upper bound on the surprise, minimizing the free energy must ensure that the divergence between the assumed model and the model under exact inference gets smaller and the free energy approximates the surprise.

$$\textit{Free energy} = \textit{Perceptual divergence} + \textit{Surprise} \tag{2.1}$$

The perceptual divergence here is between, on one side, the probability of the assumed model and the observations, and on the other side, the exact inference of the model, given the observations; technically, this is a Kullback-Leibler divergence (or relative entropy) between two probability distributions, which is either zero, when they are identical, or positive (and never negative). Working to have a small divergence means that one's assumed model is beginning to approximate (i.e., get closer to) the model that one would have by doing exact inference—this is *approximate inference*. If the divergence is close to zero, then free energy must be close to the surprise. Overall, it can be seen that free energy must always be larger than or equal to the surprise since the divergence is always zero or positive; free energy is thus a bound on the surprise. Therefore, if the organism can minimize free energy, it will be able to assess the surprise (as depicted in figure 2.8). But how can it do that in a self-supervised way?

Free energy minimization is conceived in terms of the fit between the two quantities that the organism does have access to—namely, the assumed model and the actual observations. That is, free energy increases when the observations are not providing evidence for the assumed model, and it decreases if the observations can be brought into line with the assumed model. So, if the organism is able to organize its model of the world to fit with the observations or select observations to fit with the model, then it will be minimizing its free energy. This is a kind of surprise minimization since the better the fit of the observations to one's model of the world, the less surprised one will be.

As I mentioned, minimizing free energy, understood in this way, is still based only on an assumed model, which might be inaccurate. However, in equation (2.1), this is the one thing that the organism can assess and change. Crucially, minimizing free energy (by ensuring a good fit between the assumed model and the observations) minimizes the divergence to the exact posterior, meaning that the assumed model will approximate the model that an ideal reasoner would have. This is a beautiful notion because it means that by manipulating something that the organism does have access to, its model of the world will get closer and closer to something that the model does not have access to (namely, the exact posterior).

The process for doing this is variational inference, which is a standard tool in machine learning and theoretical neurobiology. As a simple heuristic for how variational inference works, we can depict it as a kind of model-constrained, trial and error process. To illustrate this, simply, assume first that the organism we are looking at has organized its internal dynamic processes into a manageable number of distinct, causally dependent nodes (rather than messy, high-dimensional space), which will capture the parameters of the assumed model. Then, if there are, say, 10 parameters, the organism may vary the first parameter while holding the other nine fixed and noting how the free energy changes. Then the organism moves to vary the second parameter while holding the other parameters fixed, and repeats this process until the best free energy gradient is found. This is model-based in the sense of being described in terms of parameters of an assumed, parameterized model, and I regard it as trial and error because of the seemingly improvised process of "trialing" variations of one parameter after another, hoping for little ensuing "error" (in the sense of increased free energy).

This simple process of varying and holding fixed parameters against the overall objective function of *variational free energy* is guaranteed to minimize the divergence in equation (2.1); since the surprise itself cannot change as a result of adjusting free energy, the divergence must get smaller, as depicted heuristically in figure 2.9. The mathematics here relies on the convex function of the logarithm of probability distributions together with Jensen's

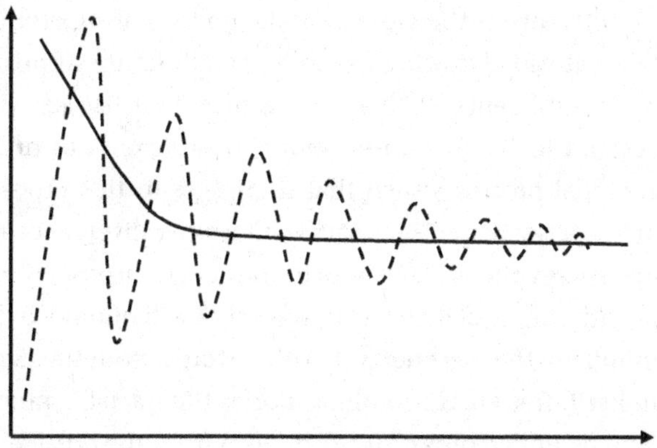

Figure 2.9

Heuristic depiction of variational inference. The solid line is the parameter estimation in exact inference of posterior probability, given some observations; the dotted line is the parameter estimation in variational inference, as parameters are varied. The x-axis is inference steps, the y-axis is parameter value. Once the variational inference is close to exact inference, the free energy is close to the actual surprise, as in equation (2.1).

inequality, which describes that for a convex function, the function evaluated at some chosen point is less than or equal to the average of the function at those points (imagine a secant line between two chosen points on the bowl shape of the convex function, which can never go below the bottom of the bowl shape). This means that the value given by the chosen points provides a bound on the actual convex function. Then maximizing that value—understood here roughly (and skipping many steps) as varying the parameters of the assumed model relative to the free energy—will approximate the actual function. In our case, this approximation minimizes the divergence in equation (2.1), which means, first, that the assumed model comes closer to the exact posterior; and second, that as that divergence gets smaller, the free energy approximates the actual surprise.[9]

The beauty of this formalism is that the normative outcomes of getting closer and closer to exact inference, and then being able to evaluate the real surprise, happen as a result of a process that is tractable to the organism that we are describing. This organism is exclusively operating to make sure that its assumed model parameters and its observations fit with each other—the organism does not care about the normative outcomes, and yet it tracks them.

Variational inference is therefore a candidate for self-supervision since it does not rely on already knowing which observations are surprising or how to engage in exact inference. Rather, an organism that has access to its own internal states and to its own observations in principle can organize itself through variational processes such that it engages in this process of minimizing the free energy bound on surprise. Once the divergence is relatively small, and free energy therefore is close to the actual surprise, we can endow the organism with the capacity to act in ways that will minimize the surprise, again by minimizing the free energy. In other words, from the perspective of the organism itself, free energy minimization is the tractable process that an organism could be undergoing, if it exists. Seen from the outside, the organism's behavior will approximate Bayesian inference over the long run and minimize surprise, given its model.[10]

The free energy principle, formulated by Friston, puts all this together to say that if an organism exists, then we must be able to describe it as minimizing its free energy. The notion of free energy is borrowed from statistical physics but is here used in an information-theoretical context; there is a vibrant research program underway investigating the free energy principle in the context of physics, where a free energy minimizing system would be understood in terms of a system in a nonequilibrium steady state. This is an area in rapid formal development; for a state-of-the-art treatment, see the development of Bayesian mechanics and path integrals by Friston and colleagues.[11]

Free Energy Minimization and Self-Evidencing

We can understand free energy in terms of the fit between the observations expected under an assumed model and the actual observations, such that the better the fit, the less the free energy. Given some assumptions about the normal shape of the probability distributions and the organization and structure of the particular organism in question, free energy can be considered the long-term average of precision-weighted prediction error. It can thus help to think of free energy minimization as minimization of such prediction errors, which connects the free energy principle to the predictive processing framework. However, to stick with the formal developments in this area and move away from previous predictive coding schemes, we will mainly use the notion of free energy minimization in this book, and relate this in a broader setting to uncertainty reduction and self-evidencing. This helps us understand how a wider range of processes than just predictive coding (such as complexity reduction or exploration) can minimize free energy.

Recall from the previous discussion that self-evidencing is what happens when a model explains an observation and the observation then becomes evidence for the model (as depicted in figure 2.2). A self-evidencing system is described in terms of a model and some observations, such that the model can explain the observation. The more a model is self-evidencing, the more the observations will have an explanatory fit with the model, and the stronger the evidence for the model will be. These elements capture an essential element of the free energy principle—namely, that when an organism minimizes its free energy and thereby tends to occupy unsurprising states, we can describe it as maximizing the marginal likelihood (i.e., increasing its model evidence). The free energy–minimizing organism does this just by operating on its own internal model and the observations that it encounters or seeks out. This operation is captured in the varying of the internal model parameters in variational inference, such that each variation in effect corresponds to a new attempted explanation of the observation (see figure 2.9), the success of which is assessed through minimized free energy or maximized model evidence. These aspects of the free energy principle mean that we can now conceive of organisms that minimize their free energy as self-evidencing organisms. Any organism that exists must be describable as minimizing its free energy or maximizing its model evidence; that is to say, as self-evidencing.

Hence, from our initial, austere analysis of the existence of an organism, via Friston's free energy principle, we get the core notion of this book, which is that existing organisms are self-evidencing organisms. This implies that the only job for the organism—the only thing that it can feasibly do, given the quantities that it has access to—is to accumulate evidence for the assumed model of its own characteristic states. We have defined the existence of the organism itself in terms of this model, so now we can move back and forth between talking about the model and talking about the existence of the organism. Then we can say that if the organism exists, then it is by and large encountering observations that its assumed, internal model can explain, and which in turn provide evidence for this model. Importantly, the free energy principle ensures that we can understand this self-evidencing process as self-supervised. Chapter 3 will argue that self-evidencing is a distinctive perspective on the free energy principle, delivering a new philosophical conception of predictive processing in terms of self-supervised abductive processing.

Concluding Remarks: Toward Self-Evidencing Organisms

This chapter began with simple motifs for things and for self-evidencing, which will be central to the rest of the book. Things and self-evidencing were then connected through Karl Friston's free energy principle. To exist is to minimize surprise (relative to a model); to minimize surprise in a self-supervised manner is to minimize free energy (in approximate, variational inference); and to minimize free energy is to maximize model evidence, which is what self-evidencing captures.

I propose self-evidencing as a useful and distinctive notion for navigating philosophically how the free energy principle aims to account for perception, attention, action, and many other topics. I see self-evidencing as preceding the broad notion of predictive processing, particularly the focus on predictive coding in many predictive processing treatments. Self-evidencing is a more apt notion because it allows several different ways that model evidence could be maximized, apart from through predictive coding. This will provide a more realistic method for describing and explaining what diverse existing agents actually do.

Self-evidencing also provides the conceptual link, via free energy minimization, to existence, which predictive processing did not encompass directly. Importantly, self-evidencing creates a link to existence while respecting the imperative that the agent in question must be conceivable as self-supervised—able to operate without external supervision. That is, this bootstrapping property of self-evidencing ensures that we can truly take the perspective of the organism when describing how it perceives, attends, decides, acts, and so on.

It is philosophically appealing to operate with self-evidencing as a method for inquiry. This is because self-evidencing is formulated within epistemology and the philosophy of science, bringing to the fore a host of tools for addressing debates about epistemology and rationality. These tools will be described and deployed in the following chapters, where we will see that they line up with the free energy principle. Self-evidencing is therefore nicely placed as a method for philosophical inquiry about self-supervised agents. On the one hand, self-evidencing precisely captures the technical notion of accumulation of model evidence at the core of the free energy principle (and indeed, self-evidencing is used in this sense in the technical literature). On the other hand, self-evidencing opens up a wide range of substantial philosophical debates.

This chapter has focused on the claim that self-evidencing provides the natural understanding of the existence of self-supervised agents under the free energy principle. A key task for the next chapter, then, is to substantiate the claims about the philosophical potential of self-evidencing. I have noted that self-evidencing organisms should be described in terms of the causal dependencies between external states, sensory boundary states, the internal states, and active boundary states back into the external states. With self-evidencing, I have variously labeled some of these states as a "model" that "explains" "observations," which become "evidence" for the model. I also have implied, but not yet explained, that self-evidencing agents can plan, make decisions, and act to maintain themselves in their characteristic states. These labels and implications will be the topic of the next chapter, which also will dive deeper into the implications of self-evidencing for normative matters around self-supervision.

3 Self-Evidencing for Active Organisms

In chapter 2, we arrived at Karl Friston's free energy principle, which says that an organism that exists must be describable as minimizing its free energy. We then tied free energy minimization and the idea of increasing model evidence back to the basic notion of self-evidencing that was presented at the beginning of that chapter. This chapter will expand on the notion of self-evidencing for active organisms. It will demonstrate how philosophically substantial properties emerge from the simple notion of self-evidencing, around themes of explanation, evidencing, agency, and normativity. Importantly, we will see how the overtly epistemic notion of self-evidencing can be brought to bear on planning, decision-making, and action, linking those processes to the existence of agents.

I begin this chapter by drawing out aspects of what it means to self-evidence, appealing to findings in the philosophy of science. I ask first where the models doing the explaining come from, and then I consider in what sense models of self-evidencing organisms can be said to "explain" observations, aligning self-evidencing via inference to the best explanation. I argue that this philosophy of science perspective perfectly matches the implications of the free energy principle. I then put those elements together to articulate how self-evidencing can help avoid a dichotomy between descriptive and normative accounts of behavior, which is a main advantage of this account. Self-evidencing is then equipped with a natural forward-looking aspect, such that the agent can selectively sample expected observations; this will underpin action, and here we bring in the notion of active inference from the free energy literature.

Putting Self-Evidencing to Work

When placed in the context of the free energy principle's account of the existence of a thing, the "self" in "self-evidencing" acquires a pleasing dual meaning. First, "self" captures the distinctive circular epistemic move, where the observations that an explanation can explain become evidence for itself. Second, "self" alludes to the analysis of existence, where the explanation, or model, that does the explaining can be used to designate the existing thing itself, so its increasing model evidence becomes evidence for the existence of the thing. Keeping these two first notions in mind, we can begin to put self-evidencing into a broader context. (These are valid notions of "self," but it is worth noting here at the outset that they do not directly imply that self-evidencing organisms are self-aware agents, nor that they have selves in a more substantial, psychological sense. In chapter 7, we will discuss a third notion of "self"—namely, the human sense of self, and whether a real self can be said to underlie the sense of self; there, we will see that the self in this psychological sense is best conceived as a self-model, which is a part of the overall model.)

As I mentioned in chapter 2, "self-evidencing" is a term borrowed from the philosophy of science, coined by Carl Hempel and discussed influentially by Peter Lipton. A prominent illustration of self-evidencing, devised by Lipton, is when you observe footprints outside your window and, to explain this novel observation, form the hypothesis that a burglar has been there. The burglar hypothesis is self-evidencing because the footprint, the occurrence of which the hypothesis explains, now becomes the evidence for the hypothesis. When asked "Why do you believe the burglar hypothesis?" you can answer, "Because I have evidence for it," and to the question "Which evidence?" you can respond, "The evidence I designed the hypothesis to explain." This may sound epistemically misguided and circular, but it is a benign circle often used in science and everyday reasoning. That is, we can rationally form beliefs on the basis of self-evidencing.

There are some dangers with this tight circle of explanation and evidencing. Intuitively, the self-evidencing circle will be valid, but of poor quality, if the chosen explanation is too far-fetched, such as "The footprint-like patterns are made by mechanized alien mini-ants"; somehow this feels like a poor explanation that should not be able to yield quality evidence for itself. We will explore the topic of what makes some explanations better than others later in this chapter, and it will loom large in later chapters too. A further danger occurs if someone doubts the occurrence of the evidence—for example, by saying, "The footprints might be caused by birds accidentally trampling around in that kind of pattern." In that case, you cannot reasonably answer by saying "No, because I have evidence for my belief that a burglar has been around" because then the circle would turn vicious. To be able to counter doubts about the occurrence of the evidence, further evidence that is independent of the burglar hypothesis must be found (such as "The birds have long since migrated"). At the limit case, where no independent evidence can be found, one must throw in the towel and acquiesce in the doubt (this skeptical theme will recur in chapter 5). If those dangers can be avoided, we should allow that self-evidencing is a rational process of belief updating.

Here is a further example, elaborated to begin to connect self-evidencing with our previous discussions. A fairly standard city-dweller hears a siren on the street, and to explain this, they assume as their model that an ambulance is nearing, which explains the auditory observation because the model can be used to generate expected observations about sirens (i.e., it is a generative model), which fit with the observed siren input. The siren input in turn becomes evidence for the belief that an ambulance is nearing. Further evidence may then be selectively sampled as the person acts by turning their head and acquiring visual observations to bolster the model, since that offers new evidence that is also explained by the ambulance model. For example, they might spot the flashing lights of the ambulance and may vary a number of parameters of the model until they have pinned down the speed of the ambulance as it gets nearer ("If it is going fast, I should see the lights over there right about now").

Through these kinds of processes, the fit between observations expected under the assumed model and the incoming evidence improves, and the model evidence increases. We saw that the way that parameters are updated in such a variational inference will ensure that the parameterized beliefs of the assumed model (about ambulances, lights, speed, and other factors) approximate the posterior beliefs that an ideal Bayesian reasoner would have for the model, given those observations (recall here the sour milk example from the previous chapter; see figure 2.3). The person's updated beliefs now have a good fit with the incoming observations, and they can then assess the surprise of the input ("This is the tenth ambulance today—something bad must have happened"). The person can use these updated beliefs to help plan, in accordance with their preferences, how to move out of the way and seek safety so they do not get run over by the ambulance or otherwise damaged by any unfolding events.

The examples of self-evidencing so far are given in terms of overt, intellectual reasoning, as in the initial examples of the milk going bad and the burglar, or habitual information-seeking and preference satisfaction, as in the ambulance case. However, nothing hangs on the particulars of the examples, as any case where a model is brought about to fit with observations in this benignly circular way will be a matter of self-evidencing. Processes that we may not want to describe as conscious or intellectual, or as overt reasoning, may also count as self-evidencing. This allows all organisms, if they exist, to be self-evidencing. This wide scope of self-evidencing is not surprising, given that we have aligned the existence of organisms with the epistemic properties of variational inference under the free energy principle.

Conversely, the argument is that, strictly speaking, a process is exclusively self-evidencing only if it is an expression of variational inference. Self-evidencing requires that the explaining of observations and the increase in model evidence happen only within the tight model-observation circle (as in figure 2.2), and without appealing to external sources of evidence. For example, we do not have self-evidencing if doubts about the occurrence of the evidence are resolved by appealing to external sources of evidence rather than just more self-evidencing. This internal, bootstrapping perspective makes self-evidencing a distinctive epistemic concept, requiring the assurances of self-supervision promised by variational inference. Self-evidencing then relies on an assumed model, just as we saw for variational inference.

Since we are taking the internal perspective of the self-supervised organism, which assumes a model to explain observations, two questions arise: Where do models come from? What sense of "explain" is in play? The next section will address the first question, and the section after that will appeal to the notion of inference to the best explanation to address the second.

Where Do Models Come From?

It would not be plausible, in either science or everyday life, to assert that models and parameter ranges are assumed afresh every time a new observation needs explaining. Perhaps there are occasional instances of such novel model construction, as in inspired creativity, or perhaps in some kinds of delusional reasoning. But in the everyday examples of self-evidencing that we have used in this discussion, the person in question would have been appealing to the broad fount of prior beliefs of their model of the world (e.g., about the sounds and speeds of emergency vehicles in urban environments, as in the ambulance example). Strictly speaking, these examples are not examples of self-evidencing when considered in and of themselves. But when a person's set of explanations of all their observations is considered in its totality, the process acquires the benign circularity characteristic of self-evidencing, where no external resolution of uncertainty is available and explained observations become the only evidence for the model. This corresponds to the perspective of the self-supervised organism. The question, therefore, becomes: Where does the entire model of the self-supervised organism come from?

The obvious answer for biological organisms like humans is that assumed models of characteristic states are acquired through evolution, emerged in development, learned from conspecifics, and shaped via learning over the organism's lifetime. The internal states of an organism would be organized in some manner through genetic expression in the context of a given environment. It is, however, unlikely that the entirety of an organism's model is completely specified since this would preclude learning about changing causes and states, which interact and conspire to cause the varied time series of observations that we receive at our sensory boundary. In principle, the states given in evolution could include some observations caused by relatively distal causes and might actually do so in some organisms. This could include input from conspecifics, such as "safe from predators" calls in some species of animals. But although it can be expedient to be able to process parts of one's world in advance, this imparts an inflexibility that is not always consistent with the state transitions in the world. For example, an organism with a predetermined understanding of a "safe from predators" call might be vulnerable to a state transition wherein a predator evolves the capacity to mimic this signal.

It may be more common that some more immediate and proximal part of the model was given to us, from which we could begin constructing and updating the wider model of distal causes and state transitions among them, which explains the observations that we have. A good candidate for a proximal part of the model that could be given to us in evolution is the observations of our *homeostatic* states such as blood pressure, glucose levels, and body temperature. These are the observational states that have limited, characteristic ranges that underwrite living, which are proximally caused by hidden states in our bodies. It makes sense to treat these as given or inherited in evolution, such that the organism can anchor the inference of other parts of the sensory input in these observations. Then the organism could endeavor to build a model of distal causes that interact with those ranges of homeostatic states. For example, they can learn which (nutrition-poor) distal states are associated with low glucose levels and act to avoid those states in favor of nutrition-rich states.

The idea of using a given model of some homeostatic, proximal states, which can be used to anchor the modeling of distal states, speaks to the distinction between modeling proximal states of the sensory boundary to the world versus more distal causes of the sensory input (as depicted in figure 2.7). If the organism is allowed this initial model, as well as the freedom to form hypotheses about how distal causes contribute to generating observations, then the organism could engage in variational inference for its entire sensorium. For example, if observations are surprising in the sense of being outside homeostatic ranges for body temperature, then the organism might initially move around randomly, either finding warmth or risking death; they could then begin to form hypotheses about how frigid observations are caused, such as snowstorms, and they could then move around less randomly, seeking shelter from the weather, and potentially at some stage, model more deeply hidden models of seasonal patterns and the presence of materials that could be used to build shelters and sew clothes. Slowly, a world-model would appear for such an organism, orchestrated around its characteristic homeostatic states but capturing the wider, more deeply hidden, and interacting latent causes in the world. This would be a self-evidencing process in which the model is updated and model evidence increases as new observations are encountered in both the interoceptive domain of homeostasis and the exteroceptive domain of distal causes.

We can now pull these things together. The generative model is a probabilistic specification of the joint occurrences of (latent or unobservable) causes and (observable, sensory) consequences, which can be expressed in terms of a likelihood (the probability of the observation, given the cause) and a prior (the probability of the cause before receiving the observation). The latent causes can be relatively proximal or distal. The preceding point is that, before learning a generative model, there are no latent states that can support prior beliefs. This means that it is possible to inherit prior beliefs over observations, but not their causes. The organism can come into the world with prior beliefs about receiving this or that type of sensation (which might well be homeostatic observations related to proximal, bodily latent states). But the organism cannot inherit beliefs about what caused those sensations; to arrive at those posterior beliefs, it must always invert the model, working from observations to causes. Once it starts this process, anchored in inherited priors over observations, it can begin to learn about the priors (i.e., empirical priors) over the causes under a hierarchical generative model.

Now we get to a crucial observation about the notion that models can be given to the organism in the ways that we have discussed, such as through evolution. The idea is that such pregiven parts of models alleviate the need to impossibly build explanations from scratch for every sensory observation. The question that I want to focus on here is how this part of the model is given to the organism. Models are described in terms of probabilistic, Bayesian beliefs about characteristic states (encoded in the causal dependencies of the organism's internal states), which are what the model evidence acquired via self-evidencing is accumulated for. Therefore, if evolution constrains the model, it does so by endowing the organism with certain beliefs. For example, evolution might structure the causal dependencies among internal states of the organism, such that it has a very strong belief that its observations of homeostatic ranges are thus-and-so. This means that the organism in effect will be nearly unable to believe that it could ever observe other homeostatic states. Hence, if an observation is encountered of a homeostatic state outside the expected range, then the organism will not revise its beliefs to accommodate this new observation. Instead, it will actively solicit new observations more in range with its expected homeostatic states (such as drinking a glass of water when thirsty or putting on a coat to stay warm). Crucially, these given beliefs of the model are of a piece with the less strongly held and more revisable beliefs about latent states that the organism are not given but may form themselves—it is all just about beliefs. The only difference is that the beliefs given in evolution are highly precise and therefore facilitate and structure the formation of other, more transient beliefs. (From now on, I will begin to call strong beliefs "precise" beliefs to indicate that strength of belief describes the variance of probability distributions, where *precision* mathematically is the inverse of variability; thus, a precise belief describes a very narrow and pointed probability distribution with little variance; precision is a key concept to self-evidencing and will be the focus of chapter 4.)

The reason why it is crucial to notice that even the given parts of the model are beliefs is that then we can stay true to the imperative of describing the organism as self-supervised (and only engaging in variational inference). Due to the high precision of the organism's given homeostatic beliefs, it is unlikely that the organism will ever vary much in its estimation of these parameters, but nevertheless these beliefs are still in the self-evidencing game. The free energy principle smoothly moves from notions of existence to epistemic notions of surprise reduction and model evidence maximization, which means that the organism's existence must be describable exclusively in terms of beliefs and evidence for beliefs. Self-evidencing can cash out this commitment, even for parts of the model given to the organism through evolution. Since beliefs are what is given, it remains the case that the organism can make do, via variational inference, with access only to its own internal states and observations.

Whether what evolution endows us with should be conceived as beliefs might seem a somewhat arcane point. But it matters for the first-principles aspirations of the free energy principle and self-evidencing. If something other than beliefs is somehow given to the organism, then we would need an additional account of what it is. Then the free energy principle would be only part of the story, which would imply that its analysis of existence is flawed. This would be bad, not only because the analysis and the underlying mathematics are convincing in themselves, but also because we would again be in the dark about how organisms can distinguish their characteristic states under their own supervision. The trick in this proposed solution is that the causal work is done in the processes underpinning variational inference, and the case for the justificatory side takes care of itself as the inferred state approximates exact Bayesian inference via the Kullbach-Leibler divergence introduced in equation (2.1) in the previous chapter. This duality is evident in figure 2.9, where what the organism can be described as actually doing approximates what it should rationally be doing.

What we should take from this is the pleasing duality of self-evidencing, simultaneously allowing a descriptive perspective (i.e., this is what the organism is actually doing, as the organism varies its parameters in variational inference) and a normative perspective (i.e., this is what the organism should be doing if the organism is rational, operating in the realm of reason). This is what allows us to understand where models come from, how they are shaped, and how they remain within the realm of self-evidencing. As we will discuss later in this chapter, since it can be difficult for any account to accommodate both the descriptive and the normative perspectives, this duality is a unique and welcome aspect of self-evidencing.[1]

Self-Evidencing and Inference to the Best Explanation

The previous section discussed where models come from, suggesting that some parts of models may be given as precise beliefs about expected observations, through evolution or development, and other parts more flexibly learned as the organism traverses its environment. I pointed out that all parts of the model, even the given ones, should be conceived in terms of probabilistic beliefs, apt for the self-evidencing process.

This section considers the next natural question: In self-evidencing, the model is said to explain the observations, but what does this mean? There is much discussion in the realms of philosophy of science and psychology about the concept of explanation, and it is not obvious which notion would apply to the specific case of self-evidencing, especially given the rudimentary link between the existence of any organism and its self-evidencing. I will propose that a rich conception of explanation arises naturally from self-evidencing, which aligns neatly with the free energy principle.

Self-evidencing comes in degrees. The degree of model evidence self-evidencing gives rise to will depend on the model's quality; that is, on how much or how little of the surprising observations are left unaccounted for by the model. This implies that to self-evidence is to adopt models that as best as possible make observations unsurprising. Philosophy of science has a long-standing (and much-discussed) notion that making observations unsurprising—predicting them—is a key component of explanation. If we assume that notion as a starting point, then we get the idea that to self-evidence is to explain as best as possible given the resources at hand. In self-evidencing, the best explanation will provide the best evidence for the model, allowing belief in—inference to—that model. Putting this together, we should be able to conceive self-evidencing organisms as engaging in what the philosophy of science knows as *inference to the best explanation*. Accordingly, in the following discussion, I will draw out the characteristics of inference to the best explanation as they emerge when guided by model evidence. It turns out these characteristics align with reorganizations of the free energy principle, supporting the claim that self-evidencing organisms engage in inference to the best explanation. This argument then shows how self-evidencing is about explanation, it clarifies what is distinctive about self-evidencing processes, and it sets out the challenges for self-supervised organisms.[2]

A good model, suitable for the real, noisy, and changeable processes in the world, is typically honed and maintained over time to achieve a reasonable fit with all sorts of observations that the organism may encounter within the environment. The *best-makers* of inference to the best explanation at play in a self-evidencing process, I suggest, are analogues of typical explanatory and scientific virtues around predictive power, comprehensiveness, simplicity, generality, and integration with other beliefs.

One best-maker relates, as remarked earlier, to establishing a good fit between explanation and data. The model delivers good explanations only if it can accommodate—explain away or make unsurprising—much of the observations encountered. Most accounts of explanation agree that making observations predictable from the model is an important condition for being the best explanation. It would be odd to say that an explanation is best if another explanation better predicts the observations in question. And having a good fit, of course, captures the idea of maximizing model evidence, which is central to self-evidencing.

However, establishing a good fit of model to observations is a balancing act. A good fit is not maximally tight because that risks overfitting the model to noise that is specific to one environment; which can make the model generalize poorly to new environments. To avoid overfitting, the model needs to be simplified, sidelining some parts of the observations so the model can be generalized. But overly simple models should also be avoided because they will tend to miss important patterns in the observations (i.e., they will be inaccurate). For an explanation to be best, then, it must obtain a balance between *complexity* and *accuracy*. Notice that this implies that the traditional explanatory virtues of predictive power, comprehensiveness, simplicity, and generality are intricately related. For example, simplicity and complexity need to be considered in the light of predictive power.

Since fitting observations to models is what self-evidencing achieves, therefore, we should expect a self-evidencing organism to maximize its model evidence in ways that balance complexity and accuracy. That is, we should expect self-evidencing to align with these best-makers in inference to the best explanation. Maintaining this balance would be helped by a capacity for adjudicating the trustworthiness (or noisiness) and relevance of observations in order to determine whether they are worth trying to explain away, and it would be helped by forming beliefs about which kinds of contexts a given model should be able to generalize to. Some organisms, such as humans, may have developed a capacity to form explicit beliefs about the explanatory best-makers to better guide their long-term self-evidencing.

Another best-maker in inference to the best explanation relates to how well a given model integrates with what is already known. All else being equal, it would detract from the quality of an explanation if it stood alone, apart from and perhaps in tension with existing knowledge. Wielding this best-maker again calls for balancing, this time between kinds of *closed-mindedness* and *open-mindedness*. In overly strong integration, new observations that do not conform to prior beliefs are discounted entirely, making the model too closed-minded. In overly weak integration, all sorts of new observations are allowed to shape the model's beliefs, making the model too open-minded. Another way to put this is in terms of the *learning rate*. The rate with which we update our beliefs from new observations should be sensitive to what we already know, and also to the quality of the evidence; if the rate of learning is good, then we do not throw out well-evidenced knowledge too quickly and on the basis of poor evidence, and neither do we ignore important new observations.

Maintaining the balance between closed-mindedness and open-mindedness (and finessing the learning rate) would be helped by a capacity, which humans may have, to form explicit beliefs about when to rely on what priors, and beliefs about what contexts may indirectly affect the occurrence and trustworthiness of observations. To harness different kinds of prior beliefs about the case at hand, as well as contextual beliefs and beliefs about trustworthiness and robustness of prior beliefs, it is useful to have a structured *hierarchical model* of causes and states in the world, which can capture how observations may relate to prior learning and be affected by contextual factors.

Notice that integration, understood as balancing closed-mindedness and open-mindedness, conceptually lines up with the best-makers for complexity and accuracy. Excessive open-mindedness would tend to show up in models that are too complex, as new beliefs are formed in response to all sorts of (possibly incidental) observations. Excessive closed-mindedness would tend to show up in models that are too simple, where important observations stemming from freshly emerging patterns of events are ignored. This suggests that one could indirectly keep track of one's open-mindedness by ensuring that one's beliefs are not becoming too complex, for example. This alignment among the best-makers implies that something may be able to unify them. As we will see later in this chapter, the free energy principle can provide this unification.

A further best-maker is more forward-looking; namely, that best explanations should be able to generate predictions for fruitful future selective sampling of observation—best explanations should be fecund. This is akin to how a good explanation should be able to be tested in a planned and well-controlled experiment in a lab, resulting in strong information gains. It would detract from the bestness of an explanation if it were predictively inert or if a competing explanation produced more fruitful predictions. The ability to plan and selectively sample informative observations is interesting because it builds planning, decision-making, and action into inference to the best explanation. We anticipated this for self-evidencing organisms in the previous chapter (see figure 2.3).

For predictive fruitfulness, another balancing act is needed. An explanation might have great promise for information gain, which is the preferred outcome, but it might have low feasibility because the actions that would deliver those outcomes are highly shaky. For example, a scientist might know what observation would strongly confirm their explanation, but the experiment to deliver this observation requires several complicated steps and challenging measurements, introducing much uncertainty about what will actually be observed. Conversely, an explanation might be easy to translate into actions leading to unambiguous outcomes, but those outcomes will not deliver much information gain. For example, a simple experiment might be using a well-tested, reliable measurement instrument, but the observations that one can hope for are fairly boring and incremental. In decision-theoretical terms, these two elements relate to *ambiguity* and *risk*. Low risk for an action is when it has a high probability of delivering an expected outcome, such as high information gain in an experiment. Low ambiguity is when the action maps onto outcomes neatly rather than messily, whether those outcomes are expected or not. When comparing the predictive fruitfulness for two competing explanations, therefore, one should assess how well each one balances risk and ambiguity. For a self-evidencing organism, we should expect some attempt to balance the risk and ambiguity of its possible actions.

We noticed previously that complexity and accuracy line up conceptually with open-mindedness and closed-mindedness, and risk and ambiguity line up with those pairs as well. If an explanation does not constrain predictions much, then it leads to highly risky action, meaning that one is open to various outcomes, and therefore complex revision of one's model. If an explanation can generate only murky predictions, then updates in the light of whatever observation comes in will likely lead to inaccuracy. This alignment among all the best-maker balances further underscores the prospects for a unifying underlying principle.

There is a key difference for this action-oriented best-maker, however—it is concerned with future, as yet unknown observations, whereas the other best-makers were concerned with current, known observations. So even if the free energy principle can unify the best-makers, it must somehow be able to distinguish between the current and future perspectives of the best-makers. We will see that the principle can indeed do this.

I will add a further forward-looking aspect, which is not directly a best-maker but rather has an affinity with a persistent concern about inference to the best explanation. Say that we have two alternative explanations for a particular observation, and the first explanation is best. Then inference to the best explanation says that we should update our beliefs to include that explanation. The persistent concern is that both explanations might be poor, even if one scores slightly higher on one best-maker balance (e.g., it is a little less complex). It is not attractive to rely on a method of belief-updating that is entirely blind to the quality of the explanations that happen to be available. I want to suggest that the best-makers listed here are informative about the quality of the explanations, this is brought out clearly by self-evidencing, and the remedy is baked into the free energy principle. Consider the case where an explanation generally fares poorly on best-makers, such that it generates risky and ambiguous predictions and generalizes and integrates poorly (i.e., it is overly open-minded/closed-minded or complex/simple). Then we could exclude the explanation from the set of explanations considered for inference to the best explanation. In other words, if an explanation is not expected to yield much model evidence, then we should not risk basing our beliefs on it. A self-evidencing organism can keep track of bestness because the quality of the explanation translates into increased model evidence. Some organisms might even be able to go offline and subjunctively assess what the model evidence would be if it were to test a certain prediction. If model evidence begins to deteriorate, then that is an indication that some of the beliefs under the model should be eliminated from consideration in inference to the best explanation. These considerations somewhat mitigate the problem of poor explanations, although of course, we can never be certain that no better set of explanations is available for any set that we are currently considering. For this strategy to work, a presumption is that organisms can keep track of, and form expectations about, the magnitude and rate of model evidence in different contexts.

Intuitively, if an explanation fares poorly on the best-makers, then it cannot reliably be used to harvest fruitful evidence. Then the explanation could be dismissed entirely, or we could switch to a less directed strategy and begin to explore the domain in question, hoping to find nuggets of information that could bring the explanation back in the fold of good-enough explanations. That is, an ability to manage the *explore-exploit* balance is useful for a scientist interested in information gain, or for an organism engaged in maximizing its model evidence by self-evidencing. This balance is familiar from decision sciences and other fields of behavioral science, where agents may exploit a resource for its utility so much that it depletes and they must explore in order to expand their map of where new resources can be found. Here, we are simply considering model evidence as the "utility" from the resource in question such that the organism should go exploring when it depletes. As we will see later in this chapter, the free energy principle captures the explore-exploit distinction. We will also see how rewards or pragmatic value itself can be conceived as model evidence.

This section began with the question how self-evidencing can be said to explain observations. The answer is that self-evidencing processes should be expected to look like inference to the best explanation, and therefore to explain observations in the sense that best explanations explain. Self-evidencing explains observations based on an accurate (often structured and hierarchical) model that generalizes and integrates well and generates fruitful predictions for future observations. This is captured in the left-to-right direction of figure 3.1.

Free Energy, Abduction, and Active Inference

As I will now explain, the best-makers of inference to the best explanation, derived from consideration of self-evidencing, neatly map to properties of the free energy principle, as the principle is mathematically organized in various ways. This is captured in the right-to-left direction of figure 3.1

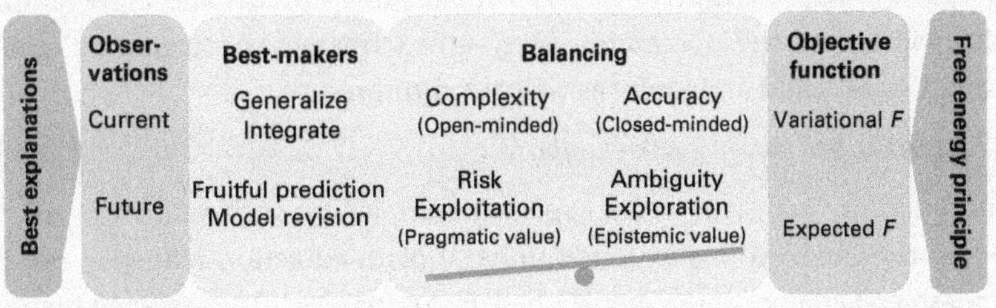

Figure 3.1
Unpacking self-evidencing processes. A self-evidencing organism accumulates model evidence (i.e., exists) through processes that resemble the best-makers in inference to the best explanation and maintain the model explaining current and future evidence by balancing best-maker characteristics (left to right). This matches the formal aspects of the free energy principle, where the objective functions of variational and expected free energy, *F*, guide the same balancing processes (right to left). This picture also makes room for model revision, exploration, and pragmatic value.

As we saw in chapter 2, the free energy is the divergence between the assumed model and what exact inference says, plus the surprise; that is:

Free energy = perceptual divergence + surprise (3.1)

These terms can be reorganized so the free energy equals complexity minus accuracy:

Free energy = model complexity − model accuracy (3.2)

In equation (3.2), complexity is the Kullbach-Leibler divergence between the approximate prior and approximate posterior probability, such that complexity is high if the beliefs have shifted a lot given the new observations. Minimizing free energy by minimizing complexity puts pressure on the organism to shift its beliefs as little as possible. Accuracy is the likelihood of the observations that the organism encounters, which suffers if observations are surprising, given the assumed model. This puts pressure on the organism to shift its beliefs to account for observations. Together, this requires a capacity for finding the balance, just as we set out previously, where enough observations are explained while retaining enough of the model. This is also what maps to the balancing between being open-minded and closed-minded. What guides this balancing act is the fluctuations in free energy (the fit between observations generated by the model and actual observations) as parameters in the model are being varied.

Shifting to the future fit between the assumed model and observations that will occur—the *expected free energy*—the terms can be reorganized again, capturing risk and ambiguity as well as fruitfulness:

$$\textit{Expected free energy} = \textit{risk} + \textit{ambiguity} \tag{3.3}$$

In equation (3.3), risk is the expected complexity, such that high risk is when the observations expected under a planned action may give rise to widespread change of the model's beliefs. Ambiguity is the expected inaccuracy, such that high ambiguity is when future observations given a planned action may not be well accounted for by the model. Organisms who minimize their expected free energy will be driven to minimize risk and ambiguity in tandem, capturing the predictiveness best-maker as before.

By reorganizing the expected free energy again, *exploration* and *exploitation* can be captured. This is often expressed in terms of planning to act for pragmatic value (exploitation to maximize utility) or for epistemic value (exploration):

$$-\textit{Expected free energy} = \textit{pragmatic value} + \textit{epistemic value} \tag{3.4}$$

In equation (3.4), pragmatic value is determined by the divergence between the organism's expected states and the states that it can expect, given a planned action; epistemic value is the information gain expected, given the planned action (and given that we want to maximize value, we now have the negative expected free energy). This means that, guided by the expected free energy, if the expected pragmatic value is low, then the organism will look for epistemic value, and vice versa. Intuitively, if the model is pretty good (i.e., the perceptual divergence is low and accuracy is high), then it is a good thing to exploit because there can be confidence in reducing uncertainty efficiently. But if the model is uncertain, it is a good idea to go exploring instead, to reestablish an accurate model. The epistemic value splits into two terms if there can be uncertainty about the model's own parameters as well as about the states, where organisms will be maximizing the epistemic value by planning actions leading to novel observations (such as in highly informative experiments in science). The pragmatic value in this version of the free energy principle brings classic utility (or reward) into the picture as part of the self-evidencing processes. In the analogy with scientific practices, this captures the intimate relation between explanation, exploration, and discovery, as well as the pragmatic point of doing science, which is to help us get close to our preferred states (note, as we will discuss in chapters 6 and 8, that knowing the pragmatic value of an action is not straightforward, as organisms often have to infer their own preferences).[3]

From expected free energy and equations (3.3) and (3.4), we have *active inference* to the best policy for action, where the best policy minimizes risk and ambiguity and maximizes pragmatic and epistemic value. Thus, planning and decision-making are conceived analogously to perception and learning: to perceive and learn, we ask which state of affairs best explains current observations; to plan and make decisions, we ask which state of affairs best explains future (expected) observations, where those states of affairs include our own actions. One can think of variational free energy as a special case of expected free energy that pertains to the present (i.e., when observations are known and do not depend on future actions). Next, I connect active inference to the notion of allostasis, and active inference will then be central to all the subsequent chapters of the book.

Now we can put together the informal considerations about inference to the best explanation for a self-evidencing organism and the formal reorganizations of the free energy principle (as in figure 3.1). If we begin by considering what processes would be in place for a self-evidencing organism that accumulates model evidence, then we get to the traditional best-makers of inference to the best explanation. We also notice that tending to these best-makers requires maintaining various modeling balances with regard to actual and expected observations, which line up under notions of complexity and accuracy. If we instead begin from consideration of the free energy principle and its reorganizations, we get to those balances directly, and we can notice that maintaining those balances optimizes the best-makers of inference to the best explanation. This all happens guided by the model evidence, which is accessible to the organism via the objective functions for variational and expected free energy.

We now have a distinctive analysis of self-evidencing, where the circular movement between explanation of observations and accumulation of model evidence anchors self-evidencing in both well-known epistemic reasoning patterns and in the free energy principle. The next sections unpack some of the underlying properties of self-evidencing and further connect the expected free energy to allostasis and active inference.

Self-Supervision as Model-Constrained Trial and Error

In a self-supervised organism, there is no external supervisor or rule book independent of the self-evidencing process, which an organism can consult for how to wield and balance the best-makers in inference to the best explanation. The organism must rely on what I conceive as a kind of trial and error process as it struggles to get these best-maker balances right. The only guidance, or objective function, is how the organism manages the long-term average fit of the generative model and the observations—how it minimizes free energy. This is the quantity that the organism has access to, and the claim here is that the organism expresses the best-makers as it "trials" various beliefs while monitoring the effect on this known quantity—minimizing the error from the trial. For example, if the model has become very complex, accounting for every observation, then the long-term average fit between predictions and observations will begin to suffer (and the free energy will increase), as the now-overfitted model will fail to generalize to new observations. The organism may then need to cast around for another approach, perhaps simplifying the model or actively seeking more predictable environments.

Notice that the element of "trial and error" here is not in the sense of blind (i.e., model-free) guessing, hoping for the best as errors happen. It is rather a kind of model-constrained trial and error, where the trials—understood as varying the values of the parameters of the model one by one, while holding the rest fixed—are constrained by past learning and beliefs handed down in evolution and captured in the internal model. The organism just does not know in advance exactly how to shift its parameter values or which parameter to focus on. So it tries various settings until it finds one that seems to maximize the fit between the model and the observations. In the end, whatever trajectory the organism takes will reflect the unfolding dynamics of its internal states as they vary and are buffeted by the observations encountered on the way (i.e., its probabilistic beliefs as reflected in the parameter values of the model at any given time). Distinct organisms of the same phenotype, with the same model structure, may well end up having different trajectories since they may exist in different contexts in the same eco-niche, have different parameter values from past learning, and happen to vary their parameters in a different order. Thereby, these organisms may encounter a different sequence of observations and update their beliefs differently along the way. It is these cascading trajectories that are described by the free energy principle, where each trajectory is determined by the states of the boundary and the dynamics of the internal states of the organism. We should thus expect self-evidencing organisms to display significant diversity.[4]

In other words, the inference to the best explanation must emerge from the organism's own organization, understood as a model-constrained trial and error process, as it stays out of uncharacteristic states as best it can. This ties self-supervision and self-evidencing to self-organization, a key concept in dynamical systems theory and theoretical biology.[5]

Notice that such self-evidencing organisms would do well to form and continuously update beliefs about the best-maker processes themselves. For example, they could have beliefs about the best balance of complexity and accuracy in particular contexts, and update these best-maker beliefs when they turn out not to be quite as good as expected (such as "My belief that my model of the bus schedule can be very simple turned out to make me wait too long, so I missed my appointment; better aim for more complex models in this domain"). Maintaining and adjusting these beliefs about the best-makers themselves are then also a trial and error process where the organism tries a certain best-maker belief, tests it against the objective function, and tries variations of the beliefs if the variational or expected free energy is not decreasing.

In general, such model-constrained trial and error processing is central to self-evidencing and will be the focus of several arguments in this book. It is a process of varying some parameters while holding others constant, and all the while checking the outcome against the objective function. Model-constrained trial and error is thus my heuristic conception of the computational variational inference processes of free energy minimization.

Descriptions, Norms, and Self-Evidencing within Bounds

If I am correct that the processes involved in self-evidencing map onto the best-makers of inference to the best explanation, then we have an argument, via the free energy principle, that inference to the best explanation captures elements of approximate inference. As described earlier, one outcome of trial and error, self-evidencing processes is that the generative model will approximate the exact Bayesian posterior. Simply by self-evidencing as best it can, without any inkling about what Bayes' rule requires, the organism's model will get closer and closer to the model of an ideal Bayesian reasoner. This is the beauty of the mathematics underlying the free energy principle.

We can now unpack some of the earlier remarks about how self-evidencing allows a synthesis of descriptive and normative accounts of existing organisms. On one side, we have a process that seems feasible and tractable for biological organisms: self-evidencing. The aspiration is that self-evidencing provides a good descriptive account of what organisms actually do; in a model-constrained, trial and error process of self-supervised inference to the best explanation, organisms form models on the basis of prior experience and given beliefs and minimize uncertainty about sensory input with respect to those models. On the other side, this trial and error process, unbeknownst to the organism, takes them closer and closer to what an ideal reasoner would infer. This duality is promising because typical accounts of such processes are either descriptive but without tractable normative direction or normative but without descriptive feasibility. For example, as discussed previously, exact Bayesian inference is intractable for biological organisms, whereas inference to the best explanation might be descriptively accurate but on its own lacks normative import; for more forward-looking decision-making accounts, Bayesian decision theory is normative but unrealistic for normal reasoning agents, whereas a heuristics and biases account and notions of bounded rationality can be descriptively adequate but has no formal link to normative reasoning. I know of only a few approaches other than self-evidencing that promise to address the tension between the descriptive with the normative.[6]

The promise to reconcile the normative with the descriptive matters because we should aim for an account of our perceptual and cognitive processes that we can feasibly execute as the limited creatures we are while simultaneously aiming at truth and reason. Self-evidencing provides this unified account because it bases perception and cognition in a conception of the everyday existence of organisms rather than on imbuing the organisms with aims, ideals, or norms that they must know and intend to comply with. We have operated with two objective functions, the variational free energy and expected free energy, and even the latter is not asserting independent goal-states because the preferences are given as just additional beliefs in the assumed model, equally subject to self-evidencing. Action via expected free energy minimization then essentially comes into play when the organism is equipped with a generic prior belief that action sequences are more plausible when they minimize expected free energy. This amounts to a generic, universal optimism, to the effect that actions will happen that fulfill the organism's preferences; importantly, this is just another belief, characteristic of the organism and subject to self-evidencing.[7]

As I see it, the ability both to describe what we actually do and to let us, as it were, blindly get closer to truth and reason fundamentally changes our conception of ourselves. Instead of characterizing humans and other organisms as "I am acting this way because I am aiming for these rewards and this knowledge and these truths, and this allows me to exist the way I do," self-evidencing casts organisms simply as "I exist, and this implies that I perceive and act in these ways." I regard this as an almost Copernican revolution in our self-conception because it grounds both what we *actually* do and what we *should* do in our existence, rather than the other way around. Karl Friston playfully conveyed this revolution by turning René Descartes's famous *Cogito, ergo sum* on its head; instead of "I think, therefore I am," it should be "I am, therefore I think." This sentiment will infuse many of the arguments to follow in this book.[8]

I mentioned earlier that whereas self-evidencing is benignly circular, the circle can turn vicious if doubts about the evidence are batted away by appealing to the self-evidencing explanation; for a self-evidencing organism, like a human, doubts about the evidence should be dealt with by appealing to independent evidence. For example, if the person from the earlier example doubts that the sound was really a siren of any sort, rather than just tinnitus, then they can appeal to a couple of bystanders who also heard it, together with the contextual knowledge that tinnitus is unlikely to exist in several bystanders at once (unless the event happens at a loud music festival, perhaps). If no contextual, independent evidence is available, then the doubt about the evidence cannot be resolved. This points to a limit case, where the totality of the evidence is doubted—somewhat like Descartes's evil demon case, where it cannot be ruled out that the demon deceives us, so that everything must be doubted and nothing can be trusted. For a self-evidencing organism, this kind of limit case would be equivalent to the totally doubtful question of why they have any real evidence rather than no evidence at all. That is, it would be the question of why they are a particular thing at all, which cannot be answered when appeal to evidence is precluded. To me, this makes good sense and might help explain why some philosophical and existential questions are so hard to get straight. At some level, all I can say is "Well, I believe I exist, and that's all there is to it." In chapter 5, we will discuss Cartesian skepticism in more detail.

I have proposed that the trial and error processes driving self-evidencing can be characterized by inference to the best explanation. The organism "flies blind" in a sense, as it tries a parameter setting of its assumed model as an explanation of the evidence and then learns how to update the model from the error that it receives, via the objective function that we saw earlier, when the explanation is less than best. When arriving at a model that generates little error, it may then begin to exploit it confidently in decisions that selectively sample for input of pragmatic value, such as getting food or building machinery. If it still generates significant error, then the organism may act for epistemic value, such as seeking information about a new patch in the environment that might have more food, or investigating better ways of making precision machinery. This was captured in equation (3.4) and motivated in the context of inference to the best explanation. In chapter 6, I will more fully unfold this idea for decision-making specifically, and for the forward-looking objective function of expected free energy under the label of inference to the best decision.

As I mentioned, the notions of inference to the best explanation and self-evidencing both originate in the philosophy of science. This does broadly assimilate the working of biological organisms to descriptions of what scientists do. There are similar concepts, however, that do not appeal to the science analogy, occurring in philosophical pragmatism via the notion of abduction, as well as in approaches such as dynamical systems theory, autopoietics, and enactivism, where the active part is emphasized more than the inferential part. In some respects, the idea that human perception works this way predates many contemporary ideas of scientists' practices. The Islamic scientist Ibn al-Haytham (Alhazen) treated perception as a kind of inferential judgment, Immanuel Kant's framework of concepts and intuitions got transformed into an idea of unconscious perceptual inference by Hermann von Helmholtz, and many ideas in psychology and cognitive science since have been close to the inferential framework (although they have mainly been without the guarantee of self-supervision stemming from the elegant mathematical underpinning from the free energy principle).[9]

The analogy with scientific inference is useful. Indeed, biological organisms are much like stubborn scientists, who will not give up on their most cherished theory (or model) no matter what evidence they encounter but will rather go back to the drawing board repeatedly to adjust peripheral parts of the theory, leaving the core sacrosanct. This makes sense with respect to the existential question alluded to earlier, where it is just not viable to doubt the totality of the evidence for one's existence. It also directly picks up on the best-maker theme, discussed earlier, of being able to balance complexity and accuracy—open-mindedness and closed-mindedness—where a recalcitrant scientist refuses to be so open-minded that nothing will be left of their theory. Of course, where the analogy to science runs out of steam is with the falsification of a theory and the happy retirement at old age of both theory and scientist. A biological organism ceases to exist altogether when the theory, or the highly precise belief that it exists, is given up.

Another useful picture here is a hill-climber traveling up and down a fairly narrow, winding path, charting a sometimes wobbling, zigzaggy route, making sure to stay within the bounds of the path. As the hill-climber traverses this winding path, they continually maintain their model, adding and pruning parameters and finessing expectations about varying precision of observations and about irreducible noise. The hill-climber will in many cases learn about the terrain they inhabit by inferring its hidden causes, and also learn about the even deeper hidden causes that interfere with known patterns of observations. They build up and finesse a deep, hierarchical model of the world that they inhabit, generating expectations of what they will observe and when. (We will return to the hill-climbing analogy in chapter 6.)

An existing organism charts such a hill-climbing path, stretched out over time and changing due to the dynamics of the world, but remaining within its characteristic states. The organism's states must be bounded, as discussed earlier, to signify that it is not disintegrating and spreading fragments of its internal states over all other states with increasingly equal probability—the path might be broadening sometimes, but it is never infinitely broad. Over the long-term average, the organism's states bounce about within those undulating bounds, filling the state space but in a way that is restricted to characteristic states. It is unlikely that one could sample from the organism's states at any time and find it deviating hugely from the long-term average—which mostly would signify that the organism is ceasing to exist (there are some nice middle cases to think about here, such as the metamorphosis from larva to moth).[10]

In this section, I have underlined how, now that we can see its connections to inference to the best explanation as well as to the free energy principle, self-evidencing provides an attractive synthesis of descriptive and normative accounts of the dynamic path of the behavior and cognitive processes of organisms.

In the next section, I will focus on the forward-looking, planning, and decision-making aspects that I have begun to include in the self-evidencing story here. It is time to shift more fully to full-fledged self-evidencing agents.

Allostatic Self-Evidencing and Active Inference

We have allowed that organisms might be endowed with highly precise homeostatic beliefs about blood pressure, glucose levels, and body temperature. Due to their high precision, homeostatic beliefs are practically unrevisable—it would take an inordinate amount of evidence to dislodge them. But these beliefs are still in the game of self-supervised self-evidencing. The organism can react to challenges to their homeostatic beliefs by performing an action, such as lying down if blood pressure begins to drop. That is, they are closed-minded with respect to their homeostatic belief, and they would rather selectively sample new observations conforming to their beliefs.

Precise homeostatic beliefs can constrain self-evidencing for other beliefs, as discussed previously. For example, the organism might form beliefs, with varying precision, about which other states of the environment affect glucose levels (say, where the good patches of berries are at what times of year). Such a wider and causally deeper model of the world would help the organism react better to challenges to their homeostasis. But homeostatic control of this sort is still somewhat precarious, in the sense that the organism is only reacting to observations that challenge homeostasis. If the observations deviate dramatically from the homeostatic beliefs, then it may be too late to react and select new observations; the organism may perish if it is taken too far from its core homeostatic expectations. For humans, being bitten by a venomous snake or succumbing to a tsunami might be examples. For this reason, some biological agents might begin to build in an anticipatory element to the model, allowing them to plan ahead. For example, officials in countries like Australia have ready stocks of antivenom for many snakebites, and throughout the Asia-Pacific, there are now elaborate tsunami warning systems. In both cases, anticipation is based on hard-earned lessons and ever-more-elaborate, hierarchical models of the kinds of latent causes of sensory input, such as the biochemistry of snake venom and antivenom, and the relation between marine earthquakes and massive bodies of water. In these cases, we, as the particular organisms we are, have a chance to preemptively act, helping us avoid the too-dramatic insults to our bodily integrity. This shifts us from a reactive homeostatic frame to an anticipatory, allostatic frame. *Allostasis* is ubiquitous among biological organisms and requires internal models that allow the requisite prospective modeling. For example, a hill-climber nearing the foot of a hill will increase their heart rate before taking it on, in anticipation of their increased energy needs and wanting to avoid a more dramatic crash as heart rate fails to keep up with energy demands halfway up the steep hill.[11]

Both homeostasis and allostasis require the basic idea of characteristic states, against which the surprise of new observations is adjudicated. Allostasis builds a further, more forward-looking component into the model. This enables the model to represent new states, such as tsunami early warnings, related to actions that it needs to undertake now to maintain homeostasis in the future, even if those warnings are not in themselves dangerous. Allostasis improves an organism's ability to exist with less deviation in its ongoing states, as it can preemptively rather than reactively avoid harm.

The allostatic expectation horizon can indeed be very long for some organisms, such as humans. We seem to have fully bought into allostatic control in a broad sense, with our numerous schemes for anticipating future insults to our existence (think insurance policies, investment portfolios, education, societal organizations for antivenom production and tsunami warning equipment, and others). There are costs to this, through the metabolic demands of maintaining huge internal models that try to anticipate the probability of insults to homeostasis over often very long time spans. Other organisms may exist in other ways, such as *Escherichia coli*, which more reactively swims up the nutrient gradients and simply tumbles when its expected nutrient threshold is violated but before its supply of nutrients depletes entirely. Different existences will have different sets of characteristic states, with much scope for divergence among organisms.

For allostasis to work, the organism may be looking ahead in time, not only for things like the good patches of berries when it is hungry but also for how future hunger might be avoided by not overly exploiting the patches that it is foraging. It models its own future homeostasis conditioned on the model, including its beliefs about what actions it will undertake. This calls for more elaborate planning and decision-making. Now the organism needs to model intricate causal interactions operating over several interacting timescales, which can bridge present and future (and be interpretable based on past learning). Rather than assessing just what to do to maintain itself in its characteristic states in homeostatic or reflexive responses to occurrent surprise, it needs to represent future surprises, given its own actions, so that it can act now to maintain itself in low-surprise states. For example, a walker packs snacks and water in anticipation of sugar lows and dehydration on a long bush walk.

We have already talked about self-evidencing in terms of expectation and surprise. But with this allostatic perspective, we shift to talking about self-evidencing relative to expected surprise, given a model and actions. This brings us to active inference, which will be the focus of several discussions in the following chapters. As explained earlier in this chapter, in active inference, a policy for action is inferred, based on the now forward-looking objective function of expected free energy. This self-evidencing perspective is similar to the homeostatically apt objective function of variational free energy, as the organism, when it enacts its selected policy, is still acting to provide evidence for its own model. In other words, the organism optimistically believes that it will likely act to occupy unsurprising (preferred or characteristic) states in the future. With this enactive focus on active inference, we will shift from self-evidencing organisms to self-evidencing *agents*.[12]

We pictured the agent as a hill-climber negotiating a winding path. Modeling the expected surprise—given an action—makes it easier for agents to stay on this path, and they can put in provisions now that allow them to stay the course in the future. For example, even though agents have plenty of resources in their current patch, they might anticipate that those resources will deplete, and therefore the agents spend time and effort exploring other patches before that danger arises. These organisms are now looking ahead on their path, with more or less confidence depending on the quality of their internal model, which generates expectations over different timescales. Taking the example of an agent walking the winding path literally, they might bring a flashlight in anticipation of having to finish the walk in the dark, which would enable them to stay safely on the path. Allostatic self-evidencing thus makes planning and decision-making central to the existence of biological agents. As agents look forward, they are considering which policy of action would serve best for self-evidencing—which policy would help evidence their characteristic states over a longer time frame? The active inference perspective is thus baked into the abductive picture of self-evidencing presented earlier in this chapter, using risk and ambiguity and exploration and exploitation as in figure 3.1 and equations (3.3) and (3.4), and allowing long-term, preemptive planning.

Concluding Remarks: Self-Evidencing Agents Do What They Do

Beginning from a simple conceptual analysis of existence, chapter 2 pushed forward to a notion of self-evidencing. This chapter expanded on self-evidencing, asking where models come from and in what sense models explain observations. We observed how this perspective captures self-supervision and points to a synthesis of descriptive and normative accounts of agency. The free energy principle was shown to entail cognitive processes aligning with a conceptual understanding of self-evidencing, in terms of inference to the best explanation. This picture was then extended to capture planning and decision-making through the important forward-looking notion of active inference. The next chapters will make good on the claim that this furnishes a surprisingly resourceful perspective, opening rich avenues for understanding everyday experience in perception, self, action, decision-making, and beyond.

I will end this chapter by highlighting what I find startling about the self-evidencing agent that we have explored so far. The initial assumption is the existence of the agent. If an agent exists, then it is self-evidencing. It is hard not to hear this as imbuing the agent with an aim—a *telos*—because we often conceive of biological functions in terms of teleology. In fact, however, as we unfolded self-evidencing, we never had to assume a goal for the agent. With existence as a starting point, the agent just *is*. Of course, with allostatic self-evidencing, there is a forward-looking element—a belief that existence stretches into the future—but this is just that: an optimistic belief in need of evidence. Allostasis is just a way of being rather than something that requires a goal that the agent aims at and against which success is measured. To be is to maximize the fit between the observations expected under the assumed model of latent causes, whatever beliefs the model harbors, and the actual observations.

This nonteleological starting point is a main attraction because it allows self-evidencing to portray truly, fundamentally self-supervised agents. There is no externally given *telos,* standing outside self-evidencing, for the organism to consult. To me, this fundamentally shifts our conception of agency, cognition, and mind. With the self-supervised, bootstrapping-like notion of self-evidencing, we do not have to somehow overcome a gulf between descriptions of what we *actually* do and normative accounts of what we *should* do. Once we exist, that gulf takes care of itself through approximate inference. In the widest perspective, it alleviates the need to think too much about the point of it all—the point or meaning *of* life. Instead, it allows each of us to focus on how we live—finding meaning *in* life. This theme will percolate through the following chapters, and meaning in life will be discussed in chapter 10.

Adopting the self-evidencing approach in our understanding of the mind can be challenging. The self-evidencing creature is oddly solitary, appearing almost depressingly disjointed from the world as it tries to finesse and evidence its model of the world and itself. Self-evidencing leads us to recalibrate our understanding of beliefs, norms, rationality, self and much else besides. I hope that the arguments in the following chapters will convey that, despite these appearances, self-evidencing is in fact both exciting and comforting. It better captures the spectrum of *everyday experience* than alternative accounts, which cannot bridge description and norms. It allows things to go wrong if our trial and error processes lead us to descend a solitary local minimum, or to go well if we engage in optimistic exploration and trial and error shaping of the world, our own selves, and others.

There is a further reason why self-evidencing is a worthwhile notion to turn to, rather than sticking with predictive processing's notion of mere prediction error minimization. The first objection that many people have when learning about predictive processing is that it must obviously be false. If agents are meant to minimize surprise, then they should just sit in a *dark room* doing nothing. Thus, since agents like us do not in fact tend to sit in darkened rooms doing nothing, it follows that predictive processing is false. The obvious, and I think conclusive, answer to this problem is that it rests on a misunderstanding of surprise. Surprise is relative to a model, so for most of us, sitting for prolonged periods in a dark room would be surprising because, given our models, we do not expect to do nothing, all the while getting, cold, hungry, and dirty. Instead, we expect to be out and about, foraging for food and earning money so we can pay our electricity and heating bills and buy warm clothes to avoid freezing when spending too much time in our rooms, whether dark or not. Conversely, if it were not surprising to be in a dark room, then that would simply reflect our model, which would harbor precise beliefs about being in dark rooms.

The only interesting debate here is how easily we explained a behavior by inferring that the agent's prior beliefs explained that behavior; this relates to something called the *complete class theorem*, which will be discussed in chapter 5, where we turn our attention to the accuracy and pragmatics of self-evidencing. My point here is that self-evidencing removes the temptation to worry about dark rooms. Self-evidencing is more open-ended and, in a sense, positive and optimistic than prediction error minimization. Self-evidencing is not about going down a gradient toward some absolute minimum; instead, it is about maximizing evidence for existence, which is hard to do by doing nothing in a changing world. Of course, explaining away prediction error is equivalent to maximizing evidence, so there is no logical shift here. Still, self-evidencing reminds us that agents can do a whole host of proactive things, actually and counterfactually, to put themselves in a good position to harvest utility and gain information, which maximize their evidence for their model of themselves.[13]

In chapter 4, we turn to the inner mechanics of self-evidencing, delving into the notion of active inference. This is important because so far, we have mainly noted that agents *must be* self-evidencing through some kind of inference to the best explanation, not how they actually do it. This will show that despite the fact that there are only a few simple core mechanisms available to the system in question to control self-evidencing, these tools can be deployed in diverse ways.

4 Self-Evidencing with Precision Control

Self-evidencing is fundamentally epistemic, as a process of maximizing the evidence for an agent's model of their characteristic states. Although epistemic, it is also an active process, as the agent acts to solicit observations that are expected under their model. This picture makes sense of agents, as they are seen from the outside. Human agents, for example, can be seen to move around the world, exploiting this patch and exploring that—and from the outside, we could say that they do that with the goal of sustaining themselves or learning about their environment. The outside perspective affords a vantage point for saying this because we imbue others with intentions and aims, assuming that they are acting under their own model of their characteristic states. But what does it look like from the inside, from the point of view of the agent? This is a key question, given the self-supervised, somewhat solitary perspective inherent in self-evidencing.

From the inside, the model is not simply known to the agent. We saw this in our discussions of the imperative for self-supervision in chapters 2 and 3. Therefore, the perspective from the outside in terms of representations of intentions and goals cannot simply be transposed to an understanding of how things look to the agent from their inside perspective. Rather, for the agent, a model is assumed, and then model evidence is sought in a trial and error manner through variational inference to selectively sort through the incoming observations. From the agent's perspective, the only thing that can be happening is a process that maximizes the evidence for the model at hand. But how does the agent actually do this?

In chapter 3, we saw that self-evidencing can be achieved either by updating the model (i.e., changing one's mind) to better explain the observations or, acting in a way that solicits the observations expected under the model's beliefs (i.e., acting to solicit observations that the agent believes they should receive, in active inference). I then suggested that the process of self-evidencing should look like inference to the best explanation and showed how this aligns with the free energy principle. But I did not say anything about how the agent engages in this process.

As we will elaborate in this chapter, this active process is basically a matter of agents self-organizing to selectively change the strength of the messages passed around within the model. "Strength" is related to the probabilistic properties of the beliefs under the model, reflecting the uncertainty or variance of a belief: little variance if it is a very peaked probability density, and much variance if it is flat. As we have seen earlier, the inverse of uncertainty is called the *precision*—the less variance there is, the more precise a belief will be. Precision provides a measure that helps naturalize notions of belief updating and active inference in terms of probability or information theory, and crucially, it explains self-evidencing as seen from the agent's perspective. This follows because to infer is to update beliefs; beliefs are inherently probabilistic and can be described in terms of their sufficient statistics, which include their precision (and mean, for normal distributions). Putting this together, if an agent exists, then it must be possible to describe them as organizing themselves by inferring—and thereby controlling—the precisions over their beliefs.

Increasing the precision of a message during inference is to increase its contribution to belief updating. A metaphor for this is trust; more trusted messages provide more reliable evidence, efficiently progressing self-evidencing. The more that a message is trusted, the more precision it should be afforded in an agent's self-evidencing. Thus, the means that the agent has for self-evidencing is the allocation of trust in light of what they already believe and what they are learning through their observations, where this includes contextually informed beliefs about what to trust and beliefs about where to look to gain trustworthy observations. *Precision control* is what this trust exercise amounts to from the self-evidencing agent's perspective. Notice that by invoking this notion of precision control, we are not making any claims about some ghostly entity being in control of precision—the claim is simply that in an existing agent, there must be selective control of the strength of flows of input through the system—if the agent simply lets all input flow freely through, then they will be fully subject to the dissipative forces of the environment and soon be indistinguishable from the environment (i.e., they will cease to exist).

As we will see, the processes for controlling precision are varied. But at heart, precision control conveys a parsimonious conception of what organisms do to self-evidence. Allowing a bit of anthropomorphizing, it is as if the agent keeps a keen eye on the sensory input that they are receiving and that they expect to receive, and they simply turn wheels and pull levers—inhibiting and disinhibiting neuronal populations—to control the flows of messages through the system, guided solely by the one thing that they know and care about, which is maximizing the fit between the input expected under the model and the sensed input. It is this selective internal mechanism—working on quantities that the agent has access to—that enables them to be truly self-supervised. In this chapter, I will unpack this sparse but also resourceful concept of self-evidencing.

First, I consider precisions for different kinds of beliefs about what happens, and then the need to build up prior beliefs about precisions (technically known as *hyperpriors*). Then I discuss the active elements in precision control in attentional and perceptual terms, moving to an understanding of action itself in terms of precision control. I next explain how precision control unfolds both in offline and online processes and offer a narrative of where precision control leaves the self-evidencing agent. The concluding remarks broaden out to consider whether this portrayal is neurocentric or disembodied in some unattractive way.

Precisions for States, State Transitions, and Policies

An agent exists when they sustain a boundary for some nontrivial period of time. This boundary is exposed to various states, ranging from benign sensory input to mortal lesions. These inputs determine the boundary's sensory states, which in turn cause changes in the agent's internal states within the boundary, where these internal states are what encode the probabilistic beliefs under the agent's model. The internal states in turn can lead some downstream parts of the boundary, the *active states*, to causally influence the states external to the agent—the latent states hidden behind the boundary or sensorimotor veil. This creates a circular causality of the perceiving and acting self-evidencing agent. For example, the agent's feet make footprints in the sand, which they can turn around and receive sensory input from and thus come to see; or the agent's hands can knead dough and bake bread, which they can then eat.

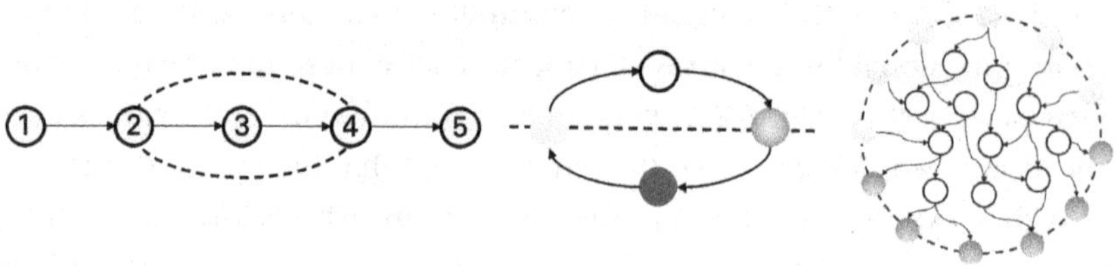

Figure 4.1
Depictions of Markov blankets (dotted lines). Left: Markov blanket for sequential causes, where nodes 2 and 4 constitute the blanket, with node 3 the blanketed state; here, nodes 1 and 5 are uninformative with respect to 3 once 2 and 4 are known. Middle: A cyclic causal net conceived for a self-evidencing agent who can close the action-perception loop, where the blanket states comprise the sensory node (light gray), and the active state (dark gray), blanketing the internal state (white), with external states outside the blanket (black). Right: A Markov blanket indicating the internal structure of causal dependencies constituting the internal model, with external states omitted.

The boundary is understood in causal terms as a *Markov blanket.* Markov blankets were briefly mentioned in chapters 2 and 3, and this extends the notion of conditional independence (or the *Markov property*) discussed in chapter 3. A Markov blanket exists, and can be drawn to represent, any case when the behavior of a group of causal nodes, by virtue of their conditional independence, is fully causally explained in terms of just their parent nodes, their descendants, and the parents of the descendants (see figure 4.1).[1]

What determines existence is whether the states affecting the agent's sensory blanket states are characteristic of it or not, given the model. If these states are wholly uncharacteristic, in the statistical sense described in chapter 2, the agent's internal states will soon begin to disperse, in the sense that this particular Markov blanket cannot be identified any more, which is indicative of inexistence.

The notion of characteristic states is the first place where precision comes in because something that is characteristic or expected in this sense is described as a probability distribution (or as we have called it, a *Bayesian belief*) with relatively high precision. The sufficient statistics for a normal probability distribution consist of the precision and the mean, and they can in principle be represented in a message-passing system (e.g., encoded by firing rates in neuronal populations of the brain). The mean tells us what to expect, and the precision conveys how certain our expectation should be. If we imagine our beliefs as a varied landscape of distributions, the more precise peaks indicate the characteristic states under that model, the states that it should be expected to be in. These precisions ensure that the organism can experience the surprise of venturing into uncharacteristic, unexpected states; for example, an agent who has high precision around the expectation to breathe air will, in this sense, be surprised to be submerged in water. This landscape of probabilities represents the agent's beliefs (in the basic Bayesian sense of beliefs as opposed to beliefs that would be consciously avowed by the agent). The greater the precision, the stronger the belief.[2]

Representing states with their varying precisions might work reasonably well in a stable and unchanging world where there are no transitions from one state to another. But the real world is rife with change, of course. There are random fluctuations in the input, there are constantly new patterns to learn about, and volatile events occur that challenge previous learning. The agent's own actions also change the input that they encounter; these self-originating changes may be more or less surprising (e.g., as in the distinction between what the agent will experience when they go foraging in a known patch versus going exploring in a new patch). The possibility of change implies that there can be transitions from one state to another. Such *state transitions* have probability distributions over them too, reflecting how likely it is that one state transitions into another. Agents who exist in changing, itinerant worlds will have models that include state transition beliefs, reflecting, "I am getting this sensory input now, but there is a high (or low) probability that it will soon be this input." This means that agents can be surprised at state transitions, given their expectations. For example, cooler temperatures occur when summer transitions into autumn, but global warming is now challenging the precision of our beliefs about that state transition.

Agents who self-evidence allostatically (as discussed in the previous chapters) benefit from knowing state transitions and keeping track of their precisions; for example, their planning rests on believing with high precision that they will transition to being thirsty when going on a run on a hot day. If state transition beliefs instead have low precision, then there is much uncertainty about what the next state will be. If all the agent's state transition probabilities are highly precise, then there are strong beliefs across the board about exactly what will happen next—and possibly well into the future. State transitions can happen over many different timescales, from tens of milliseconds (such as eye movements) to lifetimes (such as changing family life as kids grow up and grandparents pass away) and even longer (such as some aspects of climate change). State transition beliefs should be capable of reflecting all this, including how state transitions at one timescale interact with transitions on other timescales (e.g., how climate change interacts with how we organize our family lives and do our shopping); such interactions call for a hierarchical layering of beliefs over increasing timescales.

So far, we have precisions of beliefs about states and about state transitions. This yields information about current surprises and how much surprise to expect. The third application of precision relates to state transitions but is specifically about the precision of beliefs about the observations consequent upon execution of the agent's policies. Agents who act need to select among different policies for action, where policies are a kind of belief about a state transition that involves an agent's actions and observations. As chapter 3 set out, agents who select policies based on expected free energy engage in planning and decision-making through active inference. As we saw, policy selection is informed by both risk and ambiguity; recall equation (3.3) in chapter 3. Low ambiguity implies that the mapping from the states specified in a policy to observed outcomes has high fidelity (e.g., switching on this light means that I can clearly see what is going on). Risk is the divergence between outcomes expected under a policy (whether the policy is ambiguous or not) and the agent's preferences. If the agent has very precise beliefs about what should happen to it, this can translate into very precise beliefs about what to do next. In other words, the agent can be more or less confident about their selected policy.

If the belief over a policy is precise (i.e., it has a high posterior probability), then implicitly, it can be committed to with high confidence. For example, if I pick up my gelato myself, I will soon enjoy just the "Hokey Pokey" flavor I like, whereas if I adopt the policy of asking my son to go buy it for me, then I know to lower expectations about my future state. Agents gravitate toward high confidence policies because they move us more efficiently toward our characteristic states.

Notice that the observed outcomes from a policy incorporate the flow of observations as a sequence of planned subpolicies (or control states) are executed. For example, the ambiguity of the gelato policy is not just about the precision over observations of the final outcome but also the observations that I might make on the route to the gelato shop; if the route is well known, then the ambiguity is less than if I have to navigate a new neighborhood.

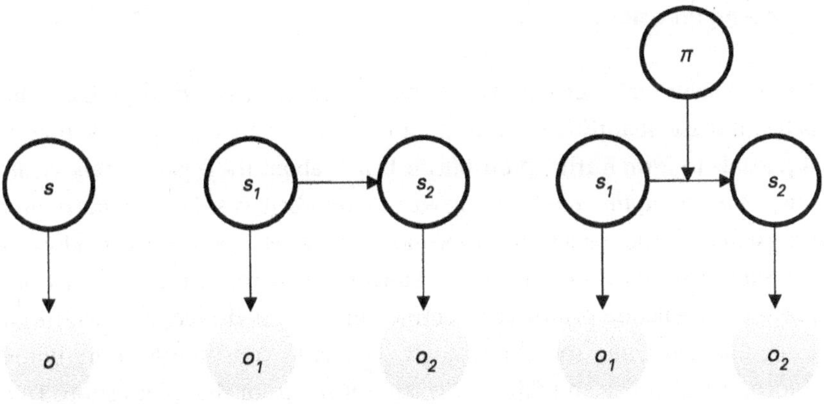

Figure 4.2
Depictions of Bayesian beliefs in a generative model under active inference. Left: Beliefs about states, $p(s)$, and outcomes under states, $p(o|s)$. Middle: Beliefs about state transitions, $p(s_2|s_1)$. Right: Beliefs about policy-dependent state transitions, $p(s_2|s_1, \pi)$. Each belief can be described by its sufficient statistics, the mean and the precision, typically captured through a partially observable Markov decision process (POMDP), with matrices for each type of belief. Models can have parametric depth, with beliefs about beliefs and hierarchical depth, with beliefs unfolding over different timescales or contextual domains.

Notice also, as I mentioned before, that policy selection, in addition to having a pragmatic value (such as getting gelato), can have epistemic value and work to increase one's knowledge, such as when I check the gelato shop's website to see if it has any new, exciting flavors; recall equation (3.4). Notice finally that, as discussed earlier, state transitions unfold over different timescales, and it is the same with policies, whose expected observations can unfold over short timescales (such as when turning the head to look left), medium timescales (getting that gelato), or long timescales (saving for retirement). Policies can also interact over many timescales, for example, the size of my retirement savings might depend on how much gelato I buy now.

We then have three aspects of precision (see figure 4.2). We should expect agents' models to encode precisions of probabilistic beliefs about states (what are the causes of the sensory input?), state transitions (how do the causes, and thereby the observations, change over time?), and policies (what state transitions involve the agent's policies and subsequent observations?).

Expecting Precisions

The sufficient statistics convey the mean and precision of a probabilistic belief. If these quantities serve as input to another part of the model, then it is possible to form further probabilistic beliefs about these parameters, creating *parametric depth*. That is, the agent can form beliefs about the means and precisions of other beliefs. Such second-order beliefs encode expectations—expected precisions—which can be violated, such that we can learn from precision prediction errors. For example, my expected precisions will need to be updated if my son (surprisingly) begins to consistently bring home the correct quantity and flavor of gelato for me from the gelato shop. This implies that the generative model of agents with such higher-order capacity includes explicit representations of the three kinds of precision.

Models with such parametric depth have an inherently "introspective" or metacognitive character because representations of precisions, as it were, "look downward" at the precisions of some of the model's own beliefs, ignoring what external states caused those distributions to have the parameters that they have. Then the agent can form beliefs, each with their own sufficient statistics, for the mean and the precision of these parameters. It may sound a bit convoluted, but it makes good intuitive sense. For example, my visual observations of the letters in the books that I read have a certain precision, which (I believe) depends on the ambient lighting wherever I sit reading. In weak light, the precision decreases, so I have come to expect a mean precision and a precision about that mean for my reading experience. As I age, I notice that the precision has decreased because I am less able to read in weak light. I then must update my expected precision, and this may lead to the selection of new policies for action, such as wearing reading glasses. Making things yet more convoluted, some models might even include third-order expectations for the second-order expected precisions of precisions; for example, I am in fact already expecting some changes to perception and cognition as I age, but I may be surprised at the rapid decline in precision in some domains.

Similar examples can be constructed for expected precisions for beliefs about state transitions and state transitions given the policies in place. Global warming and climate change induce more volatility across many aspects of life, meaning that the precision of many state transition beliefs decrease (although the high-precision belief that there is global warming makes us strongly believe that current state transition beliefs will lose precision in many domains—often cast in terms of uncertainty or unpredictability about the future). Or consider when the COVID-19 pandemic hit in early 2020, and the precisions of many of the habitual policies for private citizens and governments alike were challenged, such that we could no longer precisely expect the same old outcomes from our usual actions, such as making medical appointments, working, or going shopping.

Parametric depth implies a deep structuring, where parts of the model harbor beliefs about beliefs. This parametric structure could recur at each level of hierarchical systems, such as the cortical hierarchy in the human brain. Precisions with this structure essentially encode updateable beliefs about epistemic trust and confidence, meaning that expected precisions encode confidence in our own inferences. Agents who are able to explicitly model and predict the precisions in their environment and own internal states, thus, have this metacognitive ability, which can be regarded as a basic type of introspection (although this does not entail that the agent in question has anything like the conscious inner life that we sometimes associate with the notion of "introspection").

Notice that now we have described a kind a self-evidencing agent whose internal model has depth in two dimensions: parametric and hierarchical. First, it includes different timescales of beliefs about states, state transitions, and policies (building further levels on top of the structure depicted in figure 4.2). Second, it includes metacognitive levels, building in beliefs about beliefs. Human brains likely have hierarchies in this sense, being capable of representing states of affairs and policies evolving over very short and very long timescales for different contexts, with varying levels of metacognitive or parametric depth. Different kinds of creatures will have temporal and parametric architectures with different depths, reflecting the kinds of creatures they each are and the environmental niches they occupy. Different agents of the same species or type may have different hierarchies too, as they shape their model in response to new observations. Note that the implicit separation of timescales acquires a subtle dual aspect here. On the one hand, our beliefs about the context change more slowly—at a hierarchically superordinate level—than beliefs about the content at a lower level. On the other hand, our beliefs about states of affairs that last for years can change in a matter of seconds (e.g., beliefs about climate change can change following an alarming news report).

Precisions permeate the hierarchical, parametrically deep model on the basis of which an agent self-evidences. Learning and updating precisions are critical to an agent who exists in a changing world, where the volatility in input depends on how latent states in the environment, including the bodies of the agents, manifest over different timescales as they interact with each other. For agents like us, with deep hierarchies and extensive metacognitive processes, precision inference and learning constitute a significant burden of computational work—keeping track of a vast, evolving landscape of more or less shadowy, higher-order precisions. Learning precisions takes time because it requires identifying patterns in patterns, and it is difficult because it is an inherently introspective process to assess one's own confidence. It is a lonely task, where other people's instructions and opinions are at best only indirectly helpful because other people's beliefs themselves must be assigned precisions before the agent can meaningfully use them to update its own beliefs.[3]

Because precisions are crucial for making sense of the changing world and yet are subtle aspects of one's own internal model that are difficult to learn, it is likely that differences and deficits in precision modeling are implicated in mental illness. Precision processing in the brain is likely to involve neurotransmitters such as dopamine, which have long been speculated to be implicated in mental illnesses such as schizophrenia (dopamine might relate to the precision of beliefs about policies). This book's focus on our everyday experiences is intended to accommodate diversity across all parts of phenomenology, cognition, and belief updating, which includes mental illness. Here, we can read diversity associated with mental illness as associated with a certain kind of false inference that ensues when the learning of precisions is somehow misguided. The role of precision in self-evidencing is thus an important part of the story that accommodates diversity. (Chapter 5 looks at false inference in more detail.)[4]

To summarize, agents who exist in changeable and volatile environments may be well served to maintain and finesse internal models with parametric and hierarchical depth, laboriously building context-dependent beliefs about precisions and precisions of precisions.

Precision Control for Perception and Attention

So far, we have given the self-evidencing agent the capacity to infer and learn about the precisions of various beliefs (i.e., about states, state transitions, and policies). That is an important first step, but it does not yet explain how self-evidencing agents can actively organize themselves to stay in their characteristic states. For this job, *precision control* is needed.

Recall that by self-evidencing, the agent's inferences will approximate exact inference, as spelled out via the free energy principle given in equation (3.1). Exact inference is *Bayesian inference*, where a belief is updated optimally, according to Bayes' rule, in the light of evidence, and where the resulting posterior probability for a belief becomes the prior probability for the next inference step. Even for ground-level Bayesian inference, precision plays a key role because the amount of change in beliefs can—given new evidence—depend on the precision of two things: first, what is known prior to getting the new evidence; and second, the precision of the evidence itself. For example, if the agent already has a lot of evidence that an important meeting is at three o'clock, then their prior belief is precise, and it would take high-quality new evidence to shift it. If the agent receives evidence from someone who says that they heard someone else say that the meeting is at five o'clock, then that evidence is quite imprecise since it is hearsay, and the agent should not change the initial belief much, if at all.

If the model has explicit parametric representations of precision, then such ground-level inference can benefit from precision control. If the agent's new source provides the context that they got the information about the meeting time straight from the boss, then the agent assigns a top-down expectation of high precision to that evidence. Controlling the precision like this boosts the new evidence (e.g., by increasing neuronal gain), and now the agent is likely to update the belief about the meeting time. Precisions become a currency in the model, which can make inference context-sensitive, drawing on a wide network of contextual beliefs (such as the position of authority of a boss over the meeting schedule). We rely on boosting or attenuating the weights on signals all the time as we move from context to context. Some of this feels automatic, such as when we walk into a poorly lit room and need to reassess how much to trust our visual input from the letters on the page of a book that we are looking at. Other cases require more planning and concerted effort, such as trying to become a better listener in a relationship. In each case, inference becomes more reliant on past empirical learning, drawing on beliefs learned in many contexts and over various timescales (this is sometimes called *hierarchical Bayesian inference*).[5]

Hierarchically implemented precision-weighting of evidence in perceptual inference is one way that the self-evidencing agent controls precision, enabling perceptual inference and learning for the agent. For example, precision-weighting facilitates inference about which word is causing my visual impressions right now by drawing on contextual priors about ambient lighting, the language of the text, and the words and sentences surrounding the word in question. Precision control lends flexibility to perceptual inference, ensuring that the agent can function robustly across changing contexts. This kind of flexible, dynamically precision-controlled inference can coexist with other domains of more rigid inference, emerging in contexts that are relatively stable over time. Such stable contexts afford what is sometimes called *amortized inference*, where the belief that the agent adopts given an observation is determined by a habitual association. This can be understood as a kind of preset precision control (e.g., "I observe orange color, round shape, and orangey scent, so this is an orange"). This is consistent with self-evidencing because it is similar to how, in iterated inference to the best explanation, one explanation may have repeatedly emerged as the best one, making it unnecessary to keep considering it against alternative explanations; the observation still becomes evidence for the solo explanation. An agent's amortized inference reveals that they believe, at least implicitly, that the world is unchanging in the domain in question. Amortization is thus an efficient self-evidencing process that can work well in combination with more fluid, context-sensitive precision control.[6]

As we saw in the previous chapter, self-evidencing conveys a notion of an agent who selectively samples evidence for the model, in accordance with their beliefs; that is, the agent engages in active inference. This is in contrast to passive inference, where beliefs are updated in the light of whatever observations the agent happens to encounter, even where those observations are uncharacteristic. Selective sampling is where concerted, autonomous precision control comes in, guided not just by contextually learned beliefs but also by beliefs about the agent's characteristic states. More purposeful control of precisions could thus bias the whole agent toward visiting its characteristic states, explaining its continued existence. For example, an agent foraging through a patch containing different berries could actively preload a high-precision expectation for red berries, building on a learned association between being red, being ripe, and elevating glucose levels. By controlling precision (e.g., through neuronal gain mechanisms), evidence of red berries will then pop out and drive perceptual inference, informing foraging policies that help the agent stay in their characteristic, glucose-rich states. Now the Markov blanket describing the agent is in effect filtering the sensory input based on endogenous control of their internal states. This kind of precision control captures variations of endogenous attention; in the example, the agent has selected a policy for mental action to attend to red berries in visual search, where this decision is cashed out through precision control. Attending to something is a way of selectively sampling evidence, charting a trajectory through state space that can then stay close to characteristic states.

Self-evidencing fundamentally accomplishes selective sampling by assigning high prior probability to a belief of the generative model, making it harder for ordinary sensory input to overturn it. For example, the agent has a strong prior belief that their glucose levels are within a certain range; endogenous attention then operates to selectively collect evidence in support of that prior belief—in the end, highlighting the red berries when the agent is out foraging. It seems that humans have extensive power to control attention, on a very broad concept of attention; this includes mental actions guided by expectations playing out over very long timescales, such as an agent studiously honing their craft for thousands of hours to position themselves to eventually gain status and earning a living as a master artisan.

A capacity for sustained endogenous attention can be counterproductive if important sensory contingencies are missed as a result. If the precision weight on one sensory feature is turned up, other features may be missed. In our examples, this could be a failure to notice a hive of yellow-striped wasps among the red berries, or missing opportunities for earning quick money even before acquiring formal qualifications. For this reason, exogenous attention-grabbing also ought to be possible for the agent, such that salient contingencies are spotted. Paradigmatically, this happens through input with high signal strength, such as a wasp sting, or loud ads for earning quick money in the gig economy. However, even this exogenous-oriented process should be driven by internal beliefs about the self-evidencing value of high signal strength, in keeping with the primacy of self-supervision. These beliefs about precisions themselves will be able to control the balance between endogenous and exogenous attention, and will be of various sorts. For example, the agent will harbor a belief that inputs with high signal strength will deliver enough information gain that it is worth being distracted from the berries; or the agent might come to believe that garish ads about the benefits of joining the gig economy are not in fact trustworthy. Notice that here, we see a version at play of the balance between being closed-minded (endogenous) and open-minded (exogenous), which we listed as related to the best-makers in inference to the best explanation in chapter 3.

The agent is also likely to have developed some state transition beliefs about the agent's own allocation of attention, such that they believe that they will not stick with focused attention policies for extended periods of time; that is, the agent will believe that their changing, partially unknown environment does not afford unfettered exploitation of things like berries since wasps and darkness and other foragers will demand attention at some point in time. In terms of precision control, this means that the control is itself subject to beliefs about state transitions. In this way, the agent's self-evidencing actions can be balanced against their more open-ended beliefs about, and noticing of, sensory contingencies, which allows the agent to maintain an accurate representation of the environment. Importantly, the ability to maintain that balance is not something that is given beforehand; it must be learned in an ongoing, context-sensitive, trial and error process. A teenager might, for example, focus endogenous attention on a computer game for extended periods of time, only gradually learning the limits of parental patience.

Zooming out more, a key dynamic for the agent's trial and error efforts at turning up or down precisions across the model is the balance between what the agent already knows and what they could learn or achieve in the current situation. On the one hand, the agent will be pulled toward adjusting precisions to accommodate and learn from all the new information that comes their way—self-evidencing agents like accuracy. On the other hand, the agent will be held back by what they already know, preventing them from opening the floodgates too radically. This kind of inferential conservatism can be measured by the divergence between the prior beliefs and the posterior beliefs that result from the trial and error adjusting of precision; moving too far from prior beliefs suggests that the model is becoming overly complex—self-evidencing agents dislike complexity. In other words, the agent is constantly trying to balance accuracy and complexity, measured against the free energy. The model will try one setting of precisions that does not stray too far from its previous setting but promises uncertainty reduction, given what is happening now. This setting is assessed for expected free energy and might be tried in actual precision-controlled behavior, and if it does not work (i.e., it increases free energy), then a new setting must be tried. This dynamic constraint on precision control is indeed captured in one of the reorganizations of the free energy principle, where the free energy is complexity minus accuracy; recall equation (3.2).

In both perception and attention, expectations for precision, informed by contextual beliefs and beliefs about characteristic states, thus play a key role by inhibiting or boosting certain processes, selectively filtering input through the Markov blanket and through the system as a whole. In perception, this process approximates the optimal balancing of prior beliefs and confidence in the evidence; in attention, the process is more controlled in order to bias input to be consistent with (and provide evidence for) characteristic states.[7]

Action as Precision-Controlled, Self-Fulling Prophesying

Precision control is crucial for perception and attention, including decisions for the mental action to endogenously direct attention to some sensory feature. This active perspective generalizes to decision-making and overt bodily action. If an agent is in state s_1 and would prefer to be in state s_2, they need to adopt a policy for action to get themselves there (see figure 4.2, right). For example, if the agent is thirsty and prefers not to be thirsty, they need to figure out how to get some liquid to drink. There is an indefinite number of policies that could be enacted in principle, from turning on the tap to fill a glass of water, cupping the hands to collect rainwater, and all the way to building a desalination plant for the agent's own gratification. Traditionally, the assumption has been that agents would score all the policies (or at least the most relevant ones) on some metric and pick the best one, followed by computing the motor commands and movements needed to enact that policy. Expressed in these idealized terms, it is too much computational work to be tractable for finite organisms like human agents, and it also does not comport obviously with the imperative for self-supervision. A different story is needed, based on the notion of active inference set out in chapter 3.

For a self-evidencing agent, an action policy is a belief conveying a state transition, given a model and an action. This can be represented in the model of the agent's characteristic states: How likely is it that the agent will transition from state s_1 (thirsty) to s_2 (having quenched the thirst), given a specific policy, π (turning on the tap)? This is no different in principle from other state transitions that the model encodes (e.g., the probability that this dry weather will last into next week before the rains come). This calls for an implicit ranking of policies, with the most probable agent-centered state transitions ranked highest. For example, there is in general a high probability that the thirst will be quenched. This would be underwritten by the agent's model of characteristic states, which puts a precise prior on not dehydrating. The specific policy of quenching thirst by turning on a tap and filling a glass has high precision over it because the expected observations, seeing the glass fill and feeling thirst subside once the water is drunk, are very likely to follow just as predicted under that policy. Moreover, the policy is not too complex, so it is likely to generalize across contexts that the agent typically frequents (e.g., for taps at home and at work). This all points to the "fill glass from tap" policy to be inferred as the low-risk and low-ambiguity policy most likely to be pursued; it is the policy that is likely to afford the agent the most evidence for their model. This policy should win over the policy of building a desalination plant or of collecting rainwater by cupping one's hands.

The moment that a policy is inferred, the expected observations under that policy acquire high probability. In effect, the agent prematurely adopts the belief that they are actually making those observations. In the simple example, the agent now already believes that they are in the process of observing the water filling the glass, and this will be followed by the sensations of drinking. Of course, the agent is not in fact doing this (they are sitting in an office, parched and glassless), so now there is mounting uncertainty—a deviation from the states specified by the agent's self-induced beliefs. The agent could get rid of the uncertainty by giving up the belief that they are drinking water, but the precision of the thirsty-to-quench state transition is too high for that to happen. Instead, the agent can act in the world to make the predictions come true. However, to realize their own predictions, the agent must ignore the actual sensory evidence that they are not actually moving in the expected way. This is where precision control comes in again, in an especially powerful and enabling way: precisions can be controlled to get rid of the self-induced uncertainty. First, current sensory observations are attenuated (to help ignore the sensations that belie the fact that the arm is not currently reaching for a tap). Second, precisions on the sensory observations expected under the policy are boosted, similar to the mechanism outlined in this discussion for endogenous attention (such as the precisions over observations of the tap and the glass).

Notice how precisions must offset each other, in the sense that a precise prior belief about preferred states must be more precise than the current sensory evidence to actively realize intentions. In this example, the attenuation of sensory precision ensures that the high precision over the drinking policy will not be revised, and this enables our agent to walk to the kitchen, turn on the tap, and, ultimately, drink some water. Once the policy is enacted, the self-induced uncertainty dissipates. Here, action is not a matter of executing motor commands; it is a process of self-fulfilling prophesying realized through judicious control of precision. The agent makes the prophecy, contrary to fact, that they have water and will engage in action until they fulfill that prophecy.

The objective function—which allows the selection of policies and then the enacting of the winning policy—has nothing to do with the utility of water in itself. In the first instance, it is simply the propensity for an existing organism to gravitate toward less uncertainty (given a model) by inferring a policy guided by minimizing expected free energy. Then, once the policy is inferred, it is enacted, actualizing that expected sensory input and minimizing variational free energy. This is another aspect of self-supervision where the agent does not need to resolve an external cost function for every kind of reward (water, coffee, biscuits, or other preferred items). Instead, the agent self-evidences. The absolution from individual cost functions does not imply that self-evidencing is easy, of course, since it requires continued, internal maintenance and updating of the beliefs of the internal model. In a self-supervised agent, the cost functions are absorbed into the prior beliefs, and *preferences* are then dealt with in terms of beliefs. Active inference is elegant because it relies on these simple cost functions, but the point of the elegance is not that it makes life easier or less resource demanding for the agent; rather, the point is that the processes that we model the agent as engaging in are in fact tractable for the agent (recall here the discussions of self-supervision and free energy minimization in chapter 2).[8]

The precision control on overt action, such as getting a glass of water, is then quite similar to the precision control in covert, mental action such as endogenous attention (e.g., trying to spot the nutritious red berries). The precision of a certain kind of sensory input is boosted even if is not yet observed, and then a policy (walking to the tap; searching the berry bushes) is executed. Bodily action is not an add-on to this story, requiring complicated computation of motor commands. Essentially, the agent entrains their "body plant," forcing them to move around until the self-induced uncertainty is resolved. Such a process is just a state transition, of course, represented with a certain precision. An active state, affecting the external world, could here be a top-down proprioceptive and kinesthetic prediction observing that one's (in fact straight) arm is bending, bringing the glass of water to one's mouth. This induces a prediction error that is quashed by reflex arcs when the arm begins to bend, bringing the glass closer to the mouth. The agent need know nothing about arms or biomechanics to do this; they just need to predict things and wait for the body to do its thing until the sequence of proprioceptive and kinesthetic predictions (in this case) are realized.[9]

Controlling Online and Offline Cognition with Precision

The portrait so far is of the self-evidencing agent as a Bayesian filter, selectively letting in a quantity of sensory input via some channels and processing it through the system, where the filter's operation—its granularity, as it were—is determined by the agent's empirically shaped prior beliefs, including beliefs about their own characteristic states. This conception of precision control is apt for online processing of incoming observations. But we also engage in actions that are offline and not based directly on sensation. Dreaming, memory, imagination, and thought and reasoning all seem to require the sensory input to shut down, allowing the system to do its own thing undisturbed. This distinction between online and offline processes is a theme that will recur in subsequent chapters and addressing it is central to how self-evidencing can hope to capture the gamut of our everyday experiences.

A human self-evidencing agent seems able to engage in extensive offline processing, requiring the attenuation of distracting online input. Offline processing can be conceived as helping to finesse and explore the agent's own internal model in order to set up beliefs about state transitions so that, when the agent does actually act in the world, they will better maximize the model evidence. Memory is a way to reenact past events and associations to inform planning and decision-making, imagination helps to project plans into the future, thinking and reasoning help work through trajectories and outcomes of possible actions, and dreaming may be involved in model simplification that better enables the model to generate reliable predictions when the agent wakes up. It makes sense, then, for self-evidencing agents to be able to assess when they should be offline and when online. For example, if the agent currently exploits a resource in a certain patch but anticipates imminent depletion of resources, it may be beneficial to switch to offline mode for some period of time in order to plan the best route to begin exploring new patches.

Considering the human brain as an organ for self-evidencing, it is always in the game of organizing itself for minimizing uncertainty about current and future states; the agent is able to learn that going offline in certain conditions is better than staying online.[10]

The mechanism for adjusting the balance between offline and online processing seems to rely on precision control. For online perception, the self-evidencing model implemented in the brain reduces uncertainty, such as through prediction error minimization, which is thought to be done by inhibiting bottom-up signals accurately and precisely at prediction error nodes. For offline imagination or memory, the system should rather disinhibit (i.e., turning up the precision on) otherwise quiescent prediction error nodes and increase the internal activity so top-down units have some evidence to work on, even in the absence of sensory input from the world. It seems likely that this dual role of inhibiting and disinhibiting error units is orchestrated through the hippocampus, which in that sense sits at the apex of the predictive processing hierarchy. There is a nice, slightly counterintuitive aspect to this mechanism of precision control, where online processing is associated with inhibiting activity and offline processing is associated with exciting activity. Psychologically, this suggest that the self-evidencing agent is always, as it were, open for business and distractable to some degree by either internal thoughts or external input.[11]

The way that individual agents set up offline processing may vary considerably. In memory and imagery, for example, the bottom-up signals generated by internal disinhibition could occur at different levels of the cortical hierarchy. If they occur at relatively high levels, the resulting memory or image would be relatively removed from sensory attributes and hence be more abstract. Individuals capable of activating much lower, sensory areas may enjoy much richer, vivid memory and imagery. There are significant individual differences in reports of imagery, which may relate to preferred levels for precision control for offline processing. This could be what creates the spectrum of vividness in imagery from aphantasia to hyperphantasia, which, interestingly, has implications for mental health, relating to a propensity for intrusive memories in hyperphantasia, for example. This implies that the way that precision control adjudicates the balance between online and offline processing has implications for differences in everyday experiences, including the richness of imagery, and also implications for mental health.[12]

When the self-evidencing agent engages in offline processing, this amounts to a kind of mental action focused on internal representations. This complements other kinds of mental action, such as endogenous attention focused on online input and metacognitive representations focused on inferring the precision of beliefs. All these kinds of mental action are made possible by precision control. In the case of offline processing with perceptual characteristics, like memory and imagery, precision control changes the normally transparent nature of perception such that it is no longer felt like seeing through experiences to the properties of the world. For example, when an agent perceives a tree, the experience is transparent in the sense that it seems to reveal the properties of the tree itself; in contrast, when imagining a tree or when metacognitively assessing the precisions of a perceptual belief about a tree, the experience is made opaque, no longer pointing to the world.[13]

Mental action can seem a peculiar type of mental process, which is neither aligned with perceptual nor active inference. However, mental action is in fact not very different from overt, bodily action, at least from the perspective of the agent. Overt action for online perception is driven by precision control, such as when the agent moves their eyes from an observed nose to selectively sample the region of space (up and either left or right) where eyes would be expected to be, testing the hypothesis that it is a face that the agent is looking at. Here, the face itself is conceived as a hidden, latent cause for which inference is conducted. In offline imagery, some level within the cortical hierarchy is treated as a latent cause for which imagistic inference is conducted, simply because that level is passing signals upward, stemming from the disinhibited prediction error units at that level. Higher levels now receive input, and they do what they do, which is to infer what caused that input, even if the input originated from a lower level of the system itself, not from latent causes in the agent's environment.

I end this section with a consideration of Markov blankets (see figure 4.1). The distinction between offline and online cognition serves to illustrate how Markov blankets, relative to which inference is understood, can be identified within the agent itself, as well as at its customary organismic boundary, such as around the nervous system or at the skin. Which blanket is identified for investigation depends on the explanatory target in question. Similarly, to explain aspects of metacognition, Markov blankets are identified between levels of parametric depth, allowing inference of the sufficient statistics of lower levels (treated now as latent states) by higher, meta-awareness levels. This perspective is useful in two respects. First, it reminds us that Markov blankets are not rare since they just reflect a common causal structure (depicted in the left panel of figure 4.1), and therefore, there is unlikely to be a principled "master" blanket that we must focus on. Second, and relatedly, it reminds us that inference demands that something is doing the inference and something else is being inferred; in particular, the internal states within a given Markov blanket are inferring the latent states of the generative process external to that blanket. There can be no question of any kind of inference without this requisite internal-external structure. Hence, if we want to say that mental action is a kind of inference, internal Markov blankets must be assumed.

Further, if we want to identify a blanket further from the nervous system or the skin (such as a master artisan and their tools), then things within that blanket cannot themselves be the direct subject of inference and, conversely, things beyond that blanket are still to be inferred. In other words, the self-evidencing framework constrains the explanatory perspective, such that any talk of perceptual or (covert or overt) active inference must come with something like a Markov blanket and a distinction between internal models and external processes; if these constructs are rejected, then there can be no talk of inference. Together, this tells us that self-evidencing needs to countenance a multitude of Markov blankets and at the same time cannot get rid of the internal-external divide once a blanket is identified. We return to some of these themes at the end of this chapter, and again in chapter 5 (where a case is made that there is no tension between considering Markov blankets real and allowing that it is our pragmatic, explanatory interests that guide us about which blanket to focus on).

Solitary Precision Control in a Causal Nexus

The parsimonious tool of precision control now transpires as surprisingly resourceful. An agent who weighs precisions optimally (i.e., through the inference to the best explanation mandated by the free energy principle) and who actively controls precisions relative to their characteristic states can be self-evidencing. And, by self-evidencing via precision control, the agent

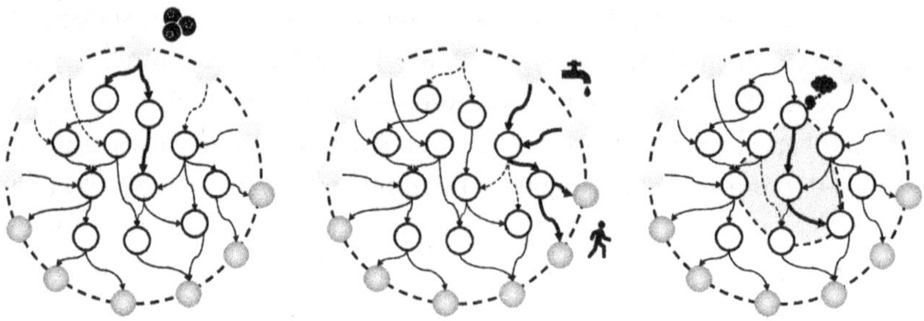

Figure 4.3
Depiction of precision control within Markov blankets (dotted circles). Thick arrows: increased precision; dotted arrows: attenuated precision. Left: Precision control for mental action (endogenous feature attention to berries). Middle: Precision control for overt action (increased precision for water tap sensory input, attenuation for other actual input; entraining body via active states). Right: Precision control for offline cognition (thought, imagination, memory) by an inner Markov blanket (gray).

is perceiving, attending, deciding, and acting. Figure 4.3 depicts examples of how precision control within the Markov blanket underpins attention, action, and offline cognition.[14]

The agent is immersed in intense, internal, self-organizing activity, allocating precision to different parts of the internal system. But the agent is also eerily passive: all the action happens within the boundary set by the Markov blanket in question. The body with which overt actions are performed is not truly in on the action—it is merely entrained through the causal interface with the blanket's active states (focusing on a blanket around the central neural system). There are just the interplay of observations at the sensory states and what happens at the active states of the blanket, conducted through the internal states.

So far as the agent is concerned, they will believe that they are acting in the real world, but they might as well be being deceived by an evil scientist hooking them up to phony sensory input. We will discuss this skeptical, evil demon scenario further in chapter 5; for now, notice that when we refer to an "agent," the real action—the doing—is exclusively found in the internal precision control. This is a crucial aspect of the self-evidencing story because this internal action, entirely performed on quantities the agent knows, cashes out the fundamental imperative for self-supervision. "Actions," as we normally understand them, in terms of eye movement, limbs moving, and bodies on some trajectory through space, have nothing directly to do with the internal action of agents. They are just parts of the generative process in the world, which must be inferred like everything else.

Being an agent is then just to control how input is filtered at the sensory nodes of blankets and through internal states (and internal blankets), and how those filtered inputs eventually set up active states. Mechanistically, self-evidencing is just filtering input through the model and back into the world. To exist as a perceiving and acting agent is just to be a filter. This seems almost unbearably reductionist. The agent is a mere model, strongly convinced that it exists and is thus self-evidencing, but condemned to infer everything beyond its boundary, from its own body and actions to the states of the world, and in other people. In later chapters, we will delve into the tools that the self-evidencing agent can use for optimizing and engaging in this kind of internal precision control. This will reveal that although there is truth to the reductionism, the full, diverse, and rich swath of everyday experience can be recovered.

The self-evidencing agent is marked out by its boundary, itself being one of the many causes making up the fabric of the world. As such, the agent has the power to affect latent causes out in the world in ways that affect what the agent may in turn come to observe. Indeed, the very idea of a filtering process assumes a continuous causal impact on the boundary from causes around the agent. The boundary is causally affected by the things that happen, and precision control lets in some of these effects while filtering others. This filtering depends on how precision control is set up, and what is allowed in is then further processed through precision-controlled layers of the internal hierarchy (subject to parametric checks and constraints) until it moves back down again, hitting the active states and affecting the world. Self-evidencing is thus inherently causal. To me, this duality of epistemic, internal precision control and enmeshment in the causal nexus is particularly attractive.[15]

Concluding Remarks: Being as Neural Filtering?

With this chapter's focus on precision control, a compelling juxtaposition of the self-evidencing agent emerges. On the one hand, agents are engaged only in internal processes, inferentially secluded by a Markov blanket from the things beyond. On the other hand, the self-evidencing agent is inextricably part of the causal flux of the world, selectively permeable to its effects. The self-evidencing agent is engaged in a solitary precision-control effort, as one node among all the other nodes in the causal fabric of the world.

Yet from this austere picture of an agent's existence just in terms of controlling inhibition and excitation of processing pathways, we can build up to who we are and how we experience the world. We saw that from precision control, we get perception, attention, overt action, and covert mental action. We extended that with offline experiences through memory, imagery, thought, and reasoning. Later chapters will add sense of self, volition, value, consciousness, and meaning and wisdom. I believe that the felt textures of life, in its full spectrum, emerge from this juxtaposition; things feel the way that they do because we are both secluded from the world, trying to make sense of it from within, and exposed to, and inescapably connected to, the world.

Here, I have described what the self-evidencing agent does in neurocentric and disembodied terms. In humans, precision control appears to be a process of the neuronal system upon itself. It happens within the Markov blanket and would not change if, instead of input from the body and the rest of the external world, it received input from other sources, such as a nefarious, evil demon. This picture contains much truth because most states of the body are beyond the blanket (depending on how wide a blanket we are considering) and as such are inferred.

Of course, it often feels as if the way that we engage in bodily action or perceive the world through engagement of our senses is not inferential, but rather fluidly enmeshed with the body and directly connected to the world. The research field of embodied, extended, and enactive cognition highlights in various ways these roles of the body in cognition and the unmediated way that we seem to perceive the world. This field also casts doubt on the neurocomputational ways that cognitive systems are conceived in the majority of cognitive science research, advocating for the inclusion of other bodily states, such as the adrenal cortex, the heart, or neuro-gastro-enterological structures, or even wholly external states such as notebooks or mobile phones. Further, it is often argued in these debates that since perception and cognition are inherently enactive, the core concept cannot be representation, sandwiched between input and output, but rather ought to be a notion such as "structural coupling."[16]

My view is that there is much to learn from these debates, but none of them threaten self-evidencing as I have presented it here. Chapter 5 will portray the self-evidencing brain as dealing in representations, but these representations arise through something akin to structural coupling. The current chapter has sought to describe the internal, solitary, epistemic mechanism of precision control but also how this picture makes sense only if we view the agent as a filter letting through some of the causal impact from the world, slowing it in parts and funneling it back to the world. Self-evidencing is liberal, in principle allowing cognition, perception, and action to happen in all sorts of nonneuronal structures; here, I focus on everyday human experience, for which the brain is obviously important and a valid focus of research attention. Crucially, causal structure and explanatory pragmatics together allow different more-or-less expansive Markov blankets to be identified as the focal point of scientific exploration. It is thus entirely plausible to include, say, the hypothalamic-pituitary-adrenal axis within the Markov blanket (if the Markov property is satisfied). In that case, as I have argued here, that combined system is in the business of inferring the latent states of processes beyond its particular Markov blanket and including those nonbrainy elements as part of its precision control mechanism. The cost of identifying a wide blanket is that the states of those parts cannot be considered as directly bringing sensory input into the self-evidencing system since they are now part of the generative model, not the generative process. There are some interesting questions here, since we are attributing a generative model to part of what we, from the point of view of the observer, would consider the generative process—namely, the human brain as a physical object of study; recall figure 2.5. There will be more on this in the next chapter and in chapter 7.

I think that often, but not necessarily always, it is more illuminating to go narrower than such extended systems and consider the signals from the heart, immune system, adrenal glands, and other organs as input to the central nervous system, leaving this system to infer and then act on those other bodily systems. To illustrate, in work led by my colleague Andrew Corcoran, we explained the role of the heart during perception of bistable stimuli in these terms, indicating how the brain's belief that there is much uncertainty in the visual input (due to the bistability) entrains the heart to deliver a less variable heartbeat, which helps to regularize input to the cortex, allowing it to engage with more unfettered attention to resolve the uncertainty. To me, this is embodied cognition, which makes best sense by placing the heart outside the Markov blanket, as a thing to be engaged with in active inference. This is all consistent with allowing more peripheral systems themselves to be cognitive in some sense, engaged in self-evidencing in their own right. Interesting questions then arise about what kind of systems these might be, and whether they do or do not have the kinds of capacities for self-evidencing as brains have.[17]

The self-evidencing framework is also consistent with a situation where agents appear to have shifting Markov blankets over time, such as changes during pregnancy that include various constellations of the placenta, the fetus, and the woman's body, or the metamorphosis seen in some insects that radically dissolve their bodily boundary, within the transitory boundary of the pupae. The notion of existence in self-evidencing and the free energy principle is purely statistical and accepting of all sorts of trajectories, whether stable, changing, or ethereal, so long as the beliefs of characteristic states do not flatten out entirely, spreading through all possible states.[18]

My overall position is thus that self-evidencing is mostly, but not always, best seen as quite neurocentric, representational, and at one remove from the body and the environment, although entirely causally coupled to both. I think that it is a kind of polarizing mistake to cast self-evidencing in diametrical opposition somehow to embodied, extended, and enactive cognition. Rather, with self-evidencing, we can get everything we want in a new synthesis, which goes beyond both traditional cognitive, representational approaches and embodied, extended, and enactive approaches.[19]

The final point to make in this chapter relates to the notion of control. If the underlying process for the self-evidencing agent is precision control, then perhaps we should pose the question, "Who is doing the controlling?" This way lies the danger of vicious regress, though, since if we need to posit an agent as being in control of the precisions, then we do not have a fundamental account of an agent after all—an endless series of nested self-evidencing agents might be needed. However, this regress does not arise when we take the existential starting point seriously. If some set of states is completely at the mercy of the dissipative forces of the universe, then there is no describable Markov blanket and no thing. If the set of states is somehow able to resist some part of the dissipative forces for some time period (i.e., if a thing exists at all), then it must necessarily be filtering these forces. In this case, it must by definition be controlling precision, even if only in some rudimentary sense of control where some channels or pathways are open and others are not (e.g., a bacterial spore letting only very few inputs through). Complex agents, with evolutionarily inherited, developmentally shaped, and empirically learned prior beliefs and the capacity for hierarchically and parametrically deep allostasis, are just the same. The way that they filter the input describes how they exist and is inscribed in the beliefs of their internal models. If the beliefs of the internal models had been different, then the control would be different too. No agent pulling strings is needed to tell this story about self-evidencing.

After the presentation of self-evidencing in chapters 2 and 3, and the underlying precision control mechanism explained in this chapter, chapter 5 will explore the notion of representation and how it connects to both the epistemic, skepticism-prone nature of self-evidencing and its harnessing in terms of Markov blankets. That chapter will help formulate the ways in which inference can be false, as well as other ways that self-evidencing can appear to go wrong.

5 Accuracy and Pragmatics

Self-evidencing is benignly circular, and it bears a passing semblance to *self-fulfilling prophesying*. This raises questions of accuracy, optimality, and veridicality. An agent implementing self-evidencing processes would be guided by the assumed model of their own characteristic states, selectively sampling observations believed to deliver the most uncertainty reduction. This appears to be a pragmatic outlook of perception and action, with no obvious connection to accuracy and truth. The agent should just have expectations and beliefs that concern their own existence and make decisions that serve those purposes, but they can afford not to care about having accurate beliefs, true representation, or making decisions that are optimal beyond having pragmatic, instrumental value.

Is the pragmatic framing in conflict with the idea that an internal model represents the world and can be a good or a bad model? Or with the notion that truth is something that agents can and should aim at? What happens to ideas of objectivity if everything is measured only relative to the agent's own, subjective model? Does the self-evidencing framework fall short in these respects, leading us down the undesirable path of subjectivism and relativism?

Those questions will be the focus of this chapter. I will argue that with self-evidencing, it turns out that we can preserve ideas of representation and truth, but the conceptual landscape within which we ask these questions is reconfigured.

There is a related question about the accuracy of the self-evidencing framework itself, as an account of the mind. Is self-evidencing a genuinely scientific framework, or does it trade too much in "just-so stories"? How literally should we take the idea of self-evidencing? Is it something that human brains do in reality, or is self-evidencing just an instrumentalist framework that is useful for predicting what agents and their brains do, but does not faithfully describe the actual brain processes? There is again an avenue to a compelling answer in favor of the more ambitious, realist view of self-evidencing.

First, I consider how self-evidencing must in effect be supervised by the world, speaking to the connection to truth. I then show how a notion of representation naturally follows from self-evidencing and what work representations do in self-evidencing. I then discuss the various ways that false inference can arise in approximate inference. Turning to the self-evidencing framework itself, I defend its scientific status and describe its relation to empirical process theories, and then argue for realism about it. Finally, I acknowledge that all this leads to Cartesian skepticism, but I soften the skepticism and argue that it faithfully conveys our epistemic situation.

Self-Evidencing and Truth

Where Is Wally? is a popular picture book series where on each page, a character named Wally is hidden in a chaotic scene of many other characters and images, and the reader's job is to spot him. (Wally is the Australian and British name; he has other names in other countries, such as Holger in Denmark, Waldo in the US, and Ali in Turkey.) This is a moderately fun and engaging visual search task. Described in self-evidencing terms, doing the task means having model parameters encoding the belief that Wally is observed, which generates the (contrary-to-fact) prediction that Wally input is being received, having beliefs about how moving the eye gaze sequentially to different parts of the page will bring about the expected input within some time frame, and adjusting expected precisions to favor some Wally-related sensory attributes (such as his glasses and red-and-white-striped jumper) over non-Wally input. The prediction induces a prediction error, and the agent, having inferred the eye-movement policy, begins quashing that error, which trains the movement of the eyes accordingly until Wally is spotted. At that point, the agent has garnered evidence for the belief encoded in the model—they have engaged in self-evidencing in some small measure.

It is tempting to cast this Wally process in wholly pragmatic terms. The agent would like to observe Wally, infer whatever policies would be in service of this preference, and then move around until the prophecy is fulfilled—they see Wally. There seems to be no need to cast the agent as somehow aiming for, or even being concerned with, truth. The agent would not do anything that would not facilitate achieving their own preferred, Wally-observing state. From the point of view of the agent, having false beliefs about eye movement, Wally's outfit, and the construction of picture books would be no impediment, so long as the agent can achieve their pragmatic goals.[1]

Does this mean that self-evidencing in principle is completely divorced from truth, and that the question of whether the internal model is an accurate representation of the world plays no role in understanding what self-evidencing is? In such a view, self-evidencing would be wholly pragmatic, and the only thing that matters is the instrumental value of the agent's beliefs. I do not believe that this kind of pure instrumentalism follows, although of course, it has an element of truth. Truth and accuracy come in various guises as the self-evidencing framework unfolds.

Imagine a frustrating trick page in a *Wally* book where, unbeknownst to the agent, Wally is in fact not in the picture at all, so there is no inbound Wally signal to amplify with endogenous precision control. Now the agent's self-fulfilling prophecy will not be able to quash the prediction error. This is an imposition on the self-evidencing agent where the prediction error cannot be gotten rid of with any of the policies currently being considered. The world failed to hold up its part of the uncertainty-reducing bargain. This example illustrates how the incoming signals from the world keep self-evidencing honest, nudging the agent's inference toward accurate representation—toward spotting Wally when, and only when, he is really there.

In this case, recognizing that the world is suddenly not yielding to the prophecy, the agent needs to infer a new policy. Instead of searching for Wally, they need to turn the page or put the book away. Policy selection is guided by what the world affords.

Although agents have much freedom in what they perceive and decide, the self-supervised process that they rely on is not unconstrained. The agent is in constant causal commerce with the surrounding environment, which provides the unlabeled training signals for the inferential processes, filtered with precision control through the agent's Markov blanket. This takes us back to the topics about the free energy principle and the importance of self-supervision (as discussed in chapter 2). The beauty of that framework is that the objective functions (variational and expected free energy) that the self-evidencing agent optimizes are tractable from the inside, and with a self-serving pragmatic perspective, and yet this optimization process is beholden to the world. Conversely, the agent's perception and decision-making are not causally determined by the signal from the world in a straightforward, bottom-up, "imprinting" manner. A self-evidencing agent wields its precision control in many varied ways as it meanders, via trial and error processes, toward approximating exact inference.

The worry was that self-fulfilling prophesying portrays self-evidencing as purely pragmatic, with no regard for truth. Now we see that, in spite of the pragmatic outlook, self-evidencing agents accurately latch onto truth (e.g., tending to see Wally when and only when he is actually there). However, the agent is not intentionally aiming at truth since they are solely engaged in self-evidencing through precision control. Truth comes along for the ride, which means that self-evidencing provides an attractive integration of pragmatism and accuracy. We can therefore agree with views insisting that cognitive function does not have representation of the world as its goal. But representation is still what characterizes the agent, and if the agent exists, then their internal model cannot be wholly unmoored from the surrounding world.[2]

Self-Evidencing Representations

If a self-evidencing agent is let loose in the environment, then over time, the causal dependencies of the internal states of their model will begin to recapitulate the generative process in the environment (see figure 2.5). The agent comes to harbor beliefs reflecting the structure of the world, as that structure relates to the causal generation of the observations.

The matching of model to process happens through the continual updating of the generative model as it encounters unexpected observations. Such observations can be first order, such as the surprise at seeing a green blueberry; or they can be second order, such as the surprise at the imprecision of the arrival times of the bus. The agent can learn from this (that unripe blueberries are green or that roadwork is wreaking havoc on the bus schedule). Unexpected observations and consequent learning can also be about the agent themselves, such as when they learn that the social discomfort that they have been feeling lately is due to a recently diagnosed hearing loss. Or, it can be about the agent's own mind, such as when their intention to attend to an important task is foiled by impulsive checking of a social media platform. The learning process can be fast or slow, but over time, the internal model will be optimized to the received input, so the model gets better and better (i.e., the encountered sensory evidence becomes increasingly unsurprising). The model is enriched, and the world feels more certain.

Self-evidencing indeed entails a rich representational picture, with a hierarchical internal model humming along with its precision control at multiple timescales and parametric depths. But in a changing, noisy world, the model is always playing catch-up. There is bound to be varying levels of irreducible noise, challenging the agent's ability to immediately model events well (e.g., if the bus just happens to be two minutes late today, the agent should not plan to arrive at the bus stop two minutes later tomorrow; such a model would be overfitted to today's happenstance). The self-evidencing process needs to avoid overfitting and should harbor beliefs about the irreducible noise and the variability in different contexts, to help control precisions accordingly. Conversely, an underfitted, overly simple model risks missing out on important patterns (such as believing that the bus is always running to schedule, ignoring increased variance during rush hour, roadwork, and the shortage of drivers). Self-evidencing needs to maintain a balance between simplicity and complexity, and it also needs to figure out this balance in a self-supervised manner, as discussed in chapter 3 about equation (3.2).

The world is also a volatile place. Setting irreducible noise and variability aside, there is much uncertainty caused by the injection of new causal factors interfering with known patterns. We have seen examples already in this discussion, such as when the bus is suddenly late due to that roadwork or when the new *Wally* book has trick pages. Volatility is a pervasive fact of life. There is suddenly a new social dynamic in the office due to a new hire, there are suddenly more floods and fires due to climate change, there is suddenly no toilet paper in the supermarket due to the pandemic, and so on. As soon as the generative model has settled into a good situation, predicting everything with reasonable complexity and accuracy, something volatile happens, and new learning of interfering underlying, latent states is required to enable the model to reduce uncertainty again. Here, the self-evidencing agent also needs to somehow learn to relegate some uncertainty to irreducible noise, and some to volatility from potentially learnable latent causes.[3]

As we have seen in chapters 2 and 3, rather than changing the generative model in the face of unexpected input, the agent can engage in active inference to help sample expected input. Active inference is conducted through the basic precision control mechanism available to the agent (presented in chapter 4), but it can manifest in different ways. The agent can leave the surprising states and move to unsurprising ones; for example, if the agent does not expect very hot states, they may get out of the sun and seek out a shady spot. The agent can also change the current state; for example, by building a structure to create shade from the sun. In the long run, agents can do much to change their own ecological niche so it increasingly matches their expectations. Over time, updating of models and actions in the environment can lead to profound changes in both, as they mutually shape each other.[4]

The upshot of engaging in active inference is that agents tend to occupy their characteristic states. In other words, the mutual information between the generative model and the generative process increases. By actively avoiding surprising states or by actively sampling or constructing unsurprising ones, the agents' model will present good representations of their world—until volatility creates new challenges.

Overall, self-evidencing leads to a coupling between the generative model and the generative process—the model becomes a rich, multilayered representation of the causal structure of the world. It is trivial to observe that this model will be highly selective since it is driven by the agent's beliefs about their characteristic states; the agent is not going to end up representing aspects of the generative process that have no conceivable direct or indirect impact whatsoever on their self-evidencing. The model will be keeping track of and continuously forming beliefs about unfolding events and about complexity and simplicity, and also keeping a keen eye on volatility and changes to levels of noise.

Putting Representations to Work

From the point of view of the agent, who is just engaged in precision control, it is irrelevant whether the model is an accurate representation of the world or not. What matters to the agent is how self-evidencing is going, as measured against the objective functions of variational and expected free energy—there is no additional objective function for truthful representation. Although accurate representation is in that sense incidental to self-evidencing, the representational properties of the model are constantly in play and are crucial to our understanding of self-evidencing.

In a self-evidencing agent, one key function of the model is to generate observations expected under the beliefs about the latent causes in the environment. The model must be able to parse individual observations, believed to be generated by certain latent causes in certain contexts. This happens in perception (e.g., expecting orange color input if what is perceived is believed to be an orange) and in active inference (e.g., for the action of drinking coffee, the agent generates expected proprioceptive observations of bending one arm, and then the hot sensation and somewhat bitter taste of the coffee). Generating expected observations demands operations on representations, where a model of interacting causes in the world is deployed to figure out what observations those causes would give rise to.

The agent's self-evidencing can also, as discussed in chapter 4, include extensive offline processes in advance of action, engaging the model to run through counterfactual scenarios to investigate the expected free energy under different policies. Here, the expected observations, representations of possible futures given this or that action, are internally generated and assessed. This again implies operations on a representation of interacting causes, parsing out and assessing sequences of observations.

Assume that an agent has encountered and learned a particular environment, such as a rat that has learned to navigate a maze. If the maze now changes and the rat observes a new route opening up (such as a shortcut), then the rat has to decide what to do with its model. It can ignore the new observation and stick to the existing model, or it can expand the model with new parameters to account for the new observation (e.g., adding a policy for running straight ahead to take the shortcut instead of only going either left or right). Model expansion is again an operation on a representation, like adding a new connecting subway line to an existing subway map. There is much excellent research on these basic navigational aspects in the brain, in particular the hippocampus, showing a sophisticated ability to segment a cascade of observations (such as those generated when traversing a maze or a subway system in a new city) into meaningful chunks, which can be stored, reaccessed, and recombined in distinctively representational ways.[5]

All self-evidencing agents must be able to assess precisions of their beliefs in some way. For example, they must be able to assess the precision of the beliefs over their policies in active inference. Some agents, such as humans, seem endowed with an extensive, explicit metacognitive capacity, being able to infer and represent their own beliefs (such as "I am certain in my belief that the gelato shop has my favored Hokey-Pokey flavor today"). Some agents might even counterfactually represent the beliefs that they would have were they to engage in a particular action (such as "If I were to eat yet another gelato, I would come to believe that I am weak-willed"). These cognitive functions imply metacognitive operations on representations; perhaps with the added capacity to represent representations, such as future beliefs.[6]

It is difficult to make sense of all these self-evidencing processes without regarding the generative model and our operations within it as richly representational. Thus, in self-evidencing, representations are put to work.

False Inference and Self-Evidencing

The agent's model is portrayed as coming to represent the latent causes of the world as the agent goes about self-evidencing. This could appear to imply that self-evidencing precludes false perceptual inference and irrational active inference. How could there be room for error if models are inexorably driven toward accuracy? Obviously, humans are error prone and often irrational. We misperceive non-Wallys for Wallys, we are fooled by illusions, we form false beliefs about the reasons for unexpected events such as buses being late, and so on. No one is totally free of false beliefs of some sort or other. In mental illness too, many experience hallucinations or delusions. When it comes to decision-making, we also often deviate from optimality, failing to infer the most optimal policies and letting ourselves down with impulsive decisions or lacking self-control. Self-evidencing needs to encompass all these kinds of false inference that we actually experience.

Let us first clarify what is meant by false inference. Inference for a self-evidencing agent should first be considered against exact inference or optimal decision-making. This is what self-evidencing through perceptual and active inference approximates, as discussed in chapters 2 and 3. False inference could then be argued to be any divergence from exact inference. In other words, if the agent should perceive or decide something according to Bayes, given the model and evidence, but actually fails to do so, then that is misperception, delusion, irrational impulsivity, and other errors. There is much scope for the agent to fail in this regard because self-evidencing is an essentially self-supervised, model-constrained trial and error process where errors are bound to happen on many of the agent's trials (as depicted in figure 2.9). In other words, false inference is possible in the gap, at a particular time, between variational inference and exact inference (or between the variational free energy and the log evidence in general; i.e., outside linear generative models, there will always be such a gap, but the size of it is unknowable because the evidence cannot be evaluated in the general case).

The agent's model is nudged toward accuracy, but always within the bounds and limits set by noise and volatile changes. The approximate inferences driving the model therefore in practice almost always fall short of perfect overlap with exact inference. It is unlikely that there will be a point in time where noise and volatility are so absent that the model gets a chance to achieve zero divergence to exact inference. Any time that an agent is getting close, something happens in the world that challenges the model, requiring it to update or requiring new policies to be inferred. The self-evidencing agent is always playing catch-up with the world. From this perspective, we are always to some degree misperceiving and irrational. A slogan sometimes used for the broad predictive processing framework is that "perception is *controlled hallucination.*" This is an apt metaphor if agents are always deviating from exact inference ("hallucinating") but are meanwhile constantly nudged by ("controlled by") the input from the world, enabling them to approximate exact inference.[7]

False inference, resulting at some point from the variational inference process, can arise in different ways and can be more or less recalcitrant. It many cases, it can be transient and easy to address by extending the variational trial and error process, such as when falsely seeing some red-and-white-striped items in a *Wally* book as Wally himself and finessing the visual search (e.g., by adding Wally's distinctive glasses to the target search). In other, more evidentially secluded domains, this kind of variational process may be more difficult. This would include interoceptively based affect and emotion and tracking one's intentions, bodily sensations, and proprioception, which is where many false inferences in psychosis and other mental disorders arise. In these domains, it is not always obvious how to extend the trial and error process to finesse self-evidencing and minimize false inference. That is, our inferences about our own and other people's mental states is another area of inference that can be relatively evidentially secluded, affording much opportunity for recalcitrant false inference.[8]

It seems that variational inference will be less prone to diverge in domains affording diverse avenues of precise hypothesis testing through active inference. To get a good idea of whether a given parameter in a model is capturing a particular cause of sensory input, an agent can manipulate the input in various ways, such as by turning around an object to see it from different perspectives, or smelling or touching it—or even sending a drone out to investigate how it looks from above. These things can be harder in more evidentially secluded domains—one cannot really see a bodily sensation from a different angle or touch another person's true intentions. Perhaps variational inference can be caught in cascades of such false inference—like ending up in a local free energy minimum—with no obvious way to turn inference around and begin to better approximate exact inference. This kind of situation could contribute to the more stubborn hallucinations and delusions seen in some psychotic conditions.

When variational inference occupies such a local minimum, the agent over time begins to update some of their model's empirical priors, and exact inference given these updated priors would in effect be brought closer to this local minimum, yet further from the true generative process. In such a cascade of inference flowing from an initial bad belief, the agent has difficulty lifting themselves out of the local minimum. So if the model has been updated as a result of being caught in a local minimum for some time, then selective sampling via active inference will begin to deliver observations consistent with this warped model. Once self-evidencing gets a pathological foothold like this, it can be difficult to dislodge. There will be much more to say about the processes involved in psychoses and neurological disorders, of course, but it seems that self-evidencing has some explanatory resources here. Recall that although self-evidencing is benignly circular, the circle turns vicious when doubt about the evidence cannot be resolved, which can happen in these evidentially secluded domains.[9]

Some illusions, and perhaps some behavioral biases, may result by following a different variational avenue. Variational inference would be helped greatly by reducing the number of parameters that the agent must sequentially vary and fix in the approximation process. This could lead to modeling choices where beliefs are allowed to have sizable free energy in some contexts if they facilitate powerful and simple inferences in other, more salient or frequent contexts. Some perceptual illusions could be explained like this, such as the ventriloquist effect, which leads to highly unusual inferences in rare contexts (such as an actual ventriloquist act, where a doll seems to speak) but greatly facilitates correct inference in the many cases where integration of visual and auditory input is informative (such as a movie theater, a noisy cocktail party, or a windy day at the beach). The complexity cost would be too high for modeling the rare contexts too (e.g., having to consider and rule out the contingency that one is surrounded by professional ventriloquists every time someone speaks in a crowded room). The literature on predictive processing has many examples of these types of explanations of perceptual phenomena, which include false inference.[10]

Intuitively, we have various ways that false inference can arise through variational processes. (I have focused on perceptual examples in this discussion, but in chapter 6, I will delve more into active inference and irrational decision-making.) Here, false inference occurs when there is divergence from exact inference. We can assess this from the outside perspective, when we help ourselves to assumptions about exact inference and about the current variational inference of the agent that we are observing. From the perspective of the agent, however, there is no such insight because the exact inference is not known. The only thing that the agent has access to is the series of inferences assessed against the objective free energy functions underpinning self-evidencing. In the moment, the agent will not know that the inference is false. Rather, the agent might get an inkling if they subsequently find that they are updating beliefs and need to infer new policies to resample the evidence. Again, this kind of epistemic imposition is a consequence of being self-supervised.

Crucially, even though from the outside, we can say that a given belief is a false inference, the agent is not culpable of irrationality in any common sense. To see this, consider that the parameters of the generative model serve as prior beliefs in approximate inference and these priors change as the agent varies parameter values and shifts between testing of different parameters. At each point throughout variational inference, the agent's inferences reflect their current, actual beliefs, and the agent is doing what they "should" do. It is difficult to blame the agent for doing what they are supposed to do. This also does not mean that we cannot be, as it were, exasperated by an agent who goes about variational inference in a cumbersome, slapdash, or inefficient way. We all develop strategies, based on our prior knowledge of how the world works, when finessing our perceptual and active inferences, and these strategies would be strongly dependent on our learning histories and more or less serendipitous insights about noise and volatility that we have gathered over time. Some people can be on a better trajectory to approximate exact inference than others. The label of "irrationality" might then better apply to agents on poor gradients toward exact inference than to those who fail to adhere to norms of reasoning as such.[11]

Just-So Stories and Process Theories

From the scientist's outside perspective, we can hypothesize about what the self-evidencing agent's model is like, once we observe the agent's behaviors. In fact, the complete class theorem is a technical result that guarantees that for each behavior, it is possible to specify a set of prior beliefs and likelihoods that would explain that behavior in terms of Bayesian inference. The complete class theorem highlights a common worry, where Bayesian explanations of behavior are "just-so stories"—the agent is postulated to have the priors and beliefs "just so" their behavior can be explained. Just-so explanations are in general viewed as problematic, such as in evolutionary contexts, where any trait can in principle always be "just so" explained to be adaptations. In a sense, a just-so story is a somewhat flimsy hypothesis, which should in principle be tested by trying to find independent evidence. But if a field of research trades in just-so stories and have no robust means to decide among them on independent grounds, then that field is seen as suspect (thus, there is a debate about evolutionary psychology in this regard). This would seem to indicate that there is no way to falsify or confirm theories in that field, and that is widely regarded as kryptonite to that field's standing as being properly scientific. The question, therefore, is if the Bayesian explanations underwriting self-evidencing are (kryptonic) just-so stories?[12]

In fact, the guaranteed existence of a Bayesian description can be seen as a strength, providing a principled starting point for investigation. I am deliberately using the term "description" and not "explanation" here. Something more is needed to turn the Bayesian just-so description into a proper explanation, which can put us on track to scientifically test hypotheses about an agent's models. For example, it should be possible to contrast two proposed sets of prior beliefs, which both can describe the behavior in question, and to proceed to garner behavioral or neurophysiological evidence for one over the other.

This is where the extensive, systematic work under the free energy principle comes in. As discussed in chapters 2 and 3, the free energy principle, which mathematically underpins our notion of self-evidencing, is fully generic. It is a principle, not a law of nature, in the sense that it is not a generalization made from observations of particular systems. Depending on how they are described, particular systems either conform to the principle or not, and if they do not, then the principle is not falsified—it would just not apply as a method to understand that system under that description. But under the free energy principle sit what are called *process theories*, which are informed by empirical observations about the anatomy and function of particular systems, such as humans with our central nervous systems. These process theories can be shown to conform to the free energy principle, given empirical assumptions about the shape of probability density functions, and various dimensionality-reducing aspects of hierarchical systems. The process theory spells out how the belief updating during variational perceptual inference and active inference is reflected in the neuronal dynamics of the brain.[13]

Process theories for primates like us build in the first instance on empirical evidence about canonical microcircuits (the neuronal circuitry traversing the six cortical layers of gray matter, repeated throughout the neocortex of the brain), and neural messaging pathways across the brain's white-matter architecture connecting microcircuits to each other. The process theory appeals to empirical hypotheses about how probabilistic functions are realized by neuronal machinery; this is based on neural mass models that allow mathematical descriptions of the activity of populations of neurons, which can then be analyzed in terms of the free energy gradient reflecting the self-evidencing process. The process theory can be used to generate synthetic brain activation (i.e., simulate the neuroimaging data that would be expected to be generated in the human brain for a particular task). For example, event-related potentials (ERPs), collected with electroencephalography (EEG), and local field potentials (LFPs), collected with intracranial recordings, are both taken as the time derivative (rate of change) of the normalized firing rate of neuronal populations. These synthetic data can then be compared to actual data from existing research or sampled in new experiments. Examples in this direction, which I have been involved with myself, include work led by Christopher Whyte on active inference in theories of consciousness. (We will encounter this work in more detail in the discussion of consciousness in chapter 9.)

There is also considerable work on the relation between active inference and neuromodulators, in particular acetylcholine and noradrenaline. Acetylcholine is hypothesized to weight the precision of the mapping between states and observations (i.e., s and o nodes in figure 4.2; or the precision of error units in predictive coding). Phasic noradrenergic tone is hypothesized to reflect the estimate of volatility, captured as the precision of state transition beliefs (i.e., between s_1 and s_2 in figure 4.2). The locus coeruleus is a primary source of noradrenaline, and phasic and tonic activity at the locus coeruleus is hypothesized to modulate parameters reflecting the forgetting rate, understood as the magnitude of belief change between posterior beliefs after each round of belief updating (i.e., relating to the balance between being open-minded and closed-minded; see chapter 3). Conceptually, it makes sense that the same neuromodulator is involved in volatility beliefs and forgetting rates since volatility indicates the emergence of new patterns that could be learned. Work in these areas could help distinguish between, for example, hypotheses focusing on state-observation precisions versus volatility beliefs; this would help substantiate and finesse initial (just-so-like) hypotheses about the beliefs of agents. There is significant and ongoing empirical work done to test, further articulate, and deploy process theories.[14]

Active inference models can also be used to simulate behavior, such as in decision-making tasks. With such simulations, a candidate Bayesian explanation can be tested and will be supported if indeed the simulated brain activity or behavior is observed, or weakened if it is not. Further, different models can be tested against each other to see which fits the observed data best, so we can distinguish between different parameter constellations that each could capture the target behavior. Here, one can compare classic approaches such as reinforcement learning with active inference models, as well as comparing among active inference models. Work led by my colleague Beth Fisher illustrates this approach. She is interested in optimism bias, where people update their beliefs more from good events than bad ones; this bias is interesting because even though it is a bias, it promotes well-being, possibly by making it more likely that agents engage in life despite experiencing a fair amount of negative outcomes. A number of different models could in principle give rise to optimism bias, but Fisher hypothesized a difference in forgetting rates for good and bad events, and constructed both reinforcement learning and active inference models to be tested against each other. The task was a two-armed bandit where rats needed to nose-poke one arm for sugar pellets and avoid the unrewarded arm, but they also could choose to just stay in their cage, so optimism would be behaviorally indicated by increased engagement. We hypothesized that rats given psilocybin (the psychoactive ingredient of magic mushrooms) would be more optimistic than rats given just saline. The behavioral data from the rats were fitted to the models, and the best fits turned out to be the active inference model with forgetting rates and, in addition, loss aversion, showing the biggest effect of the psilocybin; other active inference models and reinforcement learning models saw less good fits (and for the winning model, we could recover the parameters as well). In other words, once the actual behavioral data were in, collected over fourteen days posttreatment, we could begin to make distinctions among just-so type hypotheses that each seemed to have some face validity. Hence, we got evidence that psilocybin might work to increase optimism via a specific computational mechanism involving forgetting rates and loss aversion. This finding was then backed up by Fisher in further modeling of optimism bias, showing the face validity of the same active inference model across a developmental task, where the bias can be lost due to early adverse events, a belief-updating task confirming the classic version of the bias in the scientific literature, and a two-armed bandit task showing that this optimism bias model leads to increased engagement behaviors.[15]

This section began with a worry that perceptual and active inference evinces the specter of just-so explanations, which threaten the scientific testability of self-evidencing. In response, we considered process theories, where self-evidencing is empirically tested in particular agents, such as humans or rats. Self-evidencing can thus be seen to have rich scope for empirical testing in spite of initially invoking just-so-like stories.

From Falsification to Bayesian Model Comparison

Some might worry that the appeal to process theories does not suffice for falsificationism—the idea, associated with Karl Popper, that the demarcation of true science is the ability to generate predictions that would falsify the theory in question. It might be that the complete class theorem's guarantee of Bayesian descriptions invalidates genuine falsificationism?

To address this worry, we must first notice that philosophers of science have showed that Popperian falsificationism is not an independent demarcation criterion, simply because by necessity, falsificationism must ultimately rely on induction, and Popper in the first place proposed falsificationism to get around what he saw as confirmation problems with induction. This moves the focus from falsification and back to what might be the best inductive model for evidence accumulation and theory comparison in science. The framework that is looking most promising these days is Bayesian analysis, which is fast becoming the norm in many fields of cognitive science. A Bayesian approach to hypothesis testing is egalitarian, in the sense that it does not provide a demarcation criterion in itself and therefore applies to all hypotheses, even the crazy ones or the actual just-so stories. What Bayesian analysis does provide is objective information, at a particular time, about which theories scientists would be wise to devote their resources to testing. This makes the need for strict demarcation obsolete, as weak theories would gain less scientific attention and be expected to dwindle due to lack of testing, as fewer scientists would put their bets on them. Stronger theories would show up more clearly as attractive for more research and development. A key element here is the idea of science as a dynamically evolving process, as a continuous race to accumulate evidence for theories. Seen in this light, there is no reason to worry about the testability of different self-evidencing hypotheses in the domain of process theories, even if they initially are just-so hypotheses in the benign sense of being guaranteed by the complete class theorem.[16]

In general, this Bayesian perspective on science means that process theories, and the specific explanations under them, should be regarded as empirically testable and very much in the race for Bayesian evidence accumulation against other theories. Different process theories, say based on mapping computational mechanisms onto different anatomical and neurophysiological hypotheses about canonical microcircuits, neuromodulators, or neural architecture, can then be tested against each other. A change in process theory can cause upheaval in our understanding of predictive processing and self-evidencing, but this would be just another manifestation of the dynamic scientific process.[17]

If the scientific process for self-evidencing is indeed conceived along these Bayesian lines, then what is the explanatory target of neuroscience and behavioral science in this field? What is the empirical science of self-evidencing looking for? The answer can appear slightly unorthodox—namely, that the science is simply seeking to reveal which models are embodied in the causal dependencies of the internal states of agents, such that we can say that the probabilistic beliefs of those models are indeed what explains the agents' perceptual and active inference behaviors. Given that self-evidencing must explain the behaviors, there must be a model in there, and the job is to provide evidence for which model it really is, on the assumption that the process theory indeed captures what models look like in this or that kind of agent (e.g., in primates, octopuses, or *Escherichia coli*).

To me, these considerations deliver the following landscape, on the agreement that we are looking for empirical tests and confirmation rather than falsificationism. The free energy principle itself is not subject to empirical testing because it is a principle, not a law. Process theories can be tested empirically and can support specific mechanistic self-evidencing explanations of behaviors. If a process theory and the generative models for a particular domain conform to the free energy principle, then we know that we are imputing to the system something that it can actually do itself (i.e., it is self-supervised). Although this is a somewhat austere concept of self-evidencing as a scientific method, the free energy principle generates rich and diverse ways for generative models to explain behaviors; we saw this in the assimilation of the free energy principle to inference to the best explanation, where agents variously learn to balance complexity/accuracy, risk/ambiguity, and exploration/exploitation.

Models and Realism

In this chapter, we have first discussed notions of false inference through self-evidencing. This brought us to discuss the scientific process itself and how self-evidencing can engage with empirical science. In particular, science can investigate the neural implementation of probabilistic inference and assess which generative models best explain particular patterns of behavior. This leads to a discussion of what self-evidencing, as a scientific approach, might tell us about the world, about the actual machinery in the brain. Should we adopt scientific realism about self-evidencing, or is self-evidencing merely a convenient scientific instrument useful for predicting observations, but for which we can make no claim about biological reality? Next, I will trace a realist account, which I think is plausible even though it does not preclude further discussion in this area.[18]

The mathematics of the free energy principle does not directly generalize from any real processes in the brain because the free energy principle is a principle, not a law of nature in the common sense of being an inductive generalization. But this does not matter to realism because when developed into a process theory for a particular system, as we have seen, computational function is implemented by neuronal activity. This provides a fairly direct route to realism for the computational properties of the process theory, where in turn the process theory conforms to the free energy principle. Subject to empirical support as outlined in this discussion (in studies of microcircuits, neuromodulators, and behaviors), one would be able to say that self-evidencing represents real neural processes in humans. Similar realist arguments would be made for agents described in other process theories for other particular systems.

This is an unremarkable kind of realism, mediated through computational models and process theories. It is consistent with a broadly mechanistic approach to neuroscience, which is discussed in the philosophy of science for mechanisms in general, and with significant debate about neurocomputational mechanisms in particular. The claim is not that neurons somehow know math, just as one does not claim that a map of a terrain is literally imprinted on the terrain—we do not expect the river represented by a blue line on the map to be ink blue when we cross it, for example. The lack of literal representation for maps does not make us map irrealists or map instrumentalists. Maps represent reality, and of course, they can have more or less instrumental value at the same time, helping us to get from A to B, even if we have to cross a murky-brown river that is indicated on the map with a blue line. Being a realist about computational models of brain function is not the same as saying that those models are guaranteed to be true or useful. We can rationally form a belief about the veridicality of a model only if we test it, and its usefulness depends on what we are using it for. Again, this is no different from maps of terrain. Rather than literalism, there is a mapping function, which can take many different forms for different maps, including computational maps of the brain. The mathematical structure of the process theories, in combination with the empirical hypotheses about, for example, neuronal mass models and cortical architecture, provide the mapping function, and that is then what enables hypotheses about the actual neuronal processes implementing the self-evidencing mechanism. The free energy principle unifies all these process maps within one conception, with assurance that these systems can be self-supervised, and we are therefore describing agents at the proper, fundamental level (as discussed in chapter 2).

Several of these meta-theoretical themes are depicted in figure 5.1, which sets out my conception of the relations among the free energy principle, generative models, and neuronal dynamics, and the role of the free energy principle in assuring self-supervision for the described system, and, conversely, the rich abductive processes that can be expected for a self-evidencing system.[19]

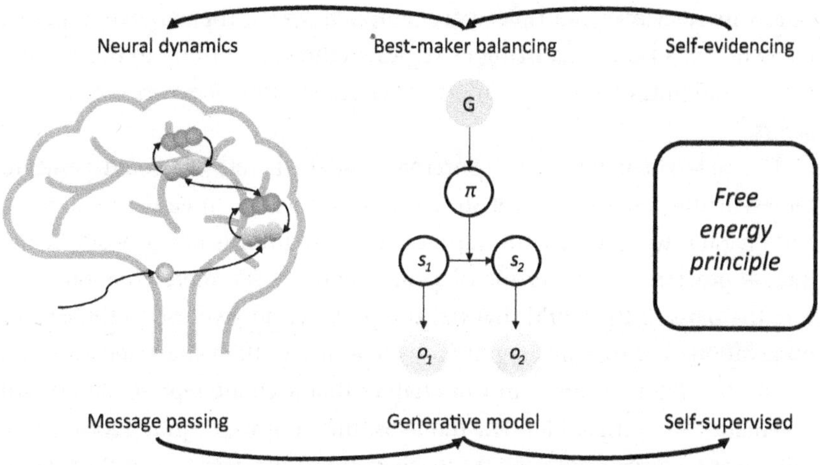

Figure 5.1
Heuristic depiction of the relations among the free energy principle, generative models (where G is the objective function for expected free energy guiding policy selection in active inference), and the neuronal dynamics described in a process theory for the brain as the particular system in question (with icons for canonical microcircuits and message passing between cortical layers and hierarchical levels). The bottom arrows ask if the behavior of the system in question can be described with a generative model, and if it conforms to the free energy principle, which would assure self-supervision. The top arrows conceive the free energy principle as a method for investigating the system in question, bringing out the abductive processes that a self-evidencing agent is expected to manifest, and hypothesizing how they are implemented in neural dynamics.

Skepticism and Self-Evidencing

Let us finally ask what role the external world plays for the self-evidencing agent, whom we have so far described in internalist terms of their internal states, within the Markov blanket describing the agent's existence. Earlier, we described the external world in terms of the generative process (i.e., the causal process external to the Markov blanket that actually generates the agent's observations and leads to updates of beliefs encoded in their internal states). We have been building up to the idea that self-evidencing implies that, when surprise is within reasonable bounds, then the internal model does a good job at representing these external causes, hidden beyond the Markov blanket, and how they interact to give rise to observations. Because causes interact over multiple spatiotemporal scales, the generative model likely benefits from organizing its representations according to these scales too, hierarchically and with appropriate constraints between hierarchical layers.[20]

The self-evidencing agent's internal model, therefore, should come to represent the causal/statistical structure of the world. In earlier sections of this chapter, we saw that this kind of representation is put to work in the diverse processes characteristic of self-evidencing. The representations capture the parts of the world that has the potential to give rise to the agent's observations and that are pragmatically relevant for their characteristic states (recall the argument made in this chapter that accurate representation and pragmatics are compatible). The model is thus, as it were, quite egocentric—full of causal arrows pointing from the world to the agent, as well as causal arrows emanating outward, from the agent to the world. Overall, this suggests that the generative model represents the generative process by virtue of some degree of structural isomorphism—matching the structure of the world to the structure of the internal model.[21]

Worries about Cartesian skepticism tend to arise when agents are described as being behind a blanket or sensory veil, left to their own devices when trying to infer the true states of the world only from the sensory input that they have available. It seems that it is always possible that one can be deceived by one's senses, as René Descartes taught us. And then it seems that we can never rule out the possibility, for any of our beliefs, that they are false—and if we cannot do that, then we do not really know anything. Descartes's famous evil demon is a nefarious, causal meddler, who conjures observations in the agent as if they were caused by real things and their interactions, such as cats, bushes, and fences, but which are in reality no such thing. The structural representation of the solitary, self-supervising agent's internal model appears vulnerable to just this kind of Cartesian skepticism.

There are broad types of reaction to such Cartesian skepticism. One can acquiesce in skepticism, accepting that we cannot rule out that we are massively deceived. One can stipulate that knowledge must be possible, and skepticism arises only due to errors in conceptual analysis or some conceptual inconsistency or inadequacy in the way that we set up the problem. The history of epistemology—the theory of knowledge—is rich with discussion about these options. One influential trend is to argue that the problem arises from the very conception of the agent as evidentially secluded behind a sensory veil, such as the Markov blanket. Here, skepticism is seen as a symptom of a pathological, internalist conception of perception and agency as sandwiched between input and output. The cure is then to get rid of the sandwich model altogether.[22]

I am in the first camp, acquiescing in skepticism, although with a significant caveat, as we will see. The Markov blanket neatly conveys a causal sandwich model of perception and agency, and we cannot give it up without giving up the fundamental explanation of existence and self-supervision in self-evidencing. The price of this explanatory and unificatory power is Cartesian skepticism. Only if we are unconcerned with self-supervision can we go the other, nonsandwich route, which prevents the skeptical problem from even arising. There can be many explanatory projects, and there is nothing illegitimate about taking that alternative route. To me, however, it is not sufficiently fundamental; it is like investigating the teacher-student relationship without ever asking how the student understands anything the teacher says in the first place (as was argued in chapter 2).

I acquiesce in skepticism also because I believe that it is faithful to the human condition. We do not, and cannot, know that we know, or remove any and all possible doubt about our beliefs. The gnawing doubt of skepticism is part and parcel of our lonely, existential struggle to self-evidence for another day. To deny the possibility of skepticism is to belie the way that we exist in the world, as well as the diverse, trial and error conditions under which we perceive and act in it.

We should acquiesce in skepticism, but as I mentioned, there is a caveat. The structural representation of the world, if the agent exists, must come to recapitulate some of the causal structure of the generative process itself. If indeed an evil demon is holding the agent captive for some time, feeding them conjured-up sensory input, then the agent's model will veridically represent something—namely, the causal structure of the demon's processes of conjuring of observations. The self-evidencing agent might believe that they are perceiving the real world of cats, bushes, and fences, but they are instead mentalizing accurately about a demon. More than that, the self-evidencing agent is not powerless—in fact, the agent can entrain their evil demon through active inference. We saw earlier that self-evidencing, even though it is rather akin to self-fulfilling prophesying, needs the world to play its part so the expected observations transpire in selective sampling. It follows that if the evil demon wants the agent to keep existing, the demon must comply with the agent's self-evidencing and deliver their expected observations. This means that the causal operations of the demon upon the agent will conform to the agent's model of their characteristic states—that is, the demon is entrained. This provides a partial victory against skepticism, empowering agents to enforce their beliefs on the (demon) world. The mechanism here is just what we discussed earlier, about how active inference shapes the agent's eco-niche.[23]

Now we can hover between epistemic optimism and skepticism. Optimistically, as self-evidencing agents, we do have the potential to uncover the real causal structure of the world, or even shape it in our own image. Skeptically, there is always the possibility that the content of our beliefs diverge from some deeply hidden part of the true generative process. The skepticism exists on a spectrum from the profound, such as our inability to rule out the demon scenario, to the everyday. For an everyday example, think of having been able to operate adequately for some time under the misconception that one's boss is looking after one's best interests, only to discover (when it is too late) that he (a real-life, evil demon) took off with the pension funds.

Self-evidencing is therefore an attractive framework to explain our existence as agents because it captures our everyday experience of this dynamic interplay of epistemic optimism and skepticism. This link between self-evidencing and everyday skepticism should come as no surprise. Back when we introduced the version of self-evidencing in Carl Hempel's philosophy of science (laid out in chapters 2 and 3), I noted that the self-evidencing circle is benign unless doubts are raised about the occurrence of the evidence. Such doubt can be addressed only by appealing to evidence that is independent of the self-evidenced belief. Cartesian skepticism is simply the limit case where there are no further avenues for finding such independent evidence, and the circularity becomes vicious, with the consequence being that knowledge disappears. The answer to this skepticism is to absorb the imperative for self-supervision, instead of pining for new evidence—and the free energy principle shows how we navigate this epistemic impediment.

Concluding Remarks: Knowing One's World through Self-Evidencing

This chapter has traversed several fundamental philosophical debates, revolving around what self-evidencing has to say about accuracy and pragmatics.

Does the self-evidenced model represent, and can it misrepresent? Is self-evidencing empirically testable, and should we be scientific realists about self-evidencing? Does self-evidencing entail skepticism? I provided affirmative answers to all these questions. In each case, the concern about assuring self-supervision sits at the heart of my answers, driving a new appreciation of what those questions mean and what those affirmative answers suggest about knowledge and mind, in particular about the intermingling of pragmatism and veridicality and the balancing of epistemic optimism and skepticism. I thus believe that the ensuing picture of the accuracy-tracking but skepticism-prone agent is revealing of our actual situation.

The self-evidencing framework thus provides inroads on fundamental questions in philosophy of mind and theoretical neurobiology. It is worth asking whether the insistent focus on generative models has more practical benefits. Why is it useful to operate with and know about models? For a self-supervised agent, like you or me, the game is to work with an assumed model of the world and self-evidence relative to this model (through internal precision control). Some aspects of the model are given through evolution, providing priors for body morphology and physiology. Other parts are more directly and concurrently assumed and then finessed in development, in learning, and through exploration and action. Each of us has some individual power to seek out expected observations, or even entrain the environment according to the model's expectations. This tells us to expect individual variability, as our prior beliefs and trial and error trajectories shaping the model diverge (although of course, we also have many priors in common). Different people can have quite varied self-evidencing paths and different models of themselves. This should foster an expectation of diversity and be a reminder in favor of humility about one's own generative model and sense of self. Perhaps this perspective could help create more empathy and compassion among diverse people. It could also help better tailor the treatments of those suffering with mental illness; perhaps computational psychiatry can work toward more individualized diagnoses—precision psychiatry through generative models—enabling more humane and bespoke treatment.

Later, in chapter 7, we will delve into how self-evidencing models the self—the self-model. In chapter 6, we finally fully focus on action, decision-making, and rationality.

6 Inference to the Best Decision

Here is the picture so far. Starting from conceptual analysis of existence, self-evidencing unfolds in terms of perceptual and active inference, proceeding in a self-supervised manner within a Markov blanket through precision control. Guided by balancing of abductive reasoning processes, self-evidencing works through finessing of an internal generative model and selective sampling of observations. Those observations in turn provide evidence for the beliefs encoded in that model. As the model is progressively finessed, it comes to represent the causal structure of the environment, where individual agents' self-evidencing paths can take different trajectories, depending upon their unique experiences and ensuing empirical priors.

Now we shift our focus more directly to active inference, decision-making, and action. We ask specifically how self-evidencing relates to rationality.

Self-evidencing leads to a conception of rationality that I will label *inference to the best decision*. I suggest that this is a compelling concept that unifies descriptive and normative accounts of rationality and that makes sense of the way that we muddle through life's decisions.

I begin this chapter by linking the imperative for self-supervision to trial and error processes in action, further developing the hill-climber analogy (encountered in chapter 3). I suggest taking an abductive perspective toward decision-making, in terms of inference to the best decision, designed as a parallel to inference to the best explanation. I follow this with a list of the best-makers for this notion. I bring this closer to everyday decision-making by focusing on deliberation and counterfactual reasoning, which I then extend to exploration and exploitation. I end by proposing that rationality just amounts to inference to the best decision.

Self-Evidencing for Action

Fundamentally, self-evidencing agents are left to their own devices when making decisions. Self-supervision, as we saw in chapters 2 and 3, is the basic constraint when marshaling candidate policies for action, considering preferences, and running through the risk and ambiguity of policies with respect to how well each would contribute to self-evidencing. This implies a kind of solitude where there is no external, objective yardstick against which inference is measured. There is only self-evidencing, which is a function of the agent's own beliefs and the observations that they make or expect to make.

None of this implies that self-evidencing agents are Robinson Crusoe–like hermits or rabid individualists. Most humans live in social settings of families, communities, workplaces, and institutions. We learn from each other and depend on each other, and through development, we hone personal relationships in multiple dimensions of life. Affordances in the environment and our bodies serve as worldly connections that funnel our decisions in distinct directions. All of this is consistent with solitary self-evidencing and should be understood in terms of the beliefs encoded in the agent's generative model. For example, the agent may believe that information from teachers or mentors typically has high precision; this belief then guides self-supervised precision control to increase the gain on what the teacher or mentor says.

This all-encompassing self-supervised process is cashed out in variational inference, as described in the previous chapters. For active inference, the objective function is the expected free energy; that is, how well self-evidencing can be expected to go, given a certain model of the world, including the agent itself, under different candidate policies. The decision to adopt a policy is the upshot of this inferential process.

Decisions abound in all our lives, in all contexts and on many timescales. We decide whether to have ice cream or fruit salad for dessert, and (once we have decided on ice cream) whether to drive or walk to the ice cream shop. We decide whether to join the workforce early in our career or pursue an education that takes longer, and how to execute that decision in detail; we decide whether to stick with our current job or explore the job market, and so on. We also engage in many kinds of collective decision-making as we vote in elections or negotiate new strategies for a company, where our policies are vetted against what we know about other people's expectations and beliefs. Given a rich, hierarchical generative model, there are many available policies for each kind of decision context. Policies can be expected to interact with causal structure in numerous ways, and predicting outcomes can easily become extremely complex. Even relatively straightforward decisions often rely on having prior beliefs about many other people's beliefs, and how those beliefs will tend to change or not given the agent's own various candidate policies. Most people are familiar with trying to balance the multitude of possible factors weighing for and against decisions in everyday examples like the ways that one could achieve the goal of losing weight, sleeping better, or becoming a better person.

Decision-making can often feel perplexingly unconstrained and nonlinear, where we have trouble retracing our steps and explaining faithfully exactly how we made our decisions. Sometimes this is chalked up to poor introspection, where the assumption is that there is a clear decision process, but we cannot access it introspectively. In contrast, I think that part of the explanation of the perplexity is that the process itself is often quite messy rather than the issue merely being about our access to the process. Even if the objective function is simple—to minimize expected free energy—the generative model harbors all the complexity and has many moving parts, and the agent is challenged to vary the various parameters of the model in a trial and error fashion when considering what to do. The decision process will be marked by fits and starts, ranging from quickly learning that one type of policy is a nonstarter to settling on a course of action, only to realize a major obstacle late in the game. Often, we update our generative model halfway through executing a policy, as the policy itself provoked some volatility, causing us to reconsider or even abort the mission. Over time, we often settle on proven policies, developing *habits* designed for particular situations. Habits can range over bodily and mental action, such as walking a particular route to get to work and attending to tasks in a certain order. Habits in turn are challenged under volatility, such as when roadwork blocks the usual footpath or new software updates meddle with our work routine; deciding whether to give up a habit is in itself difficult and takes effort. Sometimes we adopt policies that we do not really believe strongly in because doing nothing is a worse option (perhaps getting married is one such policy, since about 50 percent of marriages in some Western countries end in divorce, but it is better to be in the game with those odds than not at all). Sometimes we change our preferences, understood as our high-precision beliefs about our characteristic states, at some point as we are executing the policy (e.g., stopping myself at the door of the ice cream shop, realizing that I now prefer fruit salad).[1]

The decision process is thus often messy, nonlinear, and updated in real time, making it difficult to keep track of and to report clearly on it. But the process is not unconstrained. The main objective is to optimize self-evidencing when adopting a policy; that is, to increase the accuracy of the generative model of characteristic states by selectively sampling just the observations that we expect to see under that model. We like policies that have low risk, so they are likely to bring us close to our characteristic states. And we like policies that quite unambiguously achieve the predicted observations, independent of whether those observations are what we want. The sweet spot is when a policy is low risk and low ambiguity, when it will accurately and unambiguously produce the observations that we expect, given our characteristic, preferred outcomes. Risk and ambiguity emerge from reorganization of the free energy principle, calling for self-evidencing agents to balance them as they infer their preferences, as shown in equation (3.3).

These constraints still leave scope for much variability, depending on the overall set of beliefs in the generative model of a particular agent. An agent might be torn between two candidate policies with conflicting risk and ambiguity, settling for less by picking the low-ambiguity policy because prior preferences are imprecise in this context, and therefore ambiguity trumps risk. Agents will traverse many trajectories, adopting different polices for the same decision contexts at different times and relative to each other. This has to be the nature of decision-making—"Everything I do depends on what I believe, and what I believe depends upon what I have done in the past." Each twist and turn on these decision trajectories reflect prior beliefs about uncertainty, risk, ambiguity, and volatility, which are continuously updated as the agent learns from its perception of the world and its own actions.

A useful picture here is that of a hill-climber (picking up on the example in chapter 3), making their way through a relatively mountainous terrain toward a peak that they can sometimes vaguely make out in the distance. They have an old map to go by, but it has several gaps that they try to fill in as they go. Their route will be fairly winding rather than straight, as obstacles are in the way (such as rivers, drops, or cliffsides). Often, they must explore gullies and passes only to discover dead ends, having to retrace their steps downhill for a while; perhaps they get stuck in a ravine overnight, waiting for daylight to come before they can find a way out. Sometimes they need to deviate from their course to gain a better vantage point to look ahead. Seeing a nice, straight path stretching out in front of them can be attractive, even if it does not quite go in the desired direction. Finally, spotting a path that seems to go in the right direction is a relief, even if it is littered with fallen tree trunks and big rocks. The hill-climber might take extended detours to explore further and fill in the map in case they come back for another walk. They might give up on that faraway peak as they make their way, choosing an easier target. The climber constantly tries to look ahead, wondering if this or that path will go in the right direction, and safely.

Throughout, the climber is not guided by trying to glimpse the faraway peak, which is too often out of sight, and in any case, just seeing the peak provides little help in navigating the landscape. Instead, the climber tries to gain elevation on average as time passes. Along the way, they will try to avoid very tiring extremes where they descend all the way down to sea level again or scaling extreme cliffsides. At day's end, the hill-climber will be content if the surrounding trees are a bit shorter and the air a bit thinner. Perhaps at the end of the ordeal, they are fairly sanguine, caring less about which peak they reached than about the journey to get there.

Two different hill climbers will be unlikely to follow the same path, especially if they have somewhat different maps to begin with. Some climbers will get to their peak swiftly, others more leisurely; some will explore more than others and have an easier time when they come back later. But twists and turns and detours notwithstanding, and so long as they do not succumb to the dangers on the way, on average, they will be likely to climb upward over the long term.

Of course, even within this abductive picture, there is room for differences. Although much of our decision-making is akin to relatively improvised hill climbing, it is possible to add further constraints to the process. The objective function of expected free energy considers the surprise of future observations, given the model and the policy. This forward-looking process is quite open-ended since many different constellations of external causes and states can in principle give rise to the same observations (e.g., observing increased money earnings generated from winning a lottery as opposed to working hard). To constrain this process, and arrive at something that looks like explicit, stepwise, goal-directed planning, it could be that the agent considers not just the external observations they might make, but also the beliefs about latent states that they expect to form in the future. That is, the observation that is assessed for free energy could be (introspective) observations of the agent's own future beliefs about latent states. For example, the agent puts a strong precision on believing that they have a high-paid job, not just that they will observe an income caused by some latent state. Now the agent can better work backward, considering which is the best policy to acquire this final belief, and then continue policy by policy until they reach their current state. The analogy for this more constrained abductive process would not be the rather "jazzy" hill-climber but instead the bouldering expert who has worked out with high precision how to get to the top quickest by working backward from the top, considering the best move at each step down to the ground.[2]

Self-evidencing in decision-making is fundamentally like such deliberative or habitual abductive hill climbing (or occasionally like bouldering). We muddle along with whatever model we come equipped with, making more or less informed choices, considering risk and ambiguity as we look ahead; we are content so long as we accumulate evidence for the model on average and over time.

Inference to the Best Decision and Its Best-Makers

In chapter 3, I argued that inference to the best explanation and the free energy principle are complementary—and both are captured by self-evidencing. Specifically, both imply that agents are engaged in balancing the best-makers around complexity and accuracy, risk and ambiguity, and exploration and exploitation (recall figure 3.1). Guided by model evidence (variational or expected free energy), we do our best to weigh all these properties of explanations against each other and infer to the explanation that comes out on top as the best one. What I will do here is to focus specifically on the forward-looking aspects of inference to the best explanation, building toward a notion of inference to the best decision. This notion makes sense of how decision-making is a trial and error process and how decision processes often appear muddled and nonlinear. The agent in effect assumes, contrary to fact, that they have reached their preferred state and then asks, "What is the best explanation, in terms of states of the world and my own actions, for how I got to make these observations?" The answer to this abductive question tells the agent which policy to select. This abductive process reveals a distinctive account of decision-making, founded on the free energy principle.[3]

There is an interesting parallel here between the philosophy of science and decision science. On the one hand, there are normative accounts of both science and of rationality, where scientific inference and decision-making are seen as unitary endeavors, aiming at ideals and norms for maximizing truth and reward. On the other hand, once it is realized that actual science and decision-making are far from ideal, descriptive accounts emerge, with a focus on actual practices, conventions, heuristics, and biases. The problem with descriptive accounts is that they risk leaving rationality seem unprincipled and directionless. In both science and decision theory, there is therefore a dilemma between normative and descriptive approaches, where one horn is intractability and the other is lack of a principled grounding. As argued in chapter 3, there is reason to believe that inference to the best explanation in science, when viewed as self-evidencing, provides a route to avoid this dilemma since approximate inference moves from what we *actually* do to what we *should* do. I pursue a similar project here, harnessing the messiness of a more descriptively accurate account of rationality to reach a more principled, normative form of inference.[4]

The first question to address for the notion of inference to the best decision is: What might its best-makers be? The self-evidencing agent considers various policies, trying to decide which to adopt, balancing them against each other in variational inference, assessed against expected free energy. This decision-process is both prospective and counterfactual, as the agent considers future self-evidencing, and under different subjunctive scenarios for which observations (and perhaps future beliefs) would occur if they were to adopt this or that policy.

We saw in our discussion of active inference in chapter 3 that a policy is rated on its ambiguity, how certain it is that the action will lead to its expected observations, and its risk, whether those observations are close to preferred (i.e., characteristic) states. Different policies can then issue in more or less surprising observations, given current beliefs and prior preferences. Low ambiguity and risk then make decisions better—agents should select policies that get them precisely what they want, or, equivalently, self-evidencing policies that minimize surprise, relative to the agent's model.

Risk is aversive in decision-making analogously to how complexity is aversive in inference to the best perceptual belief. If the agent infers a risky policy, then that is tantamount to expecting a divergence between what will happen under the policy and what the agent prefers to happen. This is analogous to how, in perceptual inference, a more complex model increases the divergence between posterior and prior beliefs. The riskier the policy, the more complex the observations expected under it will be. The more risk that the agent takes on, the more complex their model of their characteristic states will become. In the extreme, policies are so risky that the agent's beliefs about counterfactual outcomes are incompatible with ongoing existence—agents could figuratively (and perhaps literally) tear themselves apart. However, risk is ubiquitous, and agents need to accept some level of risk. To illustrate, if an agent applies for a new job, then the preferred outcome is to get it, and being rejected might be the acceptable risk, whereas unacceptable risk might be that the agent's current boss gets wind of the plan and (unfairly) fires the agent. Accepting that risk would be equivalent to accepting that among one's characteristic states is being unemployed. Deciding whether to apply for the job requires careful deliberation, and perhaps some exploration and fact-finding about the current boss; perhaps this includes weighing the states of being unhappy but paid in the current job versus being happier in a new job.

Taking risk into account when making decisions is therefore difficult and amounts to more than just weighing the possible rewards and punishments. Given that active inference is self-evidencing, adopting risky policies contributes to the evidential trajectory for the agent's model, shaping it for decision-making in the future. Conversely, being too risk averse might overly restrict the policy landscape, slowing down self-evidencing and preventing development and learning. A self-supervised agent trying to navigate this best-maker for decisions must draw on wide background beliefs about the decision context, trying to adjudicate how strongly their past history of self-evidencing should inform risk assessment.

Ambiguity is linked to inaccuracy, where an ambiguous policy is not tied very closely to specific observations. This is a kind of policy entropy or imprecision where there is much variance in the observations under the policy. An unambiguous policy leads to outcomes that provide strong model evidence. For example, when entering a hotel room, it is better to select the unambiguous policy of switching on the light to reveal exactly where the furniture is, rather than the ambiguous policy of stumbling around in the dark in an unfamiliar room.

A policy can be risky but have low ambiguity, such as a slot machine that either gives a \$2 win or a \$2 loss, at some known probability. A low-risk policy can have high ambiguity, such as a lottery with a 70 percent chance of winning, but one can win anything from 1 cent to $1 million. The agent might be able to assess risk quite well but still be uncertain about which policy to infer, if the policies are all uncomfortably ambiguous. An agent might also be seduced by unambiguous policies, even if they are risky; indeed, some gambling devices might exploit this, presenting very unambiguous policies that eventually undermine the sustainability of the agent's model (through heavy financial losses). As we saw for accuracy and complexity in inference to the best explanation, wielding the best-makers for risk and ambiguity calls for often-difficult balancing of considerations.

For each policy inference, the agent needs to consider how much irreducible noise there would be in that context, which will contribute to variability in the observations, making the available policies more ambiguous. Over time, the agent also must consider what mitigating policies should be put in place to counter reducible noise in the observations. For example, an agent who decides to attend to others' speech but keeps misunderstanding might mitigate this imprecision by starting to wear a hearing aid; this would in turn change the beliefs about irreducible noise in the policy to attend, where the hearing aid works better when there is only one speaker and less well when there are more (i.e., the hearing aid changes the wearer's ability to solve the cocktail party problem of picking out a particular conversation in a loud crowd). This example speaks to an important distinction between actions that are for pragmatic value and actions that are for *epistemic value* or information gain. In the latter case, low-ambiguity policies will increase information gain, and the agent will gravitate toward such policies when trying to figure out the lay of the land (as seen, for example, in highly controlled laboratory experiments).

Low ambiguity is important for decision-making, but the agent will be stymied if the expectations for ambiguity are too fastidious. There will always be some slippage between the implemented policy and the observed outcomes, and agents need to accept some degree of volatility and ensuing lack of control (even slot machines can malfunction, for example). Adjudicating ambiguity is nontrivial for a self-supervised agent who ultimately must form and rely on their own beliefs about acceptable levels of uncertainty in different contexts. Expecting too much ambiguity means that the agent's future states quickly become uncertain, given their actions. In contrast, committing to extremely low-ambiguity policies will likely constrain the agent's policy repertoire, making them forgo opportunities. To illustrate, my colleague Kelsey

Perrykkad and I argued that some repetitive "stimming" behaviors in people with autism spectrum condition (such as repetitive flapping of the hand), might relate to the low ambiguity of policies for such movements, facilitated by their very tight perception-action loop; this might suggest diversity in different people's expectations for ambiguity.[5]

Viewed a bit more from afar, the agent's overall policy landscape will be informed by both risk and ambiguity such that, over time and everything else being equal, the agent will gravitate toward enacting policies with high fidelity between actions and outcomes, where the outcomes are consistent with the agent's characteristic states. The agent's active inferences determine a self-evidencing path over time, helping to shape their model and future expectations. Just as it is hard to balance low-risk against high-risk policies and low-ambiguity against high-ambiguity policies, it is hard to balance risk against ambiguity. Getting these balances right, then, is a key element in inference to the best decision. This self-supervised work depends on constant maintenance and nurturing of the agent's broad, hierarchical network of beliefs about levels of irreducible and state-dependent noise, as well as about their own metacognitive confidence in the beliefs of the model.

In this process, agents will learn from their actual enacted policies. Often, agents are surprised about the level of ambiguity or risk in different contexts. A situation that we predicted to afford only imprecise outcomes might actually deliver precise outcomes (such as when new artificial intelligence software in a hearing aid helps solve the cocktail party problem), or a situation that we thought would be low risk turns out to harbor unexpected risk (such as the current boss's anger about the employee applying for a job elsewhere). These cases lead to learning about which policies are worth considering, like a scientist learning which set of hypotheses they should consider in inference to the best explanation. Given that we have policy predictions and policy prediction errors here, we have prediction errors for ambiguity and risk, which can guide policy inference. A best-maker for inference to the best decision, then, is whether the policy is supported by past risk and ambiguity learning, analogous to the integration best-maker in inference to the best explanation. As we will discuss more shortly, there is a difficult balancing act here too, however, since it can also be maladaptive to rely too much on past learning—namely, under volatile conditions that call for new, untried policies. As always, adjudicating how much agents should trust prior learning depends on their wider beliefs about whether the policy in question actually projects to the current situation; this relates to the generalization best-maker in inference to the best explanation (discussed in chapter 3).

Risk and ambiguity also relate to the relative simplicity of policies. Most of us will know the difference between overly complex and overly simple courses of action. A heist planned to the tiniest detail is likely to be foiled by small divergences, such as the guard being one minute late on their round; and a simplistic policy of just storming the bank is also unlikely to work. Complex policies are also unattractive because we want to learn from action in order to update our policies for future use, and very complex plans generalize less well to new situations. Simple policies with few moving parts are great for generalizing but might fail when put to the test in a concrete situation. Again, we have a candidate best-maker, policy simplicity, which is difficult to adjudicate and demands much background knowledge to optimize. Different agents might also have different appetites for complexity, depending on how each reads the context in terms of trusting that things will work out without having planned every minute step. In this way, different agents will traverse different policy landscapes, depending on their beliefs about the overall generative process and the kinds of volatilities that might be expected. Notice that, although the agent might have such a heuristic—a prior belief about how to balance the simplicity and complexity of their policies—this belief is itself updated as each policy outcome is considered for risk and ambiguity.

This section has lined up several best-makers for inference to the best decision and highlighted the need to balance them. We have focused on policies with low risk and low ambiguity, which are analogous to best-makers relating to complexity and accuracy in inference to the best explanation. The next section considers best-makers related to deliberation specifically.

Deliberation and Counterfactuals

Next to learning—in a post-hoc manner from one's actual actions—there is vicarious learning from prospective, *counterfactual* testing of policies. This is mental deliberation of policies ahead of inferring one, under various possible policies and scenarios. Agents take themselves offline and run their generative model counterfactually on certain policies to gauge the risk and ambiguity for their expected observations. Policies may be stored in memory from past experience and, as they often are, may be suggested from conspecifics. In mental deliberation, the agent will have the opportunity to play around with alternative policies and to consider if there are available mitigating policies that could be put into place. In the heist example from the previous section, the gang can go through and simulate various more or less detailed plans, building in all the available information about the layout of the bank, the alarm system, and the movements of guards, and they might even deliberate whether it is worth preventing the conveniently lazy but also unpredictable guard from coming to work on the day of the job.

Mental deliberation for active inference plays a key role in understanding the self-evidencing agent. Since deliberation depends on self-supervised formation of prior beliefs about the generative process, outcomes of deliberation will differ among agents with different generative models (which is why it is important for the heist gang members to communicate with each other). Different agents will also have different prior beliefs about which policy landscape to consider counterfactually, and how far to counterfactually go before being satisfied that a policy is good to commit to. Policy deliberation is therefore a good candidate as a best-maker in inference to the best decision, as it affords more information about risk and ambiguity, on the assumption that the generative model is in fact reasonably accurate and well suited for offline processes. It tracks well with the idea, from scientific inference to the best explanation, that best explanations should be able to generate fruitful predictions, and also that an important part of the abductive process is to consider if the initial set of candidate explanations or policies are adequate.[6]

Counterfactuals (counterfactual conditionals) have been discussed philosophically for some time. These are subjunctive formulations of the type "If (contrary to actual fact) it were that *A*, then it would be that *B*"; for example, "If there were higher taxes, then there would be a better public health-care system." Counterfactuals are semantically fascinating and can behave in different ways depending on how they are wielded. This suggests that with creatures like humans, who have temporally deep generative models—with much room for considering all sorts of causal interactions—small changes in counterfactual scenarios can create large differences in the assessment of outcomes. As a counterfactual scenario is explored in more depth, more causal interactions can crop up, and more tuning of scenarios may be needed, leading to different lessons. A counterfactual that yields one outcome in a particular scenario will not give the same outcome when adding more detail to the scenario (formally, one cannot strengthen the antecedent of counterfactual conditionals; for example, it is likely false to say, "If there were higher taxes and higher levels of corruption, then there would be a better public healthcare system"). As an example of decision-making, an agent might deliberate about moving to another city, so they counterfactually consider what their daily lives would be like if they were to live in this or that city; they might like the idea of a city with vibrant cultural life but then dislike it when considering a city that also has a dreadful public transport system, and then go back and forth in their deliberation depending on which aspects they consider. This can help them adopt a policy for identifying a city with both good cultural life and good public transport.[7]

The way that the agent is conceived in the counterfactual scenarios matters greatly too. A well-known example, attributed to Willard van Orman Quine, considers what would happen if Julius Caesar were to appear in the Korean War—would he use nuclear bombs or catapults? One way of navigating this question is to either fix Caesar's beliefs to Roman times or to update them to the beliefs that he would have acquired in that modern scenario—the expected outcomes would probably differ accordingly. Recent developments in formal active inference theories track aspects of this semantic distinction, where active inference can be informed by deliberation about which beliefs the agent would have adopted in the given counterfactual scenario. For example, someone not wanting a dog as a pet because they believe that it would be too much work might nevertheless acquire a puppy, based on the deliberation that if they were to get one, their beliefs about the effort would be inversely correlated to the experienced cuteness and emotional bond to the dog. This is a more sophisticated active inference type, which treats the agent's belief system itself as part of the causal order of things that will have changed in the scenario. Notice that here, the person is not merely relying on a model of themselves to guide their decision-making; the person is also treating their own belief state counterfactually, to be updated as part of the planning.[8]

Optimizing deliberation with respect to both variations of counterfactual scenarios and the distinction among types of counterfactuals appears to be important for inference to the best decision. Which policy comes out the winner in such offline testing depends on the scenario under which it is counterfactually tested. Undertaking counterfactual deliberation, therefore, is a candidate best-maker for decision-making. However, since the outcome is measured only against the objective function of expected free energy, the agent needs to build up beliefs through vicarious trial and error about how to do the difficult balancing of considerations. For example, the agent needs to learn which contexts call for counterfactuals with beliefs anchored in the current state and which call for sophistication, where beliefs may change. Learning in this counterfactual space is not always easy; for example, it is difficult for agents to accurately forecast what their affective states would be in various scenarios, such as experiencing trauma, winning the lottery, or indeed getting a puppy. Another difficult consideration is when to terminate counterfactual deliberation since engaging in it for too long could risk missing the opportunity to act; for example, the puppy might be sold to another person if the agent is considering too many counterfactual variations of the puppy scenario.[9]

Overall, carefully (but not too carefully) counterfactually deliberated policies form part of inference to the best decision. The exact way that deliberation is conducted is subject to many prior beliefs on the part of the self-supervised agent. In the abductive hill-climber analogy, we should expect agents to follow quite diverse trajectories as a result of this model-constrained trial and error process.

Deliberating on Exploitation and Exploration

As we discussed earlier in this chapter and in chapter 3, sometimes we act to maximize pragmatic value or utility, and other times, we act for epistemic value or information gain—we shift between exploiting and exploring. Managing the balance between exploring and exploiting is fundamental to biological agents, and it is manifested in many ways within humans and across species. Rats, for example, will explore new patches when resources in the existing patch deplete, taking the expected distance to new patches into account. Exploration can be guided by beliefs about the lay of the land, as in the case of the rats (i.e., it is based on a model of the available patches for exploration) or by naked exploration, as in the case of *Escherichia coli,* which tumbles when nutrients deplete (i.e., model-free). Humans also navigate the explore-exploit distinction across domains of behavior. This includes mental exploration of the counterfactual types discussed previously, where agents might explore counterfactual scenarios where they push exploitation to various degrees, to decide when resources would be so depleted that they should go exploring; or agents can investigate counterfactually where they should start exploring, such that the exploration is somewhat model-based as they set out.[10]

The balance between exploitation and exploration is directly related to the expected free energy as in equation (3.4), which delivers the tractable objective function for active inference. That is, free energy is the combination of epistemic value, as the divergence of the posterior from the prior, given the policy, and pragmatic value, as the expected (log) likelihood of preferred states. Crucially, both terms are assessed by expected free energy so that they can be directly compared. This means that the agent can seamlessly shift back and forth between exploitation and exploration. If there is more expected epistemic value than pragmatic value, then the agent goes exploring, and vice versa. This means there is no computational mystery about how and when agents decide between exploring and exploiting. This is an appealingly simple story, where the simplicity is paired with a need for a good generative model to ground active inference. This puts the onus on the agent to maintain and update their model and to explore model space appropriately for policies and their expected outcomes, taking due note of expected risk and ambiguity. I turn next to discussing some of the abductive processes that the agent might go through when settling their beliefs about exploring and exploiting in the lead-up to doing either.[11]

Managing the beliefs that determine a good exploit-explore balance is central to adaptive behavior because we live in a changeable world where resources are not maintained at a stable level and where many factors, often including our own actions and the actions of others, cause resource levels to change. For example, as a fishing ground is overfished, exploration of new grounds is needed, and (depressingly) once we have depleted fish stocks beyond repair, new ways of producing fish or protein might need to be explored. Exploration is often anticipatory and may have game theoretical elements. For example, even before the fishing ground is showing signs of depletion, an agent might allocate some resources to explore new grounds in anticipation that the richness of the current ground will attract a greater number of fishing vessels that will accelerate the depletion, and wishing to be the first mover on any new grounds.

Deliberation on when to abort exploitation or return from exploration is not straightforward for self-supervised agents who need to anticipate patterns of depletion or information gain, which might be nonlinear and exponential or ridden with uncertainty, depending on other actors and many volatile factors in the environment. There are several questions that the agent must ask themselves, and answer based on their own prior beliefs. They must consider what the resource levels are, how their own and others' actions will affect the rate of depletion, when to abort exploitation to go exploring, how to decide where to go exploring, and how and whether further offline deliberation can help inform their decision, as well as how much pragmatic value or epistemic value is in the offing. All these considerations relate to the risk and ambiguity of the policies on offer and will determine the self-evidencing path of the agent over time.

The world is a place to exploit or explore, and in a sense, so is the internal policy landscape itself. One can say that the overall payoff from an agent's current policy repertoire can begin to deplete when the world begins to change, making more and more policies unfit for their purpose. This happens particularly under increased volatility—and the stark uncertainty that it evinces. In principle, volatility should lead to increased learning rate or, equivalently, less dependence on a long history of learning. This makes sense because volatility is likely occasioned by some hitherto unknown cause, which the agent would benefit from learning about. This requirement to increase the learning rate extends to learning new policies too, since volatility should make us less certain in our judgments about risk and ambiguity about our usual policies. A good (if traumatic) example is the COVID-19 pandemic, which forced almost everyone to throw out many of their more or less habitual policies and seek new goal-directed policies that were better suited to the new reality of social distancing, lockdowns, pressure on health systems, and other circumstances. This is analogous to how, in inference to the best explanation, a scientist might react to hugely unexpected findings by throwing out old models and theories altogether and brainstorming to identify new theories. This inferential movement often features in the philosophy and history of science, such as in the Copernican revolution from the geocentric to the heliocentric astronomical framework. Policy exploration under pandemic volatility ranged from the small, such as how to secure sufficient stocks of toilet paper and face masks, to the big, such as how to conduct triage at hospitals and whether to invest rapidly in messenger ribonucleic acid (mRNA) vaccines in addition to traditional vaccines.

Sometimes policy exploration must be done on the basis of quite simple, underspecified models in advance of having learned much about the new, interfering cause and its downstream effects. This sentiment was expressed poignantly by the World Health Organization (WHO) executive director, Mike Ryan, in the very early days of the COVID-19 pandemic: "Perfection is the enemy of the good when it comes to emergency management. Speed trumps perfection, and the problem in society we have at the moment is everyone is afraid of making a mistake—everyone is afraid of the consequence of error. But the greatest error is not to move. The greatest error is to be paralyzed by the fear of failure."[12] This quote captures several of the challenges to decision-making that we have discussed here, such as being overly risk averse and seeking only unambiguous policies. The example tells us that volatility is an important factor that should guide our policy inference away from low risk and low ambiguity.

Counterfactual reasoning is important for policy exploration, providing a safe, offline space to test new and perhaps even outlandish policies. But this has its own pitfalls, of course. When agents do not expose themselves to the real consequences of their actions and stick to counterfactual spaces, their ignorance and biases can be relatively unchecked. For example, it is well known that people have exponential bias, having trouble dealing with things that change exponentially, such as an infection rate higher than 1. People underestimate how quickly the infection will grow and how important it is to act while numbers still seem negligibly low (as Ryan emphasized in his statement).

I will use one of our own studies to illustrate some of these themes. During the pandemic, my then–PhD student Simon van Baal invented a version of a social goods game that we called the "self-isolation game," where ten participants played with each other in a pandemic setting with exponential growth, having to decide when to isolate in the light of infection numbers. They never learned to self-isolate in a timely fashion, even though they were slapped with ruinous "lockdowns" when infections picked up exponential steam. This bias appears to be costly when it comes to learning from our past decisions in counterfactual deliberation. If the agent's counterfactual simulation of delaying the adoption of countermeasures fails to take exponential growth into account, they will act too late. Moreover, if they live in a community where strict countermeasures were put in place, successfully curbing the pandemic—at the low numbers necessary to halt exponential growth until vaccines were rolled out—then they might run counterfactuals that make them critical of those past strict restrictions in hindsight. The agent might say that the fact that deaths and hospitalizations did not increase once restrictions were lifted is evidence that having no restrictions would have been fine, failing to counterfactually compare unvaccinated, exponential growth with vaccinated, less exponential growth. Their counterfactuals might also fail to account for how the relevant context might affect exponential growth. For example, they might compare harsh restrictions in one community to less harsh ones in another community, leading them to the belief that since infections were lower in the latter, it would have been fine to have such mild restrictions in the former; but the agent might fail to acknowledge that the communities differ in geography, demographics, income equality, and health-care systems, all of which can play a role as volatility-inducing factors when considering how successful a restriction might be. This all points to how difficult it is to learn from past policies through mental, offline deliberation, especially for complex and sophisticated counterfactuals; and how strongly such deliberation depends on the agent's generative model, biases, and other factors, adding to the difficulty of balancing policy perpetuation against policy exploration.[13]

The explore-exploit balance is critical to understanding the self-evidencing agent, and it falls elegantly out of the notion of expected free energy. However, managing the balance such that it leads to successful policies depends on deliberation, which is often complex and counterfactual, with many degrees of freedom. Deliberation is a best-maker for inference to the best decision—and like all best-makers, it takes considerable skill wield it.

So far, then, we have set out the best-makers for inference to the best decision, focusing on risk and ambiguity as well as exploration and exploitation, together with counterfactual deliberation. We have noted how the objective function, although tractable, puts the onus on the agent to wield their generative model as best they can, through trial and error, from their self-supervised standpoint. I will now draw this together in an argument that inference to the best decision plausibly captures the notion of rationality.

Rational Decision-Making as Inference to the Best Decision

I have been building up to a self-evidencing account of decision-making in terms of inference to the best decision. Earlier in this chapter, I identified and reflected on candidates for best-makers for best decisions. In an active inference setting for a self-evidencing agent, what makes a decision best is that the inferred policy is low risk and low ambiguity jointly considered, is simple but not too simple, is based on past learning but takes volatility into account, considers whether, when, and how to explore or exploit, and is stress-tested in counterfactual deliberations. If a policy comes out on top for these best-makers, then that is the policy that should be inferred, rather than policies faring less well.

Crucially, the point here is not that there are additional objective functions for the best-makers, as well as the objective function given by the expected free energy (understood as the fit of the expected observation and the model encoding the beliefs and preferences of the agent, given the policy in question). Rather, the point is that if we are looking at a self-evidencing agent engaged in planning and decision-making, then, although all the agent does is assessing policies against expected free energy, a more nuanced picture emerges when we "pop the hood" and look at the actual processes that the agent undertakes in doing so. The agent will be seen to form beliefs for the best-makers about risk, ambiguity, and other factors, and trialing and updating these over time. For example, in different contexts, the agent will form, and be guided by, beliefs about how much ambiguity versus risk can be tolerated, or about how much volatility is needed before they will explore new policies, or about how much counterfactual deliberation is needed before they are confident about a given policy. These best-maker beliefs themselves are self-evidenced in ongoing inference; for example, the agent may learn to their chagrin that free energy actually increased when they failed to deliberate counterfactually for long enough about big decisions; this could include decisions about where to live and work ("the paycheck may be large, but I failed to consider the resulting poor work-life balance"), and they may resolve to adopt the policy of spending more time on deliberation the next time that a big decision needs to be made.

The best-makers for best decisions are analogous to the best-makers for best explanations (recall figure 3.1). Risk and ambiguity relate to explaining away observations accurately and without complexity; being informed by past learning relates to generalizability, deliberation relates to integration with other beliefs and to determining the set of candidate explanations, and pragmatic and epistemic value relate to fruitfulness. Overall, the best policy should explain expected observations, as the best explanation explains the actual observations. In turn, the observations, when they actually occur, become evidence for the policy, considered as a belief about the agent's own state-transitions (e.g., from not having a coffee cup in their hand to getting one by reaching out; or from having one job to getting another, with better pay). Here, we see the self-evidencing circle in play, where the policy, if it is good, will accumulate evidence for itself and thereby stand stronger for the next decision (e.g., the agent does not have to engage much deep counterfactual deliberation to reach out for the coffee cup unless they are in a volatile situation, such as reaching for the cup on a turbulent plane flight).

The claim, then, is that an agent is rational when they are engaged in inference to the best decision. This implies that it would be irrational to pick a policy that fares worse overall on the best-makers than another policy does. Put differently, it is irrational to select a policy that does not minimize the expected free energy since expected free energy, via the best-makers, tracks just the things that we commonly expect of a rational agent. Don't take too many risks to achieve your desired states or do things with ambiguous, overly uncertain outcomes; don't try really complex or overly simple plans; explore the lay of the land if your exploitative policies begin to seem uncertain, but don't explore endlessly and pointlessly either. In addition, when making these choices, also weigh them against your beliefs about resource depletion, about your state of knowledge, and about volatility in your situation, which might well change your beliefs about what is risky or certain or simple. This tells us that what we expect from the rational agent is not a deductive beeline of normatively guided reflection but rather a winding, context-dependent trial and error process, constrained by and buffeted to varying degrees by the agent's various beliefs about context and about their own preferences and goal states.

That gives the basic formulation of the notion of inference to the best decision, as an account of rationality. I believe that it is a plausible view of rationality because it is both descriptively and normatively adequate, where, crucially, it is self-evidencing that unifies these descriptive and normative perspectives in one framework. I set out my arguments for adequacy and normativity next.

Inference to the best decision is descriptively adequate as an account of rationality because it tracks well with how I tried to capture decision-making as a nonlinear, often muddled trial and error process. This partly reflects a first-person perspective where making decisions rarely feels like following a straightforward calculus. To me at least, that is how it is when making decisions: We engage in more or less conscious and explicit deliberation, and try different policy avenues in thought (and maybe in practice as well), going back and forth until we settle on a policy that, with some degree of confidence, seems like the best one, among the things that we considered. We cast about trying to take relevant facts into account and rely on diverse auxiliary policies to scaffold our decision-making (such as deciding to ask others or large language models for advice). As we engage in these processes, we are guided by, and continuously update, beliefs about the best-makers, making mental notes for our future reference. Descriptively, inference to the best decision also accords well with a large body of literature on bounded rationality showing that human agents follow all sorts of biases and heuristics, rather than utility-maximizing algorithms or Bayesian calculation, when they make decisions. Agents try various hacks, tricks, or more or less elaborate schemes informed by environmental and social cues when making decisions; they latch on to this or that fact about the situation that makes sense for them and focus their deliberation on that, and they develop habits for solving decision problems, which they might abandon in the face of volatility.[14]

My argument is that when we try to extract the underlying patterns for this description of our decision-making, it looks like agents are guided by beliefs about the best-makers that I have set out. We try one policy path in counterfactual deliberation but find it too complex; we simplify but then lose confidence that the policy will remain low risk, so we settle on a middle ground and embark on pursuing this policy—only to discover that too much uncertainty accumulates too quickly. So we halt and update our model with inferences about some assumed hitherto unknown latent causes and try the policy again in our minds, but this time, based on advice from a mentor, we adjust for the new interfering causes and then try to act again. The claim that this descriptive account of our decision-making psychology lines up with the best-makers is empirical, of course. Here, I am going by what I think is a common experience of making decisions, recognizable to many of us—and I suggest that this idea is theoretically plausible and worthy of empirical investigation.

This description of decision-making is exactly what one would expect if the starting point is a self-evidencing agent. From a self-supervised standpoint, minimization of expected free energy depends on variational inference that makes the agent's inferences, based on their generative model, approximate optimal inference. The agent is guided by just the one objective function and have all and only their prior beliefs, harnessed in their model, at their disposal to identify the policy that will do the job. From the self-supervised, inside perspective, all the agent can do is to consecutively fix some beliefs and vary others, and then check the self-evidencing outcomes counterfactually—"Would this policy, under these beliefs about risk, ambiguity, and so on, maximize model evidence?" What our discussion shows is that even with prior learning to constrain which beliefs to fix and which to vary, this variational trial and error process has many degrees of freedom where agents must throw themselves into the fray.

The next step is to argue that in addition to being descriptively adequate, inference to the best decision is normatively adequate. *Normativity* can relate to different properties, and in discussions of decision-making, it may typically be optimality in terms of utility maximization or other norms of rationality, such as Bayesian rationality. The question is whether self-evidencing can be normative in any respectable sense, given that it relies on what looks like a bootstrapping process of self-fulfilling prophecies based on possibly idiosyncratic beliefs about risk, ambiguity, simplicity and other criteria. This was discussed in chapter 2 and elaborated on in subsequent chapters, showing the ways in which inference, via variational routes to minimizing the bound on surprise, indeed approximates exact or optimal inference. Normativity comes in by making a virtue of the model-constrained trial and error processes that we actually engage in—those processes are what ensure approximation to Bayesian inference and underpin the ability to occupy unsurprising states.

Inference to the best decision, therefore, is both descriptively and normatively adequate. The last step is to show that with self-evidencing, inference to the best decision unifies these descriptive and normative perspectives. The beauty of self-evidencing under the free energy principle is that the agent can adhere to the norms "blindly" as it were, just by muddling along as best they can, given the beliefs encoded in their generative model. That is, the agent self-evidences in a self-supervised manner based only on guidance by the internally tractable measure of expected fit between the observations generated from their assumed model and their expected observations given a policy. By doing this, they must—in a hill-climbing manner—approximate what the objective norm of Bayesian rationality prescribes. Self-evidencing makes descriptive sense of everyday experiences as a decision-maker and is at the same time, and thereby, tractably normative.

The possibility and occurrence of irrationality and error are important testing grounds for both the descriptive and normative adequacy of the notion of inference to the best decision. Agents can be described as irrational or in error only against the backdrop of normativity. Part of the focus in chapter 5 were the various ways that self-evidencing agents can go wrong or be irrational, and this discussion extends to inference to the best decision. Even though agents must be approximating optimality on average in the long term, the trial and error nature of inference to the best decision often pulls them away from optimality; for example, they try a new policy under volatility that turns out to be too risky, such as placing large bets on a roulette wheel after losing their job. This can get them stuck for extended periods of time in local attractors far from the true generative process, such as gambling addiction facilitated by nefarious slot machine algorithms that create a perfect storm of risky, unambiguous policies that allow victims to sustain homeostatic function while preventing any kind of thriving, and with barely noticeable increments of exploiting their resources.

Inference to the best decision is therefore an attractive approach to rationality and decision-making. It is both descriptively and normatively adequate. It resolves the significant tension between distinct descriptively and normatively adequate theories by showing how actual behavior is exactly what should be expected of agents approximating ideal norms through variational processes.

Concluding Remarks: Rationality as Abductive Hill Climbing

In this chapter, I have argued for inference to the best decision as an account of rational decision-making. Starting from a self-supervised, self-evidencing agent, inference to the best decision adequately accounts for and integrates descriptive and normative aspects of decision-making. I think that this account of decision-making is attractive because it puts rationality within reach of ordinary agents while making sense of the everyday experience of trying to make good decisions. It does not set out impossible ideals as the measure of rationality, nor does it leave rationality as just a matter of wielding a constellation of heuristics and biases.

In chapter 3, I offered an account of self-evidencing in terms of inference to the best explanation, focused mostly on perception and belief formation. With the development of inference to the best decision in this chapter, we can see the outlines of a self-evidencing account of rationality as such, for decision-making, perception, and belief formation—namely, the combination of inference to the best decision and inference to the best explanation. This final unification is also vouched for through self-evidencing, which places perceptual and active inference within the same framework, aligned with variational and expected free energy respectively.

In addition, I offered a hill-climbing analogy for what I described as the often muddled and nonlinear process of decision-making, which also applies to the process of variational inference to the best explanation. This type of ampliative, inductive inference is sometimes called *abduction*, so we can see rationality in perception, belief, and decision in general as abductive hill climbing. Extending the analogy, we can imagine that the climate for our hill-climber is hostile—they must keep going to avoid succumbing to the cold or other dangers. This builds existence into the analogy, in the sense that the moment when self-evidencing begins to cease, the agent's states will begin to be challenged, with increasing risk of inexistence. So long as climbers exist, they will be moving. Guided by the one belief that they should be moving upward over the long-term average, they will approximate ascension, even if they are detouring at certain periods, caught in low-lying, local valleys, or occupy smaller hilltops for some time. We then add further complexities, such as volatility in the landscape (landslides, avalanches, weather events, climate change, annoying tour groups to avoid) and in the climbers themselves (fatigue, change of heart or priorities, zest for exploration)—they might say, "I can't go up that path because there was a landslide, and my legs are tired." Now their policy selection and the way that they interact with and fill in their incomplete map is highly dynamic, depending on their changing capacities and beliefs about which paths are now safe and which not. The climbers are working the map as they work themselves through the changeable landscape, making decisions and forming beliefs as best they can, and taking the evolving model of themselves into account. I think this is an apt analogy for the way that we exist, decide, and believe as self-evidencing agents.

But who are these hill-climbers? How does who they are influence what they do? How does what they do influence who they are? Chapter 7 focuses on the agent at the center of all this—the embodied, individual self.

7 Self-Models and Selves

In this chapter, we ask what it may mean for a self-evidencing agent to have or to be a self. What kind of thing is the self, if it is a thing? We also consider the role of the body for selfhood and agency. These questions are quite pressing now, after having focused through several chapters on "self"-evidencing—on agents who exist, have everyday experiences, and make decisions for themselves.

The upshot of the chapter will be that the self is, mostly, a really ordinary thing. It is just as ordinary and as real a thing as other uncontroversially ordinary and real things. The self-evidencing approach to the self that I will present in this chapter turns out to be comforting and liberating, but also in other respects perhaps somewhat alarming. This account of the self aligns well with how the self manifests in the everyday experiences of our day-to-day lives. The idea is that selves relate to relatively invariant causal factors within ourselves, which express themselves in the way that we interact in observable ways with others and the environment. Along the way, I will disambiguate on the notion of the "self" in "self-evidencing" and the idea of having or being a self.

I begin this chapter by considering how to approach this task, distinguishing between the self and the sense of self. I then unpack the notion of the self-model as a structured representation of the self and consider the reality of the self that is represented by such a model. Next, I consider the function of the self-model in self-evidencing and what it might mean to *be* (rather than have) such a model. I end by proposing a notion of humans and our selves, casting us as what I will call "agents of volatility."

How Should We Look for the Self?

There is a long tradition in philosophy for asking about the nature and metaphysics of the self. David Hume famously observed that "when I enter most intimately into what I call myself . . . I never catch myself at any time without a perception, and never can observe any thing but the perception." Most would likely agree that the self is not easy to catch in introspection, and this difficulty has led to a wide range of psychological and philosophical theories about the self. In work led by my colleague Kelsey Perrykkad, we tallied them, stopping at thirty-nine conceptions of the self. Here is a sample of fifteen of them—variously saying that fundamentally the self is an agent, bounded, a disposition, illusory, integrative glue, interpersonal, material, minimal, narrative, a person, social, spiritual, transparent, unconstructed, and unified. Not only are many of these concepts in tension with each other, many appear to stem from accounts of very different aspects of the self, and different explanatory projects about the self. There is little philosophical or theoretical consensus on what the self is or how to focus on the question.[1]

There is nevertheless little doubt that, psychologically, we all have a sense of self and engage in much self-related processing. There are therefore rich and important research programs on the sense of self in several domains of cognitive science and clinical psychology. Several mental disorders and conditions are characterized to some degree by disturbances in self-related processes, and overall well-being and our interactions with other people depend on our sense of self. Schizophrenia, borderline personality disorder, depersonalization, and autism all involve different aspects of the sense of self, such as instability and weakening; psychological measures and constructs such as clarity of self-concept, sense of agency, and sense of self-efficacy have different kinds of impact on well-being, anxiety, and compassion.

Some studies that I have been involved in can illustrate the multifaceted research on the sense of self and self-related cognitive processing. With Perrykkad and our collaborators, we showed that clarity of self-concept and self-efficacy predicts how compassionate we report ourselves to be, although not whether we actually donate to a charity. Perrykkad also did an in-depth review and analysis of self-related processing differences in autism and systematically analyzed the involvement of the sense of self in healthy people scoring differently in traits associated with the mental disorders listed here, finding that they differ along several dimensions. With my colleague Mateusz

Woźniak, we investigated the self-prioritization effect, which is where a transient association between an image and oneself leads to preferential treatment of the stimulus and subsequent target stimuli (e.g., being shown a triangle and told "this is you" speeds up subsequent detection for a target more than when shown a circle identified as "this is a stranger"); we showed that this effect depends on an intricate interplay of top-down and bottom-up processes in the brain, consistent with precision control. In work with Anna Ciaunica, we showed that this effect manifests differently in people high on depersonalization traits, suggesting that precision control differs when the sense of self differs. In turn, Ciaunica led a study showing that those who spent a lot of time in Zoom meetings during pandemic lockdowns experienced higher levels of depersonalization.[2]

The body is, of course, integral to our sense of self. There is much research on how our embodiment, our body image, and our body schema can change over time, can differ between people, and influences how we process stimuli and overall well-being and health. Our body is such an intimate and constant component of our experience that it anchors the first-person perspective, contributes strongly to feelings of self-worth, and is the go-to attentive focus in many forms of meditation. Surprisingly, our body sense is also highly plastic and subject to false inference. Simple illusions can radically make us feel transiently that we have different body parts or inhabit different bodies, or make us adopt different first-person perspectives. Some serious conditions, such as eating disorders, involve profound changes to body image, which are subject to much research. For example, it is well known that one's sense of body size causes one to approach doorways differently, since a wide body needs to turn more sideways to get through a narrow doorway. People suffering with anorexia, who are critically underweight, appear to turn their bodies sideways even for doorways they could easily pass through, suggesting that their sense of their body size being very large is so ingrained that it changes their actions. The relations between body sense and other processes are still not fully mapped. For example, in work led by my colleagues Manja Engel and Steve Gadsby, we investigated how eating disorders relate to perception of the distance between two touches felt on the body, with the idea being that if the body is falsely experienced as larger than it is, then the mapping of the points should be farther apart. We found that this effect was most pronounced when patients were given a few seconds to consider their judgment, suggesting that the effect might reflect a postperceptual process of reinterpreting the immediate, more accurate perception in the light of their distorted beliefs about body image.[3]

What studies such as these suggest is that, although the existence of the self is hotly debated, the existence of a sense of self is not up for debate. What our own research suggests, together with much research across many fields more broadly, is that the sense of self and its relation to the body varies significantly among people, in both health and illness, and it matters in the way that we process everyday sensory stimuli, as well as for how we represent ourselves and interact with others and the environment.

Some combination of cognitive and perceptual computational processes must underpin this multifaceted sense of self, just as we assume that such processes underpin all aspects of our mental lives. It would be an odd position to hold that the processes underpinning the sense of self are wholly unrelated to what we should call the "self"—the cognitive and computational machinery underlying the sense of self should be the first place to look if we want to make inroads on what the self is. Hence, I advocate taking the naturalistic approach where, if we want to learn about the self, we should go to the sciences that appear to be concerned with investigating it. There might be limitations here, for example, such that naturalism cannot capture some conceptual properties of the self beyond the psychological sense of self, arising from philosophical analysis. But naturalism does seem a promising starting point.

In the predictive processing literature, the self has in fact received much attention; it seems there is something that makes the self amenable to the cognitive and computational machinery in this broad framework. Much of this work has sought to tie various aspects of the sense of self to predictive coding and active inference, consistent with the naturalistic stance. It also has focused on how we must build self-models, as well as on the importance of interoceptive prediction errors and their precision. In the next sections, I will build on my earlier work and work done with John Michael, and specifically focus on how a sense of self might emerge in the self-evidencing agent, exploring what this might make us say about the self and its reality. I begin by setting out the modeling materials that we have available for building the sense of self.[4]

Models and Self-Models

Here is what the previous chapters of this book have delivered so far, which will be relevant for the account of the self given in this chapter. The self-evidencing agent harbors a rich generative model that abductively drives our perception, beliefs, and decision-making. Having a model is inescapable for an agent who exists because existence is a matter of evidencing the model. At the same time, perceptions, beliefs, and policies can wax and wane as the agent's beliefs about uncertainty and volatility control precision, revising their model as they traverse the changing world. The model is a structural representation of the latent causes of sensory input (i.e., of the causes that are hidden beyond the Markov blanket's sensory states). The model is shaped through pragmatics, focusing on things that the agent believes will help them reduce uncertainty now and in the future, but it is consistent with realism because reality to some degree controls what the agent can end up believing and deciding. This realism is in turn consistent with the possibility of occasional and more widespread misrepresentation and irrationality.

The generative model is what we have available to work with when attempting to understand the sense of self and to explore whether that puts us on our way to find the self. Like any other aspect of our lived experience and phenomenology, the sense of self would stem from some of the structural representations emerging from the causal dependencies of states within the model, which emerge through the dynamics of perceptual and active inference. We therefore begin the search for the sense of self by exploring the available representations of the agent's model.

Many of the observations that agents make are driven by latent causes in the external environment, such as trees, cars, and other people. Many other observations are mostly or partially due to agents' own actions in the environment. If the agent shouts loudly, they will hear their own voice; if they turn their head, that determines what they see; if they exercise, they may feel their heart beat faster; if they are kind to others, those people might be kind to them in turn; and if they make good investments, they will reap the rewards. In perceptual inference, the agent might then infer that they are indeed kind, or a good investor, from the pattern of their observations, ruling out that others are just treating them kindly no matter what, or that the windfall was just luck. In active inference, the expected observations are generated from the model, given the agent's policies, explicitly modeling the agent themselves as a cause of their own observations.

The causes of observations stemming from the agent themselves are treated as hidden, latent causes (belonging within the generative process) that can be inferred and used to generate expected observations for active inference. In this sense, self-related causes are no different from any other latent causes. There will be a rich and hierarchically layered, agent-based set of causes that contribute to the agent's own sensory input, ranging from bodily causes to new habits to character traits. Moving one's limbs typically gives rise to observations in kinesthetic, proprioceptive, visual, auditory and other sensory modalities; having an injury such as a sprained finger causes surprising results, such as not being able to grasp a glass of water; a teenager's growth spurt can change the way they have to navigate doorways; deciding to do more exercise might change several patterns of bodily sensory input as the body strengthens. Intending to be rich or well liked can change the way that the agent spends money or interacts with others; striving to become a better person can percolate to all aspects of the agent's actions over a long period of time. Perhaps at the highest level of the model's belief hierarchy sits the mere belief about themselves that they exist or, equivalently, that they self-evidence, as a deeply hidden, internal cause of everything that the agent does.

These internal causes sometimes act in seamless concert with each other, such as an elite athlete at the peak of their performance, or they can get in the way of each other, such as when self-control falters in the competition between preferences for ice cream and fitness. Each of the internal causes in principle can be inferred, just like any other cause of an agent's observations. Learning of agent-based causes and generation of expected observations given agent-based policies presumably happen all the time, and they will be subject to all the processes and foibles of inference discussed in previous chapters. For example, it might take the agent a long time to learn that, for them, ice cream tends to win over fitness late in the day and late in the week; or an agent might labor for a long time under the self-serving misconception that they are kind to others, or the self-deception that their basketball skills will take them to the pros.

To self-evidence with respect to observations stemming from the agent's own actions, therefore, the generative model will, in addition to being a model of the external world, also be a model of the agent themselves—a self-model. The agent will continually be employing this self-model in perception, learning, and decision-making, helping them make sense of the world, and thereby themselves. They will infer self-related causes, learn about longer-term patterns, explore and discover the self, and revise the

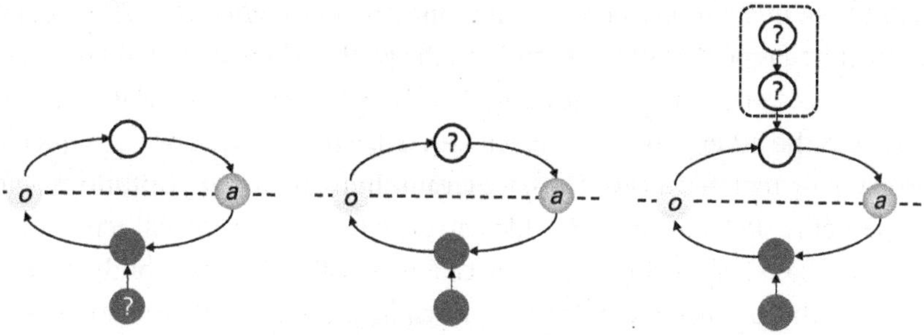

Figure 7.1
Depiction of self-models for an agent described by a Markov blanket (horizontal dotted line: Markov blanket; light gray: sensory state; dark gray: active state; white: internal state; black: external state). Left: Inference of an external cause (marked with "?") of sensory observation, *o*, mediated through a less deeply hidden cause (black). Middle: Inference of an internal cause of *o*, mediated through a less deeply hidden cause, *a* (active state of Markov blanket and external states). Right: Inference of more deeply hidden internal causes of observations, constituting the self-model (dotted square).

self-model. Figure 7.1 depicts inference of internal causes in the context of agents as described by Markov blankets.

Self-models will be a characteristic of every active agent who can infer policies in active inference. As we have seen, to exert any kind of control over an environment, the agent must represent that environment. And given that the environment includes the agent themselves as a cause of changes that happen, a part of their overall model should include a model of themselves. Inactive things, such as rocks, do not have models of themselves. Different active things will have different kinds of self-models. Some will be quite simple, others more complex; and different models can be implemented with different internal architectures, as I will discuss in later sections of this chapter.

A Structured, Real Self

The generative self-model consists of a set of inferred causes of sensory input, where those causes in fact stem from the agent themselves. The inferred causes are structured so they can explain observations relating to the agent's own agency, layering and causally associating long-term intentions, short-term inclinations, body schema, and other elements. The self-model is a structured representation of causes in the world, where that world includes those causes that inhere in the agent. In this sense, the self-model is no different from the other parts of the agent's model, which, as we saw in chapter 4, structurally represents cars, trees, ice cream shops, financial institutions, and other items. So far as the self-evidencing agent is considered, all representations are based on latching onto the causal/statistical structure of the world.

From this perspective, at least in principle, there is no difference between what the self-model represents and what other parts of the model represent. In chapter 5, I also argued that although skepticism cannot wholly be ruled out, there is no reason to be irrealists about what the generative model represents. Cars, trees, and ice cream shops are real and can be represented with more or less fidelity by the agent's model. Given the self-model is just more of the same kind of structural representation, it too should be said to be able to represent something real.

Of course, just like any other representation, misrepresentation is also a possibility for the self-model. From the fact that something is a representation, it does not follow that there is something there to be represented, let alone represented accurately. For example, I am quite sure that my representation of financial institutions is quite rudimentary, and I now know that my representation of my basketball skills was largely illusory. We even have systematic and persistent misrepresentations, such as some visual illusions (e.g., the Müller-Lyer illusion, where inward or outward-pointing adornments of lines make them seem longer or shorter than they really are) and some decision-making biases, such as our optimism bias that things will tend to turn out better than evidence would indicate. Misrepresentation notwithstanding, and setting aside classic skeptical arguments, let us agree that our representations mostly represent real things. Cars, trees, and ice cream shops exist, and the models of the world in our brains represent them.

The key questions, then, are: What does the self-model represent? What constellation of latent causes in the world does the self-model capture with more or less fidelity? There are several possible answers. I have suggested that we set aside the possibility that the entire self-model is a hallucination or delusion, capturing nothing at all about the real world. It is not plausible to hold that the agent has *no* cause that leaves a mark on the world for the agent to observe, leading to their inferences of that latent cause. It is possible that the self-model harbors some systematic misrepresentations, akin to self-illusions or self-hallucinations, such as my belief that my basketball intelligence will compensate for my lack of athleticism and send me to the big leagues. But I do not find it plausible to say that all the self-model is an illusion, systematically taking in observations from real causes in the world and always inferring something false on their basis. That would be as odd as saying that all of visual perception is illusory or all our inferred causal relations about the world are false (such as our true belief that smoking causes cancer and tobacco companies knew this but lied about it). It seems more likely that large parts of the self-model are in fact reasonably accurate, even if some parts are falsely inferred, even if there are sporadic systematic illusions or even hallucinations, and even if some internal self-related causes are unknown, not represented at all. In this sense, perhaps my self-model is not too different from my model of the financial system, where I am reasonably accurate about some surface-level pecuniary phenomena, I have adequate but imprecise representations of banking and some kinds of investment mechanisms, I probably harbor a fair few misconceptions about the stock market, and I am wholly ignorant about cryptocurrencies.

The more or less accurate self-model represents some of the structured causes residing within the agent's Markov blanket, which give rise to the agent's actions via their active blanket states. This includes the agent's longer-term traits, which causally interact with their shorter-term, more context-sensitive states. In causal terms, what is represented in the self-model is a subset of the internal nodes of this Markov blanket, which are children of some sensory nodes and other internal nodes, and parents of other internal nodes and some active states. It is tempting to label this set of self-related nodes "the self," just as it is tempting to label the set of external latent causes represented by my model of the financial system as "the financial system."

The temptation to label the set of latent causes represented by the self-model as "the self" is motivated, at least initially, by the hypothesis that it is those causes that give rise to the sense of self that I discussed earlier—the phenomenology and core beliefs about the self studied extensively in cognitive science, phenomenology, and psychology. Although I will say more about specific aspects of self-related phenomenology shortly, generally, an agent's sense of self is likely to emerge from just the set of latent causes within themselves that are represented in the self-model.

This implies that the self is real. That is, if we want to withhold reality from the self, we should also withhold reality from cars, trees, and the financial system. From this self-evidencing point of view, Hume's mistake when he failed to find the self in introspection was as rudimentary as looking inside a bank and complaining that one cannot see the financial system. This realism about the self is not very momentous; it just says that the self is whatever set of causes plays the self-role as extracted through consideration of the sense of self. Nothing much follows from this realism, except that it demystifies the self, putting it on par with other real things; and perhaps it can help temper some of the more antirealist ideas about the self.[5]

Sometimes realism about a domain has failed, such as in cases where nothing has played the causal role in question. Consider, for example, phlogiston, the element with negative weight hypothesized in the seventeenth century to explain the reduced weight of things after burning, or consider *élan vital,* the force hypothesized to explain life as late as the early twentieth century. In both cases, subsequent theoretical and empirical development helped restore realism for more scientifically sound concepts. It is, of course, possible that our story about the self will go that way, although I doubt we will throw out our concept of self altogether.[6]

Earlier, I subscribed to the idea that existence requires that a Markov blanket persists, setting a boundary for the thing in question, indicating that its internal states are not dispersing. I also argued that the agent is self-evidencing relative to its Markov blanket. This leads to the question where realism about the self sits with respect to existence, understood in Markov blanket terms. Markov blankets are not metaphysically "heavy" notions because they just require the Markov property of conditional independence and a certain organization of parent and children nodes, and they can be found at multiple scales, nested within and connected to other blankets (recall figure 2.1). When we model Markov blankets within a causal net, each blanket in principle can stand out as a causal node playing a role in the larger net, and with its own internal mechanisms (e.g., think of organs comprised of interacting cells). Given the way that the self causally behaves, it seems likely that conditional independences would allow us to draw a Markov blanket around the set of internal causes comprising the self-model (perhaps somewhat like the financial system exists within the larger economy). This might make perfect sense from a modeling perspective, even if the self is not a particularly paradigmatic, thinglike thing (just as universities and the financial system are not neatly thinglike things either; recall here our liberal concept of things from the beginning of chapter 2). The Markov blanket lending existence to the self might be less like a cell membrane and more like the blanket for a university or the financial system, which encompasses a dynamic system with different parts more fluidly changing their connections to each other and to external states over time.[7]

Given the self-evidencing framework and its reliance on internal models, including models of the self, there is nothing untoward about being a realist about the self. This is a humdrum realism, not much more exciting than being a realist about any number of things around us. The realism is also consistent with the Markov blanket concept of existence that we have used to explicate self-evidencing.

The Role of the Self in Self-Evidencing

So far, I have pointed to the way that the self-model forms through self-evidencing, as latent causes of self-generated sensory input are inferred. Just as surprise arising from receiving sensory input from any other latent cause is explained away by inferring structured relations and transitions between them, surprise arising from self-related causes is explained by inferring a structured self.

The self-model is not a mere curiosum, doing nothing much once modeled—it is not like the inference of the existence of a planet circling a distant sun. The self-model is crucial for figuring out the expected free energy in active inference and is therefore finessed and wielded in distinct ways by the self-evidencing agent. It is important to active inference that the agent has an accurate model of themselves. Key to policies in active inference is the transition probabilities informing the agent how one state is likely to transition to another, given the intervention of the agent. For example, the agent might give a high probability to transitioning from being jobless to having a job conditional on writing a good, polished job application, and that belief depends on several fine-tuned predictions about the agent themselves (ability to write well, conscientiousness, and other characteristics), as well as how those aspects of the agent will operate in the given context (knowing what company the application is going to, what the deadline is, and other elements). Many further beliefs about the self play into this simple example. The ambiguity of the policy (the expected fidelity of the mapping from action to outcomes) depends on the agent's belief in their self-efficacy, their ability to precisely affect outcomes by their actions; if the agent generally believes that they have low self-efficacy, then they might not give a high probability to getting the job by writing a good application—why bother honing one's writing skills if the jobs always go to the kids from the posh schools?

Maintaining and finessing the self-model are therefore important, and of course, this would be an engaged and active self-supervised process where the agent seeks to have a model that garners as little surprise as possible over the long term, and which can robustly guide policy inference.

Maintaining and finessing the self-model involve quite difficult and intricate learning processes where the agent looks backward to replay events to figure out why things happened the way that they did. This speaks to a self-evidencing version of what in reinforcement learning is called the "credit assignment problem"—when an agent has acquired a reward after a sequence of actions, they need to learn which of the actions caused the reward to occur. For example, a basketball player might consider whether the game was won because of their own hard training regime, their teammates, or their wearing their lucky bracelet. For an active inference agent, there might have been a policy predicting the rewarding outcome, but after the fact, the agent may learn certain facts that make them doubt what actually caused the outcome. So they might mentally go back in time and replay the events, relying on episodic memory, to check if their confidence was in fact misplaced (e.g., the basketball player might recall making several decisive moves in the game, which they had practiced beforehand). This can help shore up the policy for future occasions. The agent can also go back in time and mentally play out the event in several counterfactual scenarios, shifting both contexts and policies. Here lies the risk of rumination and regret, when agents consider what went wrong and what they could have done otherwise; for example, the agent might counterfactually consider how much effort would have been involved in writing a more polished job application and play out the events of the job interview to see if it would have gone better. Of course, replay in counterfactual thought experiments can also lead to important lessons, which can guide inference of better policies for the future. As the agent repeatedly replays the events with slight modifications, some invariant features might stand out, such as the reaction of the job selection committee to a slightly inappropriate joke or when the job candidate's lack of critical skills became apparent. These lessons can help the agent adjust precisions over policies to simplify their self-model ("I might not be so funny after all"), making it more applicable for the future.

The self-model is thus revised in episodic and counterfactual replays to assign credit for outcomes and finesse the model for future use. The self-model is also central to prospective and counterfactual "preplays" for planning and decision-making, as discussed in chapter 6. In planning a future course of action, the agent is centrally involving their model of their own capacity for making things happen. The agent can project how future events will unfold and assess the expected free energy, given their own actions, or compare or sequence actions in different ways; or they can condition the projection on different skill levels, such as to ascertain if they should undertake training first; or they can engage episodic future thinking, where they project what experiences they would have if they were to undertake certain policies in order to figure out if they would subjectively prefer the outcomes or what beliefs they would acquire. In each case, it will be important to have a relatively accurate self-model because otherwise, there will be ambiguity or risk in the course of action. For example, it is no good deciding to invest one's life savings if one is too lazy to learn how the financial system works, so it would take too much time to maintain the investment and the rewards would be uncertain and the risk of losses too high.

In some of the examples given here, agents are considering how to exploit their own causal capacities. In other examples, agents were exploring their self-model, seeking to learn more about themselves. Agents can take a quite proactive approach to exploring themselves, both counterfactually and actually. An agent might ask themselves if they are the kind of person who would help a person in need and preplay scenarios in which they are; they may then discover to their dismay that when actually put in the situation, they failed because they had underestimated situational factors (evidently being in a bit of a hurry was enough to make the agent ignore the person in need, throwing doubt on the hypothesis that they are kind). They might then episodically replay the situation and revisit their relative preferences between getting to their destination in time and being kind. Here, agents are exploring and discovering, and even changing, who they are.

Notice that in this example, agents are also inferring a set of preferences for themselves; they are figuring out what kinds of outcomes they prefer (being on time versus being kind). This inference engages the circular causality of self-evidencing, with its air of self-fulfilling prophesying. The agent forms the belief that they prefer to be kind, which informs their policies and makes them selectively sample some observations (stopping to help the next person in need and consequently being late for an appointment), and these observations then provide evidence for their beliefs about their preferences. This is active inference directed at the self—*self-shaping*—where beliefs of the self-model causally shape the internal causes of action. Active inference bends back on itself when agents shape themselves, but otherwise, such cases are similar to the way that active inference can change external causes to drive different observations (such as using bricks to build a house that shields one from directly observing the effects of the weather). In chapter 5, I argued that self-fulfilling prophesying is constrained by the extent to which the world plays along in delivering the expected observations, and this holds for self-shaping too. The agent might well discover that it was false to infer that they preferred kindness since they in fact never stop to help a person in need; that is, they infer that the kindness policy is in fact too risky, making their self-model too complex.[8]

When discussing inference to the best decision in chapter 6, we looked at the exploit-explore distinction, where agents need to decide how much to exploit a resource and how soon in advance of the expected depletion they should go exploring other patches. There is a kind of exploration that has been relatively ignored in this debate—namely, exploring the self. In work led by Kelsey Perrykkad, we call this type of action "foraging for the self"—when uncertainty about some aspect of the self-model begins to accumulate and the agent engages in actions that try to shore up or restore the self-model. We think that this happens in particular when the agent becomes uncertain about how well they control states of the world through their actions. In our studies, the agent uses a computer mouse to control one of several squares on the screen, with varying levels of noise added to the movements, making it hard for the agent to know exactly which one they are controlling. We know which square the agent is hypothesizing that they are controlling because they look at it and we track their eye movements; and we know how much uncertainty is accumulating by calculating the difference between their mouse movement and the square's actual movement. We find that when uncertainty accumulates, agents change their hypothesis—they look at a different square to explore if that is the one they are controlling. In a study where agents either occupy a "beach" or a "water" environment, with different underlying statistics, we find that when uncertainty mounts, agents will leave one environment for the other, hoping to learn better and acquire more certainty about their control.[9]

Interestingly, in Perrykkad's studies, we repeatedly find that healthy individuals with high scores on questionnaires for autism traits tolerate less uncertainty about their control before they jump to new hypotheses or new environments; in contrast, individuals scoring lower on autism traits accept more uncertainty before they jump. Different people thus manage their sense of agency differently and expect their control of a situation to be within different ranges. We hypothesize that people with more autism traits prefer less ambiguous policies, expecting precise transition probabilities for their actions, and perhaps this is driven by less informative prior beliefs about contextual factors that may interfere with the execution of a policy to make it more ambiguous. In other words, the way that different people go about their daily business (such as getting out of the currents and crashing waves of the ocean quickly) can be related to their sense of self and how they interact with and control their self-model, not with seeking rewards or information about the external world.

In this discussion, I have conveyed the rich and varied role of the self-model in self-evidencing. Not only does the agent infer the states of the self as they encounter observations that are best explained by positing internal causes, but they also work with the model retrospectively, prospectively, and counterfactually. They finesse, revise, expand, and simplify the self-model, and they constantly consider how their model of their self translates into the observations that they might make when they decide to do things in the world.

My claim is that all this processing over the self-model is what gives rise to the sense of self. For example, the agent's sense of self-efficacy may reflect the overall ambiguity of their policies as explored in replays and preplays; their sense of agency may relate to their execution of policies in various contexts; and their sense of clarity of self-concept may be driven by the precision of the beliefs about their traits, and perhaps their ability to infer their preferences.

The sense of self is with us all the time that we are conscious, more or less clearly, waxing and waning according to the context and the activity that we undertake. Much of the time, we do not pay much explicit heed to it; and at other times, we focus intently on who we are. When deeply engulfed in an attention-demanding task, we seem to lose ourselves. But when we withdraw from the task, we occupy a holding pattern where we often engage in self-related mentation. There is rich neuroscience suggesting that distinct brain networks are engaged in these different phases of daily life, and of course, the kinds of processes listed here make sense of the waxing and waning of the sense of self. For example, in the squares task in Perrykkad's controlling-the-squares experiment, there is little temptation to question the agent's control if we do not add the noise and they can just get on with the task.[10]

This section has traced the many diverse ways that self-evidencing agents engage their self-model and how this plays into their sense of self. The self-model is inferred, real, and is put to work in varieties of active inference. In the next section, I consider some broad features of the self and discuss model architectures that could sustain them.

Models for the Minimal and Narrative Sense of Self

Philosophers have identified some invariant features in the phenomenological sense of self, which can be underpinned by self-evidencing for the self-model. These are the minimal self and the narrative self

It seems like all our conscious experiences are accompanied by a sense of "mineness," considered a minimal sense of self. Try to think the thought, "*Seven Samurais* is a classic movie," and, although I am the one who asked you to think it, you will sense the thought as yours. The same holds for all your perceptions, thoughts, emotions, and experiences, all of which seem to include such a sense of subjectivity or minimal self.[11]

The narrative self is a richer, invariant feature accompanying the minimal self, captured in Daniel Dennett (1992)'s well-known metaphor of the self as the "center of narrative gravity." There is much work on the relevance of narrative for the sense of self, but also dissenting voices arguing that sense of self is nothing like a narrative. That is, perhaps the self is not so much an epic narrative as, at best, short stories or episodic snippets loosely strung together. Sometimes this debate is presented as an either-or, but I think there is a spectrum of narrativity for different people, and perhaps for the same person over different phases of life. Some people have a very strong narrative sense of self; others have a strong sense of separate snippets of self with no substantial connective thread (and can lead their lives just fine). Different degrees of narrativity may relate to different habits and capacities for episodic replays and episodic future thinking; those with much episodic capacity may find it natural to string together lived episodes in longer, more or less epic narratives, whereas those operating more with conceptual representations may be content with a more abstract, staccato sense of self.[12]

The self-model can sustain both narrative and less narrative senses of self since the cohesiveness of the generative process as represented by the model depends on how the model is put together structurally, and the agent's habits for putting the model to work in replays and preplays. The shape and depth of the hierarchical self-model and the precision of its parameters matter here. For example, as we saw for autism traits, perhaps higher-level contextual beliefs can be less informative, leaving less general connective or narrative threads to tie more concrete parts of the self-model together; perhaps in other cases, very informative abstract, higher-level beliefs systematize beliefs about the self in a way that makes it obsolete to care about how more concrete, low-level individual elements hang together. Or an agent who only rarely finds uncertainty accumulating for their sense of self may be less inclined to go exploring to discover their preferences, so they may be less likely to challenge and change their sense of self, including the extent to which it is narrative or not. The core idea here is that "narrative," for the sense of self, is not about fiction and telling stories, but just about the structural way that higher levels of the generative self-model subsume lower levels, informing their state-transition beliefs, and how that subsumption plays out when the self-evidencing agent maintains the model in more or less rich and deep episodic replays and preplays.

In the previous section, I described the self-model as being actively involved as the agent perceives and acts in everyday life. This implies that the self-model is causally interacting with the rest of the model. The model architecture underwriting this arrangement could be understood as distinct internal Markov blankets, coupled with each other. The self-model represents dynamics of the agent changing relatively slowly—asking "Who am I?"—and other parts of the model, which we can call the "world-model," represent dynamics that change faster—asking "What do I perceive, attend to, and do?" There is top-down causation from the self-model's beliefs that constrain the beliefs of the world-model; for example, the belief that the agent is kind may help control precisions for attending to those in need. In turn, the evidence sampled by the world-model determines the input to the self-model, constraining its slower dynamics. The coupling between the two blankets is constrained by the overall system, the agent themselves, such that there are limited degrees of freedom in how they can interact; they can interact only in certain top-down and bottom-up ways and do not uncouple.

This means that functionally, the self-model is "learning" at the slower dynamic scale, while the world-model is "perceiving" at the faster dynamic scale. The combination of the two coupled blankets with these functional profiles lends a nice robustness or resilience to the overall agent, where beliefs about the agent's self can be updated in light of what is actually observed, and in turn can constrain what is actually observed. If one part of the system begins to get things wrong or encounters volatility, the other part can help the system correct its course. The architecture that I have described is of course yet another instance of the self-evidencing motif, where the top-down causation is seeking to obtain certain observations consistent with the self-model, and the actual observations of the world-model in turn become evidence for it.[13]

We can speculate that the phenomenologically invariant features, the minimal and narrative selves, can be accommodated with this kind of model architecture. The minimal mineness that accompanies all experiences might relate to the belief that the perceptual and active workings of the world-model are constrained by the self-model, and in that sense are "mine" (we will elaborate on this kind of subjectivity in chapter 9, which focuses on consciousness). The narrative self is not just a story—it is also, in Dennett's term, something that requires a "point of narrative gravity." The narrative self requires a thread (however stable, coherent, and epic or piecemeal, broken up, and novella-like) through all the experiences minimally labeled as "mine." This thread might be provided by the slower dynamics of the self-model constraining the faster dynamics of the world-model, such that there is something relatively stable, anchoring the sequence of everyday experiences. A schematic for this proposed, sparsely coupled architecture for the self and the overall model is set out in figure 7.2.

Presenting the self-model as a set of internal nodes with its own Markov blanket, sparsely coupled to the world-model, implies that the self is literally a thing since Markov blankets are used to individuate things. This is consistent with my discussion of the reality of the self, employing what I believe are humdrum (even if liberal) meanings of the terms "thing" and "reality." I will return to the proposed kind of coupled blanket architecture for the self-model in chapters 8 and 9.[14]

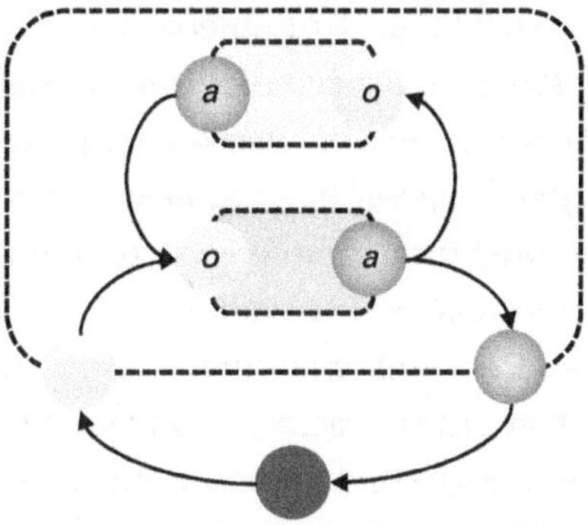

Figure 7.2
Schematic for the self-model as a nested Markov blanket (light gray internal blanket), sparsely coupled to other internal world-models (darker gray internal blanket), such that the (potentially slower-acting) self-model can selectively sample and act on the (potentially faster-acting) world-model.

What Is It Like to Be a Model?

So far, this chapter has argued that the self-evidencing agent must have a self-model as part of their generative model, which represents the causes of their sensory input arising from within themselves. The self-model underpins the agent's sense of self (i.e., their experiences and beliefs about their own self) and constrains the world-model. Agents thus rely on good models of themselves for efficient inference to the best decision and can maintain their self-models only from their self-supervised vantage point. Self-supervised maintenance includes active exploration of the self and of its agency, together with replays to assign credit to self or other factors for outcomes, episodic future thinking, and counterfactual reasoning for prospective agency. This multifaceted set of self-related processing is what in creatures like us could explain our varied phenomenology of self, with multiple more or less rich and cohesive elements that can wax and wane and can differ between and within individuals. Now we ask what it is like to have, or be, such a model.

There is an old joke invoking a problem about introspection for psychological behaviorism. Two behaviorists meet each other in the street, and one says to the other, "You're well today, how am I?" The joke trades on the problem that, for behaviorism, the best arbiter of mental states is behavior, and other people may well be the best observers of your own behavior, and thereby your mental states. In behaviorism, mental states are inferred from behaviors, not experienced from a first-person perspective. A worry might be that self-evidencing about the self has a somewhat similar problem: Do we know about ourselves only through inference, beginning with observations of the effects of our own actions? For example, it is not very plausible to say that I know that I want ice cream only when I observe myself with ice cream in my hand. I think there is a satisfactory, multifaceted answer to this worry.

On some occasions, I might find out about the desire for ice cream ahead of buying it from internal observation of my own covert actions, such as mental planning through active inference; however, on other occasions, I fail to introspect and might indeed wantonly discover my desire as I actually buy the ice cream. Further, if we change the ice cream example to an example about what it takes for me to come to believe that I have the character trait of self-control (or not), then it seems that our self-model does indeed rely at least partially on behavioral observations since my belief that I have self-control is sorely tested by my repeated observation of myself eating ice cream. Here, we do not have the silliness conveyed by the behaviorist joke because self-related inference is not done from scratch every time we make an observation. Inference is always a precision-weighted compromise between the strength of the prior belief and what is being learned from the current observation. Strong empirical priors are based on extensive past observations, and they reduce the weight allocated to new observations; given my history, I fully believe, in advance of the next observation of my own behavior, that I want ice cream and that I have no self-control when it comes to ice cream.

Although self-evidencing does not fall foul of the worry highlighted by the behaviorist joke, there are some other potential problems concerning the idea that the self is real and is inferred in the self-evidencing agent's model, which gives rise to the sense of self. The first thing to address is a potential confusion about the occurrence of "self" in "self-evidencing," in the context of the discussion of the sense of self and the reality of the self in this chapter. "Self-evidencing" is a term from the philosophy of science (see chapters 2 and 3) concerning the way that a model (or theory or hypothesis) that explains some observations turns those observations into evidence for itself—the model exists and is evidenced to the extent that it can explain the observations itself. Self-evidencing in this classic sense is only about evidence and explanation, with no direct conceptual relation to the notion of an agent's self. Self-evidencing in the sense that we have discussed it in the book so far is also primarily about evidence and explanation, but the model is now understood to be a self-organizing system—an agent endowed with some autonomy to organize its internal states (via precision control). We use the notion of self-evidencing to conceptually understand the existence of such an agent, and it is not unreasonable to say that an existing agent in some sense is an existing self. However, in this discussion, we have limited the concept of self to whatever gives rise to the sense of self, understood as experiences and beliefs that in some way have the self as their content. This conceptual distinction between the model that defines the whole agent and the self-model that captures the self is useful, given the functional role of the self-model, as discussed in the previous section. Conceptual distinctions can be negotiated in various ways, but in my proposed way of carving out the concepts, there is no confusion between "self" used as the prefix of "self-evidencing" and "self" used to pick out the self as inferentially revealed through the self-model.

Now a somewhat obscure-sounding issue should be addressed. It feels natural to say that someone can "have" a model, which they use for some purpose. For example, a hill-climber can have a map and use it to navigate, or a computational modeler can have a mathematical model that they use to understand the financial system. It feels less natural to say that an agent "is" a model, existing and operating as a model. Yet this is how we should understand the self-evidencing agent. For expository ease, I have not been overly strict in my usage so far, mostly talking about "the agent's model," but this should be understood as the model embodied by the agent and enacted in their behavior, not as the agent having a model in the way that a hill-climber might have a map.

As I see things, it does not gel well with the self-evidencing framework to say that the agent is one thing and the generative model something else, and the agent "has" the model and uses it. In previous chapters, we have cast the existence of the agent in the causal terms of a Markov blanket, with sensory states, active states, and internal states. The internal states will come to model the environment in the sense that, as the agent goes about self-evidencing and accumulating evidence, the structure of the internal states will begin to recapitulate the causal structure of the world (as discussed in chapter 5). Putting these elements together, we should say that the agent models the world but does not have a model of the world.[15]

To be a self-organized and self-supervised model, then, is to be the kind of thing that undertakes the processes associated with self-evidencing. Through precision control (see chapter 4), the model predicts, selectively samples, infers, attends, plans, decides, and learns. These are all processes that make sense as operations on the representations appealed to in chapter 5. Although it may at first sound odd to say that the agent "is" a model, once we flesh this out in terms of self-evidencing processes, with the understanding of both agents and models in terms of structural dependencies among internal states, it is not a mysterious notion. The self-model as we have discussed it here also fits into this overall picture of the agent as a model. The causal structure of the world that is modeled include the causes emerging from within the agent themselves, which is therefore part of the model. I said earlier that for the self, as part of the conditionally independent cause of the agent's own sensory input, we can carve out its model as one part of the agent's internal states, separated from other parts by a Markov blanket (see figure 7.2).

Viewed as a self-organizing biological system, then, the agent is a model where agency arises in self-organization through learning and active precision control. This can still give rise to behaviors that, from the outside perspective, look as if the agent has a model that they consult before they act. For example, when the agent comes back to their old neighborhood and walks around, they might stop occasionally, consult their inner, memorized map, and then decide where to turn next; or the aspiring good Samaritan might pause, in spite of being in a hurry, consult the preferences represented in their self-model, and decide to stop and help a person in need. In these cases, it feels natural to say that the agent has a model, but it is still the agent as a model that is in play. For example, in the navigation case, the agent has inferred a set of latent causes (e.g., segments of the terrain in terms of left and right turns) and can prospectively line up policies for getting to their destination and assess these against expected free energy, given the precision of their beliefs about the order of the segments. We should also leave room for a hierarchical arrangement where higher levels explicitly represent the policies of lower levels. In the Samaritan example, the agent might be able to represent the extent to which they follow through on their policy of attending to a certain task (such as keeping an eye out for people in need), and can in turn ask themselves if this is the policy that they would prefer to follow. This can again look as if the agent has a model that they consult, but it is just about the model embodied by the agent having parametric depth that allows nested policies to be inferred and enacted.

In this account, there may still be some apparent oddity in the idea of a model inferring a model of itself. It seems to be an unusual situation because our standard idea of a model is that it models something other than itself—a map models a terrain, but rarely itself. But the model of the self-evidencing agent bends back on itself, representing its own states. In fact, for systems that actively control states of affairs in the environment, it is necessary that they model that environment, including the ways in which that system intervenes in the environment to exercise its control. The self-model simply picks up on this necessity and makes a virtue of it, prodding and probing the self to arrive at the best possible policies for control (by its own, self-supervised lights). This self-modeling plays out in the cascading causality within the Markov blanket, as nodes influence each other from sensory states to internal states and on to active states, before circling back through the environment with new inputs to the sensory states. The states represented by the self-model comprise some subset of all these blanketed states.

Here, the self-model is not an epiphenomenon, idly representing things. The beliefs encoded in the self-model (about preferences, dispositions, or traits, for example) themselves can be part of the causal flow leading to the inference of policies that, when enacted, cause new observations, which in turn further shape the self-model. Another analogy here is the way that a scientist may discover that they are themselves introducing confounds into their data gathering through their own actions (e.g., by selecting only a particular type of study participant or by asking participants questions that give away what they are interested in hearing from their participants). Self-evidencing in active biological organisms like us is (benignly) epistemically circular and, at the same time, characterized by circular causality. In that context, it is not odd that the model represents itself.

Agents of Volatility

I will end this chapter with a speculative proposal about the kinds of selves that we humans are. I will base this on the ideas developed up to now, trying to pinpoint a peculiar aspect of our self-models.

We noted in chapter 5 that self-evidencing agents have the capacity to accurately represent states of affairs in their eco-niche, and they can misrepresent them too. This point was not very profound, merely reflecting that the agent's internal model will pick up on facts about their bodies and the environment in which they exist. It would be odd if the internal model of a social, land-dwelling agent like a human were representing causal structures that are completely irrelevant to how they go about self-evidencing, given their model. There is thus a pragmatic element to what we end up representing, wedded to our inferred beliefs about our goals and preferences, understood as our characteristic states. These states can be transient (like the desire for ice cream), constant (like the desire to earn money), or tied to our biologically determined, homeostatic set ranges. We tend to represent the things that are expedient for those characteristic states, such as the whereabouts of gelato shops, how to navigate education and job markets, and how to avoid states that take us outside comfortable bodily ranges, such as extreme temperatures or lack of water or food. Each type or species of self-evidencing agent will have an internal model that represents the characteristic causal and statistical aspects of their niches that matter to their existence.

Humans are mammals with certain physical and biological characteristics, and we occupy an eco-niche appropriate for those characteristics. We are successful (sometimes regrettably so) at inserting ourselves into the niches of pretty much every other organism, exploring and exploiting as we go along. Other creatures and invasive species can encroach too, of course, but we seem to be very good at it. There must be something about our causal model of the world, and the policies that we enact as the selves we are, which explains our capacity for encroaching on all corners of the world and the particular way that we do it. What are the beliefs about ourselves as agents that make us the particular kinds of encroaching agents that we are?

My speculative suggestion is that we occupy a niche that is not just defined in terms of variables that matter for us in a given environmental or biotic setting, but also in terms of statistics. That is, we have evolved to exploit and explore what might be labeled a statistical resource—volatility. The niche of any species will have volatile events, as latent causes interact and create more or less nonlinear, sudden changes in the observations that the organisms make. For example, humans around the Pacific will have unexpected nonlinear changes in what we observe when the climate shifts from La Niña to El Niño, nowadays further complicated by anthropogenic climate change.

Any kind of change from one state to another is modeled via state transition beliefs in the internal model, including those state transitions, encoded in policies, that pertain to our own agency. Learning about state transitions is useful for having a longer horizon for self-evidencing. Thus, knowing in advance that next year will transition to El Niño from this year's La Niña allows countries in the Asia-Pacific region, such as Australia, to prepare more for drought and bush fires than for flooding. Whenever we encounter volatility, there is an opportunity to learn, to explain away unexpected uncertainty and be better at anticipating it and perhaps controlling its outcomes.

This engagement with volatility is no different in principle from an animal whose heart rate goes up in anticipation of physical exercise. But I think creatures like us may have evolved to specifically exploit policy inference and learning under volatility at ever-broader spatiotemporal contexts. The model needs much parametric and hierarchical depth to deal with complex interactions between latent causes, and we have plenty of such depth in our brain's internal model. We also have a self-model that fits the longer spatiotemporal scales over which volatility plays out causally. We can project ourselves impressively far into the past, to extract information about how our agency has interacted with hidden contextual factors to create unexpected turns of events, and into the future, to test not just what observations we will make given a policy, but also how that policy might interact with multiple latent factors in the world ("If I were to complain about that annoying thing, then might that cause others to acquire certain beliefs that they didn't have before, which might cause my social network to turn against me?"). This suggests an ability to infer and learn about volatility, and then to exploit what we have learned ("I will make an anonymous complaint instead, which will throw a wrench into the works, and then I will build an alliance to support my cause").

The key point is that we can model our response to volatility rather than just responding to it—it seems that we have gone one more step up in the hierarchy of volatility modeling than other animals. This gives humans great advantages because we can travel through ever-new patches or radically transform existing ones, and survive there, where other animals would perish, and deplete their resources before moving to the next patch. With great advantages comes tragedy too, of course. We are so successful at controlling volatility that other species suffer, and indeed we make our own species suffer too as we run out of patches to deplete.

We have a rich and deep internal model, which we can take offline and epistemically forage through to all sorts of counterfactual worlds in search of exploitable volatility. With this ability comes both benefits and dangers. At the outer reaches of a rich model attuned to nonlinearities and interactions of hidden causes lie fantastical fiction, abstract thought, and creative arts, and also conspiratorial suspicions and intricate investment schemes and scams. Volatility is inherently difficult to learn about and to control because one needs to hold many causal variables constant while systematically varying others, keeping an eye on expected outcomes all along. This is fraught, with many degrees of freedom for positing and mentally testing deeply hidden latent causes. Maybe this way lies some of the mental distress and illness that seem prevalent in humans specifically; for example, if an agent becomes uncertain about the existence of latent causes in other people's minds, including their volatility, then they may lose confidence in their existing policies involving other people and endlessly cycle through thoughts about what the right policy could be (this could potentially contribute to some of the social signs and symptoms common to many psychiatric illnesses, ranging from borderline personality disorder to schizophrenia).

If there is truth to the notion that the human eco-niche is centrally characterized in terms of volatility, then, as a species, we do not turn away from volatility but rather seek it out and try to turn it to our advantage. I think this is the case. We mentally explore volatile scenarios and consider our own actions as new latent causes in new patches of ground, engendering often-radical state transitions in our various environments to suit us. We constantly explore mental scenarios in which we can either harness volatility or make volatile changes through our actions (such as causing revolutions or staging hostile takeovers). Sometimes this is considered and elaborate, such as when executing a long-term plan to take over power in a political cabal. Other times, it is more in the vein of Dashiell Hammett's detective in *The Maltese Falcon*, Sam Spade, whose "way of learning is to heave a wild and unpredictable monkey-wrench into the machinery" to create new unknown spaces to explore, provoking nonlinearities that can lead to solving the murder mystery. Our self-models, the beliefs that we have about ourselves, may then be centered on intricate ways in which our policies can engage with volatility.

In a slogan, humans are *agents of volatility*. We are "of" volatility, in the sense that our niche is characterized by volatility, and we are "agents" of volatility, in the sense that we cause volatility in our eco-niche.

If humans are indeed set apart from other creatures by virtue of being such agents of volatility, then that should inform our discussion about the way that we undertake self-evidencing and the kinds of beliefs about ourselves that we enact through our models. I will return to agents of volatility in the remaining chapters of this book, delving into questions about volition and will, consciousness and wisdom.

Concluding Remarks: Solitary yet Causally Connected Selves

In this chapter, I have focused on a specific part of the internal model embodied by the agent—the self-model. I argued that the self-model arises through inference of causes of the agent's sensory input originating from within the agent themselves. It is a structured, causal representation that plays a multifaceted, active role in self-evidencing, as the agent mentally travels into the past and future to maintain, revise, and finesse their model of themselves, constraining perception and action. Operating at slower dynamics, the self-model is critical for the agent's capacity to select policies of action that over shorter timescales may constrain policies leading to their characteristic states. Self-evidencing for the self-model gives rise to the agent's sense of self, which in turn depends on their more or less idiosyncratic abductive journey, as they maintain their self-model so it can play its central role in active inference. Agents are self-supervised, self-organizing models moving around their eco-niche to selectively sample their expected observations, seemingly being especially adept at dealing with and exploiting volatility. I portrayed humans, in particular, as agents of volatility, modeling the world and themselves to make the best of volatile events, in essence occupying a statistically described eco-niche allowing them to push themselves into ever-new environments, generating volatility for exploitative advantage wherever they go.

The self-evidencing agent is, as we saw in chapter 4, confined to conducting its self-evidencing through self-supervised precision control, obsessing only over the association between internal and sensory states, mediated through active states and external states. In this process, the agent's self-model keeps track of their own ability to constrain this circular causality. This portrays the self-evidencing agent as essentially solitary, secluded behind its sensory veil, just trying to organize its internal states to filter its observations to fit with its expectations. But, as we discussed in chapter 5, accompanying this etiolated, skepticism-prone portrait sits a wholly environmentally enmeshed and embodied agent.

This is the core duality of existence as self-evidencing agents—both epistemically solitary and causally connected. Neither side makes sense without the other. Our self-model, with its focus on our own capacity for constraining sensory flows in a volatile world expresses this duality. I think this duality is a kind of existential invariant—the tension between solitary precision control and our causal connection to the body, things, and other minds is always there. Volatility is always just around the corner, or we seek it out or create it ourselves, reminding us of the boundary between agent and nonagent. Every time that we solve a perceptual or active inference problem and things are going smoothly, changes abound, and we must explore new ways of being.

This conception of the self-evidencing agent of volatility can come across as somewhat existentially barren, epistemically bleak, and causally determined. Chapter 8, therefore, asks what self-evidencing does to volition, emotion, and value.

8 Volition, Emotion, and Value

In this book, we have begun with a simple premise of existence and built to the idea that, if agents exist and are self-supervised, then they must be self-evidencing. At various points, I have said that self-evidencing is not teleological because it begins with existence rather than having continued existence as an explicitly represented goal state. The agent believes that they exist; that is, they believe that they are self-evidencing, and they act and finesse their model as they evidence this belief about themselves. The agent can be wrong, inaccurate, or irrational by the light of their own beliefs, but there is no external, objective standard that they can measure their success against, even if from the outside, we can say that their beliefs are false or their actions irrational.

Given this picture of the self-evidencing agent, it could seem that self-evidencing is somewhat inert and desolate—devoid of volition, purpose, salience, affect, and value. The agent simply exists, and their way of existence is determined by the prior beliefs that they have and the way that they act to provide evidence for them. Different agents (different humans or different species) can exist in different ways, where the differences reflect what their models are like. One way of self-evidencing cannot be ranked as better than another since the objective function available is responsive only to the agent's own model, to their prior beliefs and expectations. The question arises, then: How can such an agent want anything, have aims, rank outcomes, have affective responses or emotions, or value things?

Related to this concern is an observation that the proliferation of "self" as a prefix in this discussion makes the whole view rather "selfy." Self-evidencing may seem to preclude anything but self-centered behavior. Is self-evidencing inescapably egotistical, ruling out prosociality and compassion?

This chapter will be a conceptual exploration of the account given so far, to see if and how volition, will, emotion, and value, and perhaps aspects of morality and ethics as well, can emerge for the self-evidencing agent. It turns out that we can locate volition, will, emotion, and value within self-evidencing in ways that make them unmysterious; self-evidencing is consistent with some approaches to morality and ethics, too. This chapter thus delves into some much-discussed topics in philosophy and public debate. I do not attempt to address these issues comprehensively. Rather, I intend to extract aspects of self-evidencing that could provide useful starting points for constructive exploration of the debates. The point here is to show that self-evidencing, its meager starting point in self-supervised inference notwithstanding, has the capacity to engage with substantial aspects of everyday experience and discourse.

First, I describe how the generative model relates to expressions of volition, and then I explore how this conception of volition can relate to debates about free will, particularly compatibilism. Next, we shift to considering how emotion and sensation might arise from inference, and what this tells us about their nature. With those elements in place, I argue that we can recover salience, value, and norms from within the austere self-evidencing perspective, and briefly indicate how the account could speak to some contemporary ethical debates.

Self-Organization and the Motley of Volition

The internal model of a self-evidencing agent is characterized in terms of the prior beliefs of the agent at a certain time. These beliefs define what states the agent expects to occupy and how those states are expected to transition from one to the other. Beliefs about state transitions codify the agent's belief that the world is a changing and at times volatile place. This implies that the agent can never really rest easy; they must always anticipate uncertainty and be prepared to infer policies that will evidence their model in the face of that uncertainty. The belief that the agent exists is then tied to beliefs that they will actively transition from their current state to another (such as seeing something else, eating an apple, driving somewhere, or getting a degree). This implies that the agent is always poised for either overt or covert action. In other words, as a self-organizing system, the agent's beliefs will drive their active states such that they selectively sample expected inputs at their sensory states, thereby reducing uncertainty.[1]

Action-initiation happens when the agent infers a policy specifying certain not-yet-actual input (such as having turned their head, holding an apple in their hand, driving along a road, or being in a classroom to receive education). The sensory input, given a policy, induces a prediction error ("I believe that I have an apple in my hand, but I don't"), which is quashed as the body plant is entrained to execute the plan (i.e., going to the fruit market to buy an apple). Action-initiation happens because active inference of the policy as it were, "drags" the agent's bodily or mental action along—our volition is in the throes of our priors.

From this fundamental perspective, there is no difference in volition between the most elaborate, deliberate scheme and a mere bodily reflex such as blinking at a puff of air to the eye. In each case, there must be some prior probabilistic belief that induces a prediction error, which entrained action then quashes. Self-evidencing then seems to present volition as a matter of the agent finding themselves doing this or that thing after finding themselves with this or that belief. This contrasts with volition understood in a more everyday sense, as an act of willing something, of oneself choosing the most preferred option in a given situation and therefore enacting it. The question, therefore, is whether inference to the best decision can accommodate our everyday experience of volition. I think that it can.

Policies are ordered and selected according to their ability to best achieve preferred states, understood as the agent's low-surprise states. This means the ordering of policies is determined by the risk and ambiguity (as described in chapter 6) of the beliefs encoding them. An agent prefers low-risk policies, which will take them to their characteristic states and not to other, less characteristic, and thus surprising states. They will consider the risk in conjunction with ambiguity, which is the fidelity of the mapping between the action and the outcomes given their model, which will (if the risk is also low) take them precisely to their characteristic states.

Agents use inference to the best decision to select the policy that best leads to the preferred states, and this is a model-constrained trial and error process that leaves much room for variability and doubt, especially in cases that are more complex than bodily reflexes. Even after some deliberation about a complex case, the agent may hesitantly (i.e., with low confidence) infer a policy that in fact fails to achieve preferred states; self-evidencing does not imply that there is always a very low-risk, unambiguous policy available since the world and our own capacities do not always cleanly afford our preferences. Often, one is compelled to infer a policy that is known to be risky—namely, if the window of opportunity is closing and doing nothing will be a worse policy than doing something risky (examples might be pandemic management, and marriage). Self-evidencing is also consistent with cases where there seems to be an unresolvable conflict between preferences. For example, if the agent's supervisor at work directs them to do some questionable task that conflicts with preferences about ethics but accords with preferences about being paid, then the agent may be stymied and unable to make a confident decision.

Across choice situations, the agent engages more or less deliberatively with their inferences to the best decision. It seems plausible that actions are experienced as more volitional, seeming more like expressions of the agent's will, when they have undergone more deliberation, exploring more deeply into counterfactual scenarios. For example, the agent instructed to do something undesirable at work might consider a range of policies counterfactually, trying to resolve the ethical-financial impasse. In the end, they might decide to do the task after careful consideration of the financial consequences of losing the job. The close scrutiny might make their own preferences more salient for the decision, amplifying the sense of volition even though the choice situation itself is foisted upon them.[2]

We should be careful, however, about aligning volition solely with the degree of deliberation. Sometimes a rash, seemingly impulsive action can appear strongly volitional. For example, the agent might initially decide to do the questionable task at work, but just as they are about to do it, they suddenly quit their job; they might in a flash, and with clarity, have come to infer with respect to their self-model that doing the task will undermine their core values. The agent might experience this as strongly volitional, although still mystified about their own thought processes. I think that this kind of case suggests that volition should reflect not only preferences but also the degree to which the policy in question is constrained via the self-model (introduced in chapter 7). If the self-model has high-precision beliefs about characteristic states and state transitions, and if volition is an expression of the will of the self, then volition would be tied to a strong self-model. Volition would then be related to the invariant features of self that we discussed in chapter 7—namely, minimal "mineness" and narrative self. The experience of a decision as both "mine" and embedded in a narrative (whether epic or more aphoristic) would reflect that it is constrained by the self-model and could underpin the experience of voluntariness.[3]

The everyday experience of the link between volition and self-model might be further amplified in sophisticated deliberation that infers which beliefs the agent would acquire under counterfactual scenarios (as discussed in chapter 6). For example, sophisticated deliberation may lead to the inference that the agent would wind up with a highly unexpected belief about themselves if they perform the questionable action requested by the boss—"If I do this thing, then I will acquire evidence for a belief that I am a bad person, so I won't do it. I'll quit instead." These considerations reflect that self-evidencing treats the self-model as dynamically updated in the light of self-generated observations, and volitional actions could be guided by anticipated revisions. In essence, this is risk assessment, remembering that risk is related to complexity—and a large revision to the self-model indeed comes at a complexity cost. If the agent's policy would entail a complex revision of the self-model, then this could indicate that it is not volitional, in the sense of being constrained by the self. Again, however, there will be scenarios pulling in the other direction. This may become evident if the agent has significant accumulated uncertainty for their self-model and indeed goes exploring for more radical changes to it. For example, the agent might become increasingly unhappy about their daily job, even though it helps them accumulate wealth, and counterfactually explore the policy of giving all their money to charity, liking what they infer about the new beliefs about their future self as a result ("I would come to believe that I'm a nice and happier person if I did that"), and volitionally decide to execute the policy in spite of the complexity cost to the self-model. In this case, there are still constraints from the self-model, under an exploratory policy that the agent believes might afford significant uncertainty reduction in the long run.

In the various examples we have considered, involuntariness lies just around the corner. The wealthy person might shy away from giving to charity and stick with their current but uncertainty-riddled self-model, which more chronically prevents their actions from manifesting a strong, authentic connection to the self. Or the worker might do the ethically questionable action but feel that they are under duress from the supervisor—that is, more constrained by the supervisor's self-model than their own.[4]

I have couched the discussion mainly in terms of volition, which is in turn manifested through a complex net of connections to other notions, such as agency, authorship, self-control, impulsivity, autonomy, authenticity, and self-transformation. The examples that I have used tap into these related notions in various ways, helping to demonstrate that volition is expressed in a multifaceted and context-dependent motley of processes. The question was if volition, understood as such a motley, can find a place in inference to the best decision, given that it portrays the agent's decisions as determined by their priors. As the cases that we have discussed suggest, inference to the best decision can accommodate several cases where everyday experience of volition is prominent. This ranges from common, low-key volitional actions that conform accurately to our preferences all the way to profound ethical challenges that require careful deliberation about our core values, and on to decisions relating to transformative experiences that question who we are. I think this corresponds well to our conception of volition, capturing a motley of cases that can be highly context-specific, or, in self-evidencing terms, specific to many varied beliefs of the agent's overall generative model.

In this account, there is however no exercise of the agent's volition in and of itself. As I mentioned at the start, if the agent believes they are in a changing, sometimes volatile world, then they will be poised for action—poised for state transitions. This means they will come to believe that things will soon be different and those beliefs entrain their agency, making them do what they do. But there is still plenty of room for volition in this picture, as it is diversely expressed simply by virtue of existing in a changing world, and the particular kind of volition that we manifest then reflects how we exist; that is, it expresses our model. A sense of volition might, in humans in particular, relate to how sophisticated policy inference is constrained by the Markov-blanketed self-model. I regard this reduction of volition to inference of state transition beliefs constrained by the self-model to be a metaphysical benefit, absolving us of finding a mystical seat of volition in the brain. In the next section, I connect these ideas to the thorny question of free will.

Causal Nets and Free Will

The problem of free will and determinism is that free will, understood as the freedom to do otherwise, is incompatible with determinism, understood as the notion that everything that happens is determined by laws of nature and initial conditions. If an agent chooses ice cream over broccoli, then that choice was determined by the causal chain leading to it from initial conditions as governed by laws of nature. The agent could not have done otherwise because they could have chosen the broccoli only if the initial conditions or laws of nature had been different, and none of that is in their control. (Adding probabilistic laws to the mix does not help since a choice being caused to have a certain probability does not allow the agent to do otherwise, and irreducible chance is not how we conceive free choice.)

The popular compatibilist answer to the problem is familiar to philosophers, but in my experience, it is surprisingly often unacknowledged in other fields, including popular science. The core idea is, as the label suggests, that free will and determinism are compatible. The key move is to offer a different conceptual analysis of free will than the ability to do otherwise. Compatibilist free will is the freedom to do what one wants, to act in accordance with one's beliefs and desires. Since the agent wanted the ice cream, they acted freely when they got it (and they can then be regarded as responsible for that choice). Compatibilism comes in many stripes, but the motivation is that free actions should indeed be part of the causal nexus because we want them to be caused by the agent's beliefs and desires.

The obvious challenge for the compatibilist conception of free will is that some beliefs and desires are not what the agent wants; for example, if they have been unduly caused to have those beliefs and desires by some external force (such as through coercion) or some pathology (such as addiction). Therefore, compatibilist accounts of free will have various further conditions on beliefs and desires that must be satisfied before an action counts as free. One popular idea is that, to count as freely acting, one must have second-order desires to have the desires that one acts on—the agent must desire to desire ice cream, for example. Of course, this view must somehow avoid the problem that even second-order desires could be the result of external control, and there are various moves in the literature to try to tell this story. Another popular idea is that free will demands that the agent is tracking norms of rationality or moral good, or is appropriately responsive to reason. The intuition is that unfree action is bluntly caused, whereas free action is controlled appropriately by the agent when it conforms to norms of reason. Of course, even when an action is responsive to reason, there is a causal chain determining it, but this chain is controlled by something the agent does, which ties it in the right way to what they want.[5]

Free will is one of the most debated of all philosophical questions, and the conceptual landscape of positions on free will is vast. Not even the most sophisticated versions of compatibilism or incompatibilism are free of serious problems. It would be foolhardy to claim self-evidencing as a new solution. What I will do, then, is situate self-evidencing within the landscape, showing that by virtue of the proliferation of "self"-prefixes that we have been operating with, self-evidencing has some promising properties for compatibilism in particular.

The self-evidencing agent is fully part of the causal nexus. We have defined the agent in terms of their Markov blanket, which is a causal net taking causal input at its sensory states from states external to it, filtering this causal flow through its internal states onto its active states that in turn have their effects on external states, the further effects of which can then be observed by the agent themselves. There is no room for a noncausal element here, and therefore self-evidencing seems to fit with the determinist side in a compatibilist position on free will.

The question then is if self-evidencing provides some properties that would help deliver the kind of control that compatibilism is after, where free actions are those controlled in the right ways by the agent's wants.

Total lack of freedom would be uncontrolled causal flow through the agent, or, in other words, the kind of entropic causal flow that tears apart the agent, making them cease to exist. If the agent exists, they must, in some sense, control or filter the causal flow from their senses, through their internal states and onto their actions. This reiterates the picture of precision control considered in chapter 4; self-evidencing is all about control, of slowing and regularizing the dissipative forces from the world. This signals a kind of autonomy or capacity for self-regulation, which is a good starting point for writing compatibilism into self-evidencing. This makes good sense in terms of the requirement for self-supervision where, in spite of being causally embedded, the self-evidencing agent must resolve its inference to the best decision on its own and without external guidance.

The mere controlled filtering aspect of self-evidencing is not quite sufficient for compatibilist free will, however, since just existing is not enough for acting freely in the everyday sense that we are after. To take a paradigmatic example from the free will debate, an addict is not acting freely, and yet they exist and therefore in some sense control their causal flow; or consider a thermostat that is actively controlling its sensory input but whose actions we would not deem free. The previous section worked through the motley of ways that we might say that an action is voluntary, which identifies types of controlled self-evidencing that can help narrow down the cases. This included acting according to preferences, relying on prospective and counterfactual deliberation, and having actions be constrained by the self-model. The addict and the thermostat should not have these characteristics as strongly as an agent freely deciding to take a drug or voluntarily deciding to control the temperature in a room.

Acting in accordance with one's preferences is to selectively sample observations that fit with one's relatively precise belief about characteristic states. Both the addict and the thermostat do this (the addict expects to take drugs and the thermostat "believes" that it will stay in states between 18°C and 22°C, and acts to make this happen). If we allow that the agent in question can infer their own preferences from their observations (as discussed in chapter 7), then a wedge might be driven between being free and unfree in a way that is consistent with second-order desire compatibilism. There is a sense in which the thermostat's beliefs about its own preferred internal states are directly driven by its actions; for example, it has a bilayer of metal that bends according to temperature and thereby controls the temperature switch such that its active states (i.e., the bending) directly drive its beliefs (the bending encoding the temperature range). If we now eviscerate the direct causal link from the active states to the internal states, then the agent must infer their own beliefs from its observations (in a human, from sensing the ambient temperature changes following their action to turn the heat up or down); that is, the agent must infer with some confidence that they prefer to occupy states between 18°C and 22°C. Inferring one's preferences—what one wants—comes closer to endorsing that preference as one's own, compared to just finding oneself with a preference. Adding more hierarchical and parametric depth can then put into place inferential processes for second-order desires too, in just the same indirect and inferential way, where the observation is generated by a pattern of behavior (e.g., taking drugs and repeatedly regretting it). The best explanation of such a pattern might be that the agent infers that they do not want to prefer drugs. The drug taking is not free, then, and may happen in contravention to their second-order desire, perhaps because of extremely low ambiguity of the drug-taking policy (the precise mapping from a simple syringe action to observed blissful outcomes) that overwhelms the low risk for the longer-term policy to not take drugs.

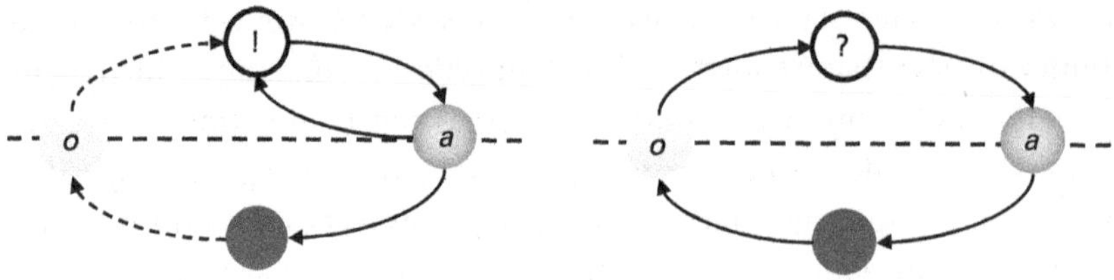

Figure 8.1
Schematic for active things with and without direct links from active states to internal states. Left: The active states directly determine the internal states, absolving the need for inference. Right: An active thing deprived of direct information about its active states, who must infer their own actions (for details of the latter, called "strange things," see Friston, Da Costa, Sakthivadivel, et al., 2023).

On this account of what endorsed preferences and second-order desires could look like for a self-evidencing agent, what allows free action is Markov blanket seclusion, where even one's own preferences and actions are inferred. Intriguingly, the key move is to deprive the agent of direct information (flowing directly from active states to internal states), forcing them to infer preferences as latent states of the world—the capacity for (compatibilist) free will thus comes about through an additional inferential burden. The benefit from taking on that burden is, indeed, a kind of freedom from direct causal determination, where the agent can form hypotheses about what their preferences are and test those hypotheses in action (as discussed in chapter 7). This affords greater flexibility in the action repertoire and allows better fit to the eco-niche of volatility—agents of volatility have greater scope for free will. The internal states of the thermostat, in contrast, are so directly driven by its actions that it has no opportunity to uncouple and explore its preference landscape. See figure 8.1 for this proposed architecture.[6]

The second main type of compatibilism centered on being responsive to reason. To accommodate this notion, it may be useful to appeal to inference to the best decision as a process that approximates optimal inference, which is a paradigm of reasoning. If inference of a policy in a variational manner tracks optimality, then it is in some sense responsive to reason. The agent is not aware of this reason because they are just engaged in the messier "reasoning" of a trial and error abductive process of testing various hypotheses and seeing how they fare relative to expected surprise. To some compatibilists, this distance to the actual, exact reasons might detract from suitability for free will since the agent is not directly appreciating and being guided by reason. I see this feature as a strength because, to me, the abductive process is more psychologically realistic than something like explicit syllogistic or probabilistic reasoning. Action that is not free, according to this self-evidencing conception of being responsive to reason, could be when habitual or context-insensitive policies are mindlessly favored over goal-directed policies with lower expected free energy requiring effortful reasoning; the consequence would be that a chronically habit-driven agent would not be acting freely—it would be an automaton. Or perhaps lack of freedom is when agents persevere with a trialed policy in spite of the increased uncertainty (such as persevering with the drug-taking policy); this might begin to capture some of the kinds of cases that reason-responsive compatibilists would say are not free.

This approach tries to capture compatibilist notions through appealing to indirect, abductive responsiveness to reason. Interestingly, it is the indirectness, where the agent is deprived of direct guidance or knowledge from objective norms, that makes this an attractive package of views. This in turn speaks to another version of reason-based compatibilism—namely, where actions are free if they track objective norms of rationality. This removes the focus from being responsive to reason in one's actions and simply requires that actions track rationality. Self-evidencing under the free energy principle is about approximating exact inference via variational means, such as inference to the best decision, where exact inference is indeed an objective, Bayesian norm of rationality. The strength of this approach is that agents can track (i.e., approximate) the norm without knowing the norm by doing something else—self-evidencing. (Note that this kind of compatibilism often has a moral inflection, in the sense that the free action should track moral principles. Such a moral aspect is not easy to capture within self-evidencing as we have presented it so far; later sections in this chapter will consider some moral themes.)

The intention with these comments about compatibilism within a self-evidencing framework is not to argue that we have a full-fledged theory of free will here. All the compatibilist theories of free will are subject to extensive continued debate. I suspect that there are elements of truth in several of the diverse versions of compatibilism, and a theoretical framework that can accommodate a range of compatibilist intuitions holds promise. My comments in this section might therefore serve as an invitation to work seriously with self-evidencing as a fruitful theoretical biology method or system for compatibilism. If there is an overall theoretical suggestion here, it is that compatibilist free will might require the kind of causal and computational seclusion characteristic of (sophisticated, "strange") self-evidencers (as in figure 8.1). Seclusion puts an inferential burden on agents but affords them the freedom required to thrive as agents of volatility. I think this is what free will feels like, too; being a responsible decision-maker is hard inferential work, and one's decisions constantly get knocked around by uncertainty and volatility in the environment and within oneself.[7]

Emotion and Sensation

Looking over the self-evidencing landscape, things can look a bit desolate, even after allowing for volition. Everything seems determined by quantitative processes relating to precision, risk, complexity, learning rates, model selection, and evidencing. All these processes depend on the beliefs encoded in the model, which are updated relative to the general-purpose objective functions of variational and expected free energy. One might wonder what kind of probabilistic inference could be involved when an agent feels a certain emotion or experiences a bodily sensation.

There is in fact a rich vein of research into approaching emotion and sensation as a roughly inferential phenomenon—namely, in well-known constructivist and appraisal theories of emotions. These theories connect to the classic James-Lange theory of emotion, which says that emotions arise as the agent seeks to explain away generic heightened arousal, such as increased heart rate. For example, the agent can sense the arousal but is uncertain about what caused it and resolves that uncertainty by inferring, with the help of contextual cues, that they are angry (someone bumped into them, so the arousal must signal anger) or happy (someone gave them a present, so they must be happy). In other words, it is not that there is high arousal because one is angry or happy; rather, one is happy or angry because there is high arousal. There are several diverse members of this family of theories, some of which (such as Lisa Feldman Barrett's work), are directly appealing to Bayesian frameworks, whereas others are more loosely consistent with an inferential aspect.[8]

Since the first tranche of predictive processing works, emotion and sensation have been the focus of several treatments in terms of predictive coding. My own first foray was to conceptually analyze emotion in terms of perceptual inference, exploratory action, and precision modulation with regard to increased uncertainty indicated by bodily arousal. In that work, I was particularly trying to figure out how inferential engagement with arousal could not only give rise to emotions or sensations but also dynamically change that arousal and thereby dynamically determine them. For example, if the agent senses thirst and suspects their arousal is caused by dehydration, then they can selectively try drinking a glass of water to see if the thirst goes away, which changes the bodily sensation while evidencing the belief; or they might hypothesize that loud music is causing their arousal, explore this by opening the window and listening, and in the process infer that their arousal signals irritation; or they might pay attention to their big toe to determine what's wrong, amplifying the inferred pain and then rubbing the toe to test the pain further. I want to briefly recount these ideas because they are incorporating an active element, capturing the action-perception loop inherent in self-evidencing. Nowadays, even though there are several theories of the James-Lange type, there has been less deployment of active inference. In this section, therefore, I explore further what happens once the inferential approach to emotion and bodily sensation is transposed into active inference and self-evidencing terms.[9]

If emotions depend on active inference, then the story revolves around expected self-evidencing, where heightened arousal is the current sensory input, which increases actual or expected uncertainty given the expectations under the model. For example, the agent's heart rate has gone up, taking the agent outside their expected homeostatic range; the agent then needs to infer a policy that would help resolve the uncertainty (i.e., help return the agent to their characteristic states) on the basis of an action that homes in on the latent cause of the arousal. This inference should be what gives rise to emotions and sensations, as the constructivist theories would have it, albeit in active rather than perceptual inference.

The examples I have given here involve policies that predict the outcomes from an action that engages with the arousal, sometimes to make the arousal go away again (as in the thirst case) or to increase the arousal (hopefully transiently, as when touching the big toe to see if the pain intensifies). The idea is that sensations and emotions arise as the policy is inferred and are then dynamically modulated as the policy is enacted. Thirst is felt as the "get a glass of water" policy is inferred ("This must be thirst because I've planned to get a glass of water") and is dynamically changed as the policy is enacted

("It was thirst because it is gone now that I've drunk a glass of water"). Irritation is felt as the "listening closer" policy is inferred, and it is modulated as it is enacted by the agent walking around trying to find out which neighbor is the source of the nuisance.

The assertion here is that emotions and sensations arise upon active inference, and then through enaction of a policy pertaining to the uncertainty indicated from arousal. The picture is not that the thirst or the irritation is the target of the inference; for example, the agent does not infer that the irritation is the cause of the arousal. The inferences are engaging with the particular situation in whatever the context is, which can be about glasses of water, sources of noise, or a stubbed big toe, for example. The particular emotion or sensation begins to emerge as soon as the policy is being inferred and then dynamically waxes and wanes as the particular policy is enacted and evidence begins to accumulate—this is what inferring and enacting arousal-related policies feels like.

Such a self-evidencing account of emotion and sensation allows considerable variability in how emotions and sensations are felt, due to the different factors in selecting and enacting policies. The active inferences spurred by arousal can, as all active inference, be exploitative or exploratory, and they can have different constellations of risk and ambiguity. The examples given here of active inference underpinning emotion can play out in acting for pragmatic value, as in the thirst case, or acting for epistemic value, as in the irritation case. As always, what matters for inference is the ambiguity and risk of the policy or the expected gain in self-evidencing under the policy and the ability to occupy one's characteristic states. If the context is clear and the agent has prior experience, then it can be easy to find a good policy, such as feeling thirst just after doing exercise or already knowing what strategies are relevant for checking for an irritating noise (e.g., open the windows to check for outside nuisances or close them to check for inside noises). In these cases, inference is strongly driven by priors, and the emotion and sensation would come into relief almost immediately; for example, as soon as the noise is there, the agent leaps up in irritation, determined to find its source. Often, there is no clear candidate policy and the agent needs to investigate on the basis of more imprecise beliefs. For example, as the agent begins to work from home more frequently, they find that they are more unsettled in the morning and the later afternoon, but they do not know why. They might have a vague feeling of irritation, but because the context is new, it could be due to any number of things, so they need to go through several exploratory steps more haphazardly before finding out that it is the rush-hour noise from commuters on the busy road outside (which they never heard when working in the office) that is creating their sense of unease.

Focusing on the time course of active inference, a dynamic landscape appears. Policies can be lined up and tested vicariously in counterfactual deliberation, where the agent's inference might come in fits and starts, favoring one policy but soon giving it up for another, when deeper counterfactual testing goes against it. As soon as a policy is entertained, even if only in a short counterfactual window, precision control begins to kick in to amplify some sources of evidence, which on this account should give rise to fledgling or fleeting emotions. When a policy is finally settled on but not yet enacted in overt behavior, covert, attentional precision control will begin to push toward the emotional state in question, which then amplifies dynamically as the policy is finally enacted over some time period. Since active inference is forward-looking, it acquires this dynamic profile, and it therefore becomes well suited to our dynamic and varied everyday experience of emotion and sensation. In contrast, passive perceptual inference presents the inferred belief much more as the end point of unconscious inference, making it harder to accommodate emotion and sensation.

The dynamic time course of active inference makes good sense of how some emotions and sensations arise quickly, such as a flash of anger, and others more slowly, such as the simmering, gradual realization that one has been feeling irritated all morning. I think it is also possible to capture anticipated emotions ahead of actual arousal, such as the dread of having to do a painful task—namely, in sophisticated inference where the agent is inferring what beliefs they would acquire if they were to engage in an arousal-inducing task (e.g., the agent might anticipate the dread-filled active inference that they would make if they were to give a nerve-racking public speech).

With the variability in inferred policies, there is also the possibility of false inference for emotion and sensation. A familiar example is the agent who reports anxiety with heart rate increases and sweating each afternoon at work around three or four o'clock, just before they leave for home, and perhaps they report unease and headaches around lunchtime on the weekends. They might infer something not being right at home as the best explanation of the unease, and they might test this belief actively by staying at work on a Saturday, where indeed the headache does not arise even though the afternoon unease recurs. They might then end up feeling emotionally distraught about their family situation. The better explanation for all this is that the agent drinks too much coffee at work and too little on the weekends. The inference to the best decision in this case would be to explore caffeine intake, reducing it and quickly feeling better on workdays and not needing as much to reach threshold on the weekends. The initial, more anxious policies and the subsequent, more accurate series of policies would both come with their own dynamic emotion and sensation trajectories. For example, the agent would shift from the emotional state of distraught anxiety to a more bodily sensation of caffeine nausea. Again, there is a wide scope for different false or suboptimal inferences to the best arousal-related decisions during the varied dynamic unfoldings of active inference.[10]

Sometimes constructivist-style theories of emotion are contrasted with the notion of basic emotions arising from midbrain structures, without the need for contextualizing, inferential processing in higher brain areas. According to this view, several core types of emotional experience, such as fear, rage, lust, and panic, arise in ways that do not seem to involve elaborate context-dependent processing. There is ongoing research and open questions in the debate between these two stances on emotion, such as about the extent to which there is in fact any consciousness involved in the behavioral responses that seem (quite convincingly) to indicate basic emotions when no higher brain areas are involved. Here, I will make the point that the existence of basic emotions would not in itself invalidate the self-evidencing account of emotions in terms of active inference. Best policies can be inferred from among a very slender set of candidates and in the absence of much contextual information; fear, for instance, is evaluated very quickly and precisely. Some policies might be so favored over developmental or even evolutionary timescales that they become amortized (as discussed in chapter 4), meaning that there is a direct association ensuring that the policy (such as fight-or-flight) is triggered as soon as arousal evidence comes in.[11]

An agent might also be afforded considerable resilience and flexibility for dealing with volatility by combining these basic and constructivist systems for emotions, suggesting that they are complementary. Thus, self-evidencing agents may have two internal Markov blankets: one for rapidly processed and relatively context-independent basic emotions and one for more fine-grained context-dependent emotions, providing input to each other but having faster and slower dynamics, respectively. We saw this kind of sparsely coupled architecture might be in play for the self-model within the overall generative model, in chapter 7 (recall figure 7.2). When two blanketed systems constrain each other in this coupled structure, then one system can keep operating rapidly on its own, while the other investigates if it is possible to fine-grain the inference in question as more evidence comes in. The basic system might infer "be afraid" and help drive a fight-or-flight policy, but also passing this to the constructivist system to work with, testing more fine-grained contextual hypotheses ("Am I watching a scary movie?"), which can in turn feed into the basic system ("Stand down—we are enjoying this fear, after all").

It seems an interesting possibility if the two main approaches to emotion processing could be understood in complementary terms of two coupled, Markov-blanketed systems in this manner. It would be consistent with this architecture that basic emotions could exist in the absence of the slower fine-graining system (e.g., in a decorticated individual). This would seem better able to accommodate empirical findings than a presumption that there is one network processing the same data, suggesting that only one of the two approaches to emotion can be true. I find it worth thinking about as well because this kind of architecture could speak to some of the emotional dissonance that we often feel, such as being thrilled and scared at the same time, as if two systems are telling stories that are not quite consistent and are awaiting reconciliation through further self-evidencing.

I have been exploring what might be said if some of the standard approaches to emotion and sensation were cast more explicitly in terms of active inference and self-evidencing. It seems that this can help capture some of our intuitive, everyday phenomenology where emotions and sensations can arise in different ways, rapidly or slowly, and with either strong policies attached or calling out for more vaguely guided exploration, or with feelings of dissonance and conflict. This varied phenomenology is perhaps tied to the fact that human self-evidencers are, as I speculated in chapter 7, agents of volatility with deep, temporal internal models, who are always on the lookout to exploit volatility to learn about new, hidden causal processes. This implies some flexibility in the policies of choice, ensuring that the agent does not always just fight or flee but sometimes hangs around precariously to explore and learn, which may give rise to fine-grained, dynamic emotional landscapes that are informative about the volatile-prone properties of their eco-niche.[12]

Emotions and Expected State Transitions

I have proposed that emotions and sensations might be captured well by a self-evidencing framework. This leads to a question about what is being represented in the active inferences through which emotions and sensations arise: What are emotions and sensations picking up, and what are they "about"?

The internal generative model is a structural representation of the latent causes in the world and within the body (as discussed in chapter 4) as these causes interact with each other over different timescales and give rise to transitions for the sensory states of the Markov blanket. This tells us that representation cannot be conveyed as a mere one-to-one representation (such as "This neuronal population on its own represents this cat" independent of other neuronal activity picking up relations between cats and occluding fences and chasing dogs and cat food). Representation comes about through a structured, causal role, implemented by the way that neuronal populations dynamically relate to each other. This causal dependency within the internal states of the system is, through some mapping function, relatively isomorphic to the structure of the generative process in the world and in the body. The question then is: What kind of structural representation is involved in active inference when, specifically, emotions and sensations are in play? This question matters because it is not always obvious that emotions and sensations have representational content; for example, philosophers argue about whether the representational content of pain could really be something like "There is damage to my body."

To answer this question, notice first that causal structure comes in many shapes and can be represented at several levels. In particular, the statistical properties of a generative process, such as its variance, can themselves be represented. For example, the agent may infer the pattern of variance in the occurrence of certain nuisance sounds in their apartment complex, setting aside what the noises themselves are ("I don't know what this noise is, but it is most probable to occur around 11 P.M. at night, and also at other times, but weaker on the weekend"). This is useful to gauge one's confidence in inference and to guide precision control (as discussed in chapter 4). Agents of volatility, like humans, with an interest in exploiting volatility, are also well served to keep a close eye on different kinds of uncertainty; this allows them to figure out if an event occurs within expected variability or is perhaps a volatile event occasioned by some new interfering latent cause (e.g., a very loud noise might be caused by the neighbor's party, which the agent hopes is a farewell party, making them happy and prompting them to put in an early application to upgrade to the neighbor's apartment).

With such representations of statistical properties in place, a plausible account of emotions and sensations would be that they arise from active inferences that include representations of statistical variability in state transitions that specifically have either direct or indirect implications for the agent's expected self-evidencing. The process begins with arousal, signaling unspecified increased uncertainty about transitions from the current state to new states—the agent needs to reconsider their confidence that their actions will lead to the expected observations. Now the agent considers policies for action that will resolve this uncertainty, and specific emotions can arise during this process. For example, the agent has so far deployed a relatively proven and unambiguous set of policies at work, but a new colleague has been added to their team, causing the agent to expect uncertainty about their own policies. Initially, there may be broad, indistinct uncertainty because the new colleague's role, skills, and ambitions are not yet known, causing arousal, general anxiety, or nervousness; gradually, contextual knowledge helps resolve this into a more fine-grained assessment with appropriate emotions such as joy and relief, if the new hire skillfully facilitates the agent's own tasks toward less ambiguity and risk, or frustration and anger, if in fact the new role is an interfering competitor, leading to imprecision or risky outcomes about the agent's future policies. Positive emotions arise when expected self-evidencing is better than before, and negative when expected self-evidencing is worse. That is, emotions track representations of expected self-evidencing. This could lead us to say that emotions do have representational content, in the shape of beliefs about the model evidence gradient.[13]

Overall, when emotion is understood as arising as the agent processes expected self-evidencing, we get a dynamic account, which computationally tracks statistical properties of the environment relating to self-involving state-transitions, and physiologically is indicated by the waxing and waning of arousal. This suggests that emotions and sensations are practically always with us, since the agent is always in some informative state of low, middling, or high arousal. Perhaps the vibrant emotions like anger and fear get most of the attention, but emotions also arise in low arousal states, such as quiet contentment. Even very low arousal states, like some forms of meditation, have emotional and affective tones that likely reflect a low expectation of volatility and therefore diminished need for abolishing current policies in favor of new ones. The intensity of affect and emotion throughout the day and night then would be driven by the magnitude of violations to expected self-evidencing in this or that context and be amplified by the precision control exercised in particular situations. I think that this captures well our experiences of affect and emotion as always in some sense present, with current states of affect coloring every moment and with the strongest emotions tending to stand out.

Emotion research is rich and varied, and what I have suggested in the preceding discussion only serves to start telling a story that is distinctively along self-evidencing lines. One immediate question might be: How could this account deal with the fact that we can feel emotions vicariously when watching other people's emotions, or even actors pretending to have emotions? If emotion represents prospective self-evidencing, then how can an agent feel vicarious joy when their aunt wins the lottery or sadness when Bambi's mother dies? To account for this, it seems likely that an agent's own counterfactual processing can be engaged in a process of vicarious inference and learning while observing others dealing with arousal, and experiencing emotions while seeking their own policies to resolve uncertainty. And if there is counterfactual processing, then it seems that the agent could feel secondhand emotions arise when considering the scenarios in question. This should perhaps not be surprising since counterfactual reasoning is itself offline, and in that sense is already fictional.

It is easy to speculate about why this kind of empathetic inference should be easy to achieve for self-evidencers. One of the greatest sources of volatility throughout everyday life is the interfering hidden causes persisting in other people's minds, and therefore, a key task for the self-evidencing agent of volatility is to infer the mental states of others. When prospectively trying to identify policies with low risk and ambiguity, agents mentally go through the reactions of other people, trying to predict if their existing mental states, or the mental states that they would acquire once they become aware of the agent's plans, would foil the agent's actions. The agent might wonder if they should go for a promotion, but then worry what executing that plan would do to the expected self-evidencing of their team members; perhaps going for a promotion would increase the risk for a colleague's policy of staying in their role, making them jittery and jealous and causing them to change tack and also go for the promotion. Human agents live intricate social lives, and it may be expedient to self-evidencing to build and maintain a model that represents not only the preferences and habits of other people but also their second-order beliefs about their policies, capturing their emotions.

Self-evidencing is all about actively sampling expected observations, given the model, and since the model can have counterfactual-based beliefs about other agents' beliefs about ambiguity and risk, it also makes sense that we actively exploit counterfactual scenarios in advance of trying actual plans. We wonder what other people feel or would feel, form hypotheses about them, and test these hypotheses by engaging with others; for example, someone else looks parched, and we ask if they feel like a cold drink. Active sampling can also play out in an overt counterfactual way, such as when a person in a relationship first proposes subjunctively to their significant other, "If I were to ask you to marry me, what would you say?" to elicit a change in their partner's policy landscape, and only then pop the question in actuality. Overall, for agents of volatility, who are tracking the influence of new hidden causes, a capacity for vicarious emotion appears adaptive.

This section has considered what the representational content of emotions and sensations might be if they were based on active inference. The proposal, which also extends to vicarious emotions and sensations, is that they would track the expected uncertainty about agent-involving state transitions.

Salience, Value, and Norms

Self-evidencing could initially seem to provide a merely causal and affectively inert probabilistic picture in which it would be hard to locate real-life agents. So far in this chapter, we have been able to plausibly locate self-evidencing approaches to volition and compatibilist free will, as well as emotion and affect. The previous section also speculatively built in our capacity for modeling other people's mental states and our capacity for vicariously experiencing their emotions. This section will explore if we can go further and begin to discover notions of salience, value, and norms, familiar from ethics and political philosophy.

When the agent, from the perspective of the precision over states and policies in their internal model, looks over their actual and counterfactual contexts, they will see a rich and varied landscape of more or less salient and attention-grabbing opportunities for self-evidencing. Some aspects of the environment will almost automatically elicit actions, beliefs, and emotions because they are associated with highly unambiguous policies in prior learning. Other aspects of the context require deep counterfactual planning and active investigation before salient expected observations are revealed in deliberation, spurring the agent to act. In all cases, it is the anticipated risk and ambiguity with respect to expected free energy that make some opportunities stand out, shaping and guiding all these choices. Salience is here tied to active inference, and therefore to the agent's beliefs about their own characteristic states, constrained by their self-model. This means that salience is more than just high-precision beliefs. Although the self-evidencing picture might be merely probabilistic, when tied to the agent's beliefs about the state transitions that involve themselves, it comes alive with salience.

Pragmatic value—utility—is part of the picture too, as we saw in chapters 3 and 6. Active inference can be in the service of utility, in the sense of delivering sensory input that is close to the agent's preferred states. This means that self-evidencing captures the common notion that agents tend to act for utility. However, utility is not the be-all and end-all of decision-making because it is maximized relative to an objective function for free energy, which is neutral on acting for utility (exploiting) or for epistemic value (exploring). Intrinsically, there is no difference in the value to the agent of exploitation versus exploration. The balance between the two is governed by the expected self-evidencing under the agent's available policies. From the point of view of the agent, the only thing that matters is maximizing the fit, over the long-term average of their own model, policies, and observations that are expected to be sampled.

Even with value-laden notions of salience and utility, it may seem that genuine value is still missing. Everything that matters is relative to the agent's prior beliefs, in a process aimed at providing evidence for those beliefs. The question is: Is this a kind of ego-centered relativism that drains the world of real value? Might self-evidencing be wholly self-serving? This could be the impression if self-evidencing is cast in teleological terms, as being the aim of maximizing utility and information gain to enable the agent's own continued existence. But as we have emphasized several times, the self-supervised nature of self-evidencing means that there is no such objective goal state to aim for. The agent, if they exist, rather forms beliefs and selectively samples observations that fit well with those beliefs. Self-evidencing does not specify anything about what the contents of those beliefs should be, or whether they should be self-centered or perhaps altruistic. In this sense, self-evidencing is consistent with the full spectrum of human behavior. An agent who tends to behave prosocially simply has prosocial beliefs, and a selfish agent has selfish beliefs. It is thus not correct to say, for example, that the prosocial agent has the prosocial beliefs because it will facilitate the ultimately selfish goal of their own existence. The objective functions of variational and expected free energy are what is tractable to the self-supervised agent, and they do not feature survival as an objective, independent goal-state.

I see this focus on existence rather than teleology as a strength of the self-evidencing framework. Humans display a wide range of behaviors, from the obnoxiously self-obsessed to the saintly compassionate. Self-evidencing should be able to accommodate this diversity. This is easily achieved by having priors explain the behavior, no matter what that behavior is. Of course, explanations that quickly appeal to priors and stops there are in some sense shallow just-so stories (as discussed in chapter 5). Why did the agent donate their immense wealth to eradicate hunger? It was because they had the prior belief that they donate their wealth like that, and then they did donate, and this provided evidence for their belief. However, behind the shallow explanations lies the prospect of deeper explanations, as we broaden the scope to wider contexts and behaviors, and when we begin to think about ways of intervening to test which beliefs agents actually have and how they may be changed. Then more unifying and powerful explanations can emerge. For example, we might discover that more subtle sets of beliefs really underlie the wealthy person's donation. Similarly, we might discover that the same behaviors can emerge from different belief sets such that, for example, different individuals with the same phobia should receive different therapies tailored to the phenotype of their particular phobic beliefs.[14]

This leads to the substantial question of how agents may have diverse prior beliefs. Why do some agents harbor prosocial priors and others antisocial priors? Even within a species occupying one eco-niche, there can be many different hill-climbing trajectories, signifying different variational trial and error inferences. The broad gamut of influences on behavior can shape these trajectories. For example, two quite similar agents may both flip from acting for utility to acting for epistemic value when their current patch is being depleted of resources, but then end up exploring quite different patches, which serendipitously shapes one agent in more prosocial ways than the other. Human agents' propensity for counterfactual deliberation can also drive otherwise similar agents along different paths; there are not very strong constraints on the choice of counterfactual scenarios or the choice of alternative policies to test in these scenarios, so the upshot can quite easily favor divergent policies for different agents.

I think there is a further factor that pushes human agents to diverge in their priors. I proposed in chapter 7 that humans are agents of volatility—that is, we fit our behavior to an eco-niche that is partly characterized statistically in terms of volatility. The thought is that we seek out volatility rather than shy away from it because, equipped with an extensive capacity for deep counterfactual deliberation, we can exploit volatility by learning about new, underlying latent causes. When an agent encounters a volatile event, they should rely less on past learning and habits and increase their learning rate while searching for new policies—a task for which counterfactual reasoning is useful. This predicts divergence in beliefs for agents of volatility, as different agents will converge less on their priors over the long term if they relinquish their priors and increase their learning rate relatively often, adopting new policies in response to whatever volatile event they have come upon and want to exploit.

Much of the volatility that we come upon relates to other agents—so there is a double meaning to "agents of volatility." We interfere in each other's business and frequently have to deal with unexpected nonlinearity in our self-evidencing efforts due to the way that others react to our actions. This suggests that much of our counterfactual exploration of volatile spaces will relate to social interactions. Human agents are wont to ask themselves, "What would happen if I act like this within this particular community, rather than like that?" It seems likely that different, more or less prosocial patterns of social interaction policies can emerge from these exercises, which could in principle be expressed in terms of diverging social mores or moral codes. For example, a policy of donating funds to charity could gain strength in one group of agents and not in another, such as if charities are well regulated in one country and poorly regulated in another, affecting the ambiguity of the donating policy.

However, self-evidencing ultimately cannot escape a kind of value neutrality. No matter how divergent agents or groups of agents are, there is no common metric against which their goodness can be assessed. The prosocials and antisocials simply harbor different beliefs, and they might well both be self-evidencing quite well, by their own lights, which is all they can do. To self-evidence is just to accumulate evidence for a model, and this is neither a good nor a bad thing in the end—it just is.

Of course, we can agree to put a standard in place (such as longevity), and we might discover that altruistic species in fact survive longer and can be selected for in evolutionary terms. Recent work on active inference over evolutionary timescales posits a coupling between two Markov-blanketed, genotype and phenotype systems that would in principle allow us to see altruism as a selected-for prior belief in self-evidencing over evolutionary timescales. This is interesting in itself, but I do not think that we can simply argue that a goal of longevity promotes prosocial behavior. Plenty of bad people and evil groups outlive good people and benign groups.[15]

We have seen that self-evidencing grounds our existence and allows diverse values and prosocial (and antisocial) priors, but it cannot directly be an arbiter among value priors. This does not imply that we must throw in the towel on disagreement about values and say, "Well, you have antisocial priors and I don't, but nothing can be done about that." Our self-evidencing happens in an intensely social setting, where we harbor rich internal models about other people's mental states, including their suffering. We are buzzing collectives of little Markov blankets constantly bumping into each other, receiving input emanating from other agents' active states, which filters into our own minds via our sensory states. We interact, communicate, and engage in discourse, and we can act more or less concertedly to probe and change other people's prior beliefs. For example, we can try to influence other people to engage in morally relevant counterfactual reasoning that may change their beliefs about the suffering of others. A charity that uses emotionally charged images of emaciated children in their funding drives operates like this, as do parents encouraging their child to see things from another child's perspective. Political, diplomatic, religious, contemplative, and philosophical discourse all operate in the marketplace of priors too, exploring and communicating ways in which to argue for, or perhaps manipulate others in favor of, different moral systems. The key is to see communication as the upshot of self-evidencing active inference, where the communicative act elicits a change in the environment (namely, in the internal states of other people), which in turn gives rise to a change in their behavior.

Therefore, if an agent happens to believe that communities should be prosocial and work to minimize the suffering of all, they will be acting to try to make this a self-fulfilling prophecy on a social scale. Even this is compatible with a diverse range of actual behaviors because inference of a particular policy for promoting prosociality depends on the agent's model of other people's internal states (e.g., considering how far antidemocratic autocrats are from common humanity beliefs), of states of the world (how climate change will affect the likelihood of prosociality), and of risk and ambiguity (whether, if the agent leads by example by donating, they will have enough to feed their own family, and whether the signal strength of donating is enough to promote prosociality in others). The agent might despair at the prospects of influencing anyone much in today's world—even the prosocial can be cynical.

I am not suggesting that moral discourse is easy or can be expected to converge smoothly on consensus via the coupling of individual Markov blankets—far from it. It is part of our everyday experience that negotiating norms of behavior is among the most frustrating and fraught of activities. It is not easy within families, within local communities, or on a national or worldwide scale. Nevertheless, norms are negotiated in this kind of active discourse, in a pulsating give-and-take of active inferences by individuals and within and among communities. It is difficult to see any higher authority than the waxing and waning over time of various prior beliefs with moral implications. I offer this here merely as a perspective informed by self-evidencing, not as a substantial position in normative ethics, meta-ethical debates, or political theory.

If there is a lesson offered by self-evidencing for morality, ethics, and value, it is that continual, broad, and participatory discourse matters for the adoption of prosocial, environmentally sound, and democratic beliefs. Beliefs strengthen when they acquire evidence for themselves, and hence if they are not continually articulated and shared, and challenged, refined, and tested, then they will lose force. As human agents, we need to deliberate and engage with beliefs, reinforce them in each other, and allow all agents to engage actively with beliefs to themselves notice the accumulation of evidence. This includes renewal, where the underlying assumption is that, although past observations often are useful guides to inference, the world may change under our feet, making it imperative that we retest and reexplore the foundations for our beliefs. For example, we urgently need to engage in continued participatory discourse about whether our democratic institutions are still up to the task of dealing with the combination of enormously powerful international tech giants, misinformation on the internet, and the rise of autocrats. Similarly, ongoing, deep, and diverse conversations need to be had about how democratic institutions can be renewed to ensure that the voices of marginalized people can finally be heard and properly inculcated into everyone's internal models. Merely presenting information to disenfranchised electorates for their passive registration, or mechanically conveying both sides' viewpoints without challenging evidential status and precision, is unlikely to lead to prosocial results.[16]

Moral Model Fitting

So far, I have argued that self-evidencing is consistent with the emergence of prosocial norms and moral codes that may be elaborated via ethical frameworks. I have also made it clear that self-evidencing itself does not push moral norms in any particular direction, neither egotistical or altruistic, and cannot arbitrate among positions in normative ethics. I want to end by exploring whether there is nevertheless something to learn directly from self-evidencing for our understanding of morality. I think that there is, once we consider that a moral code, principle, or theory, be it in philosophy, religion, contemplative traditions, or political society, is essentially a set of prior beliefs—a moral model—that informs active inference to drive certain behaviors. This means that we might draw on some of the standard tools for model maintenance for self-evidencing agents, which may help understand what a healthy moral model might look like, while acknowledging the neutrality of the content of these models.

Moral models come in many shapes. They can consist in relatively explicit rules, like the Ten Commandments of Christianity, principles of sufism, or the rules expressed in the dozens of vows undertaken by Buddhist monastics. They can be expressed in prescriptive ways, such as a religion's do's and don'ts, or they can be expressed in terms of rights, such as the Universal Declaration of Human Rights by the United Nations. Moral models can also be conceived as more generic moral normative theories like consequentialism focused on maximizing utility and minimizing harm, deontology focused on duties or rules, or ethics focused on virtuous character and developing sound moral dispositions. Moral models can be implemented differently in different communities, for example, depending on degree of orthodoxy and on socioeconomic factors, or in different areas of life, such as business ethics and medical ethics. We appeal to moral models when we make plans, such as considering a new workplace or health-care policy, or when we encounter behavior that we believe should be regulated, such as athletes taking performance-enhancing drugs. We all encounter and apply moral models as we move around in our everyday social and work environments.

Previously, I argued that the adequacy or correctness of a moral model is arbitrated in social interaction and discourse. It is interesting to see what kinds of considerations can be brought to such discourse from the perspective of self-evidencing and model maintenance. So let us consider a moral model now. We will pick consequentialism, which calls on agents to minimize harm and maximize utility.

The starting point would be to consider notions of variational and active moral inference in the abductive trial and error sense that we have developed up to now. Given the moral model and the observations, there is an optimal set of policies that would minimize expected surprise with respect to that model. The agent approximates these policies through trial and error processes, as in inference to the best decision (chapter 6). In this variational process of moral self-evidencing, there is an expectation of acceptable levels of uncertainty, as the approximation process undulates more or less dramatically around optimal inference; recall here the hill-climber metaphor (chapter 6), where the hill-climber ascends on average in the long term, but via many detours and temporary ups and downs. This means that as the agent (or community) goes about their moral self-evidencing, there will be expectations for the overall rate of uncertainty reduction, given the model. Looking backward and forward in time, the agent will learn from the observed and expected utility and harm and get a sense of the divergence from expectations.

In this deliberative process, the agent might form the belief that the moral model needs revision if the moral path indeed repeatedly has extreme divergences. Just as my model of how to drive a car is no good if I keep having near-fatal accidents, even if I do tend to make it to work, a moral model is no good if it allows huge divergences in utility and harm, even if utility is maximized and harm minimized on average. The issue here is that the same average can be obtained from measurements that vary very little around the mean, as well as from measurements that vary wildly around the mean. One way to express this is to say that the ergodicity of the moral model should be observed, where *ergodicity* (in the technical sense used here) reflects a high probability of obtaining a measurement close to the mean in random sampling. In other words, an ergodic moral principle could be that no matter when or where or whom we sample, their utility or suffering should not be too far from the mean. Bringing this kind of consideration to the assessment of a moral model can be critical, for example, if one group of people is experiencing extreme harm even if the majority in society is mainly fine. The ensuing policy could be to ensure that a participatory, autonomous voice is given to anyone in public discourse—sadly, this is a policy that we have historically been extremely bad at implementing (think of low castes, women, people with disabilities, the young, First Nations peoples . . . the list is endless).

Of course, much of moral discourse revolves around different agents' very different expectations for the acceptable variability of utility and harm. For example, there can be moral discourse about whether the extreme harm inflicted by a given war is acceptable if it brings about some expected long-term utility. An appeal to ergodicity is still morally neutral among moral codes in and of itself, and of course, antidemocratic and repressive moral models can also improve their models through ergodicity considerations, relative to their model; for example, versions of puritanism might be taking ergodicity to extremes, tolerating absolutely no deviance from moral codes for anyone, at any time.

Elements in this ergodic ethical approach speak to notions of the universality of moral principles and human rights. Simply put, those notions may align well with the way that we maintain our models through self-evidencing in general, emerging from the best-makers in inference to the best explanation and decision. Thus, it may be that ergodicity considerations are useful for negotiating moral principles because they allow a balancing of common ethical concerns revolving around risk and ambiguity.

One set of concerns relates to risk. On the one hand, there is the concern that extreme harm to some individuals is unacceptable; and on the other hand, there is the concern that in practice, some harms are unavoidable (since the world throws curveballs at us and sometimes the rights of some can be in tension with the rights of others). The first concern pulls in the direction of moral perfectionism, with exceptionless, universal principles; perfectionism could be described as inferring an extreme low-risk policy, where no harm is tolerated. The other concern pulls in the direction of a laissez-faire stance, where zero resources are expended to avoid harm; this could be described as inferring a high-risk policy. Think again of our hill-climber, who would be a stuck perfectionist if they blankly refused to ever descend even in the slightest, or who would be forever lost if they always take the route that takes the least effort; they must find a balance between perfectionism and laissez-faire.

In addition to risk, ambiguity is relevant to moral ergodicity. A very unambiguous moral policy might step out very precisely how much harm reduction would follow from an action, and a very ambiguous policy would leave much uncertainty about what the harm reduction outcomes would be. Limited resources can drive us toward low-ambiguity policies—to ensure value for money. Low ambiguity does not have to align with low risk, however, and policymakers can be so blinded by efficiency that they forget to consider the risk. Moreover, in complicated, real-world scenarios, low ambiguity can be hard to achieve since interfering factors can be hard to control. For example, town planners might precisely reduce harm from traffic congestion by widening freeways but put the population at risk of increased pollution and ever-increasing undermining of public transportation options. Picking the low-ambiguity policy might reduce harm to some people (car drivers) some of the time, but looking at others (people using public transportation and people with lung conditions) would show increased harm.

In the remainder of this section, I will briefly explore what this ergodic, self-evidencing perspective might tell us about some current debates. A controversial contemporary version of consequentialism promotes "longtermism," the view that everything should be done now to ensure utility for countless future generations, tens of thousands of years from now, whose summated potential for well-being dwarfs ours. A debated consequence of longtermism seems to be that considerable harm should be tolerated now to ensure that distant future. This might include severe hardship for some people, or it might give license to ultrarich oligarchs to maximize profits now, promising to invest in the future, but not the present (via taxes). From the ergodic point of view, longtermism could be argued to allow too many extreme deviations from the mean in harm and utility to current generations, without implying that we should forget about the welfare of future generations altogether. Longtermism could also be criticized for not properly incorporating sensitivity to the way that expected uncertainty accumulates over increasing timescales. Risk and ambiguity are much easier to assess in the short term since we can predict in greater detail, and with more confidence, over short time spans. On the other hand, we shy away from extreme long-term planning because risk and ambiguity for policies with expected observations in the far future are influenced by all the many volatile events that can happen between now and then. This timescale sensitivity can indeed make our policy selection relatively myopic; this might be contributing to the way that we continue to ignore the real dangers (which have been known for more than 100 years) of anthropogenic climate change. Careful balancing is again called for here, between beliefs about variability now and in the future.[17]

There has been debate in ethical theory about how useful universal moral principles are, given how hard it is to make them apply in concrete cases. The particularist position, for example, has argued for abolishing ethical principles for this reason, instead adopting moral actions that befit particular situations. In our terms, particularism arises from a model-fitting concern that typical moral principles are overly simple, so they fail to apply outside ideal cases. In the famous trolley problem, for example, an agent must decide whether to divert a runaway train from running over five innocent people and onto another train line, where only one innocent person will be killed. Perhaps a universal utilitarian principle applies cleanly to this trolley problem, but our intuitions begin to waver as soon as the details of the case are filled in or change in certain ways: What should we do if instead of diverting the train, we push a person in front of it to stop it, or what if we have to sacrifice ourselves to stop the train, or other options (the trolley case is now subject to countless variations)? What this suggests is that moral principles are underfitted to actually morally relevant facts, and they need to be augmented with some kind of context sensitivity. However, as I understand particularism, it has not been successful in avoiding a fairly directionless stance, threatening laissez-faire, so this position appears to cast around for new bespoke principles at each new moral decision point. Although proponents of this theory of course have things to say about these objections, stated in model-fitting terms, the danger is that they have responded to the underfitting of moral principles by constructing a model that is overfitted to particular facts, leaving no robust guidance for new cases. In this debate, then, there is a familiar element from model-fitting that should be brought in, concerning how to balance between having overly simple, underfitted models and overly complex, overfitted models. The self-evidencing perspective again offers a useful conceptual tool because active inference scores risk, the analogy of complexity, and drives the agent toward low-risk options that do not involve a radical departure from their prior beliefs. Applied to moral model-fitting, penalizing risk ensures that the moral model will do everything it can to respond to particular moral situations, but only so long as it does not constantly abandon its prior beliefs (or principles).[18]

Consider next a further, controversial version of consequentialism, "effective altruism," which suggests that we should choose the altruistic acts that create the most good in the most efficient way. For example, delivering mosquito nets directly to affected malaria areas is said to increase lives saved more than many other charitable actions, such as investing in education, sanitation, or health-care systems. Effective altruism essentially puts a premium on altruistic policies with the lowest ambiguity. Investing in education, sanitation, and health-care systems may be argued to save many lives, but that would unfold over several years, with many complex steps involved, so it would be more ambiguous, even if it could be argued to have the same risk as the mosquito net policy.

However, as we saw in chapter 6, agents can be seduced by low ambiguity, choosing policies that in the long run do not help that much in robustly achieving preferred states. For example, neglecting to invest in education, sanitation, and health care may in the long term increase uncertainty and the variance of lives saved and quality-adjusted life-years for a community, making exposure to malaria a persistent risk. Similarly, if the moral edict is to donate to efficient charities rather than pay taxes in one's affluent home country, that might have the long-term risk of undermining high-intensity, state-funded research in malaria prevention, or it might contribute to undermining tax-based welfare systems, in turn destabilizing regional peace. As before, the proposal here is not that effective altruism is an outright failed moral model, but rather that a good moral model should balance risk and ambiguity and consider policies in terms of ergodicity. Such balancing might well unravel the enticing low ambiguity of effective altruism.

Longtermism, discussed earlier, is sometimes seen as emerging from effective altruism, but now we can see that the two are somewhat strange bedfellows because the low ambiguity associated with effective altruism policies is what cannot be achieved over the extremely long timescales of longtermism. The alternative, while sticking to the utilitarian outlook, would be policies that balance ambiguity and risk over a number of parameters and over a realistic span of years for robust structural changes in afflicted regions (for malaria, for example, this could pertain to the investment and local community engagement needed to improve health and education systems over, perhaps, ten to twenty years).

This section has essentially worked through how aspects of inference to the best decision for self-evidencing agents (as described in chapter 6) can speak to elements of moral discourse. The core idea is that moral models are maintained in much the same way as any model for a self-evidencing agent, and this may in turn provide a helpful perspective with respect to some moral debates. At least it shows that there is no serious gulf between self-evidencing and moral discourse. (We return to aspects of moral aspiration in chapter 10.)

Concluding Remarks: Self-Evidencing for Reflective Equilibrium

The objective in this chapter was to address a challenge to self-evidencing—namely, that it is a merely probabilistic and causal framework, with no room for important and diverse aspects of our everyday experience, including volition and will, emotion and affect, and value and morality. The chapter therefore had a rather tall order of discussing some of the thorniest topics in philosophy—free will, affect, and ethics.

Rather than providing comprehensive accounts, I have focused on providing proof of concept that self-evidencing can begin to speak meaningfully, and perhaps even fruitfully, in these big debates. This goes to show that self-evidencing is not cut adrift from important aspects of daily life or from philosophically substantial debates. The picture that I have painted is as follows: Self-evidencing allows the full spectrum of volition and free will, cast in terms of compatibilism about free will, and allowing many diverse routes to free decision-making. Emotion and affect arise as agents engage in assessment of risk and ambiguity in their own prospects for active inference. Value and morality must come down to specific prior beliefs constituting an agent's moral model, which can be negotiated and finessed in continual moral discourse.

Across all these elements of human experience, self-evidencing emphasizes the kind of abductive balancing of considerations we saw in action in chapter 6's notion of inference to the best decision and chapter 3's notion of inference to the best explanation. This pulls away from some ideas of rational and ethical inference and decision-making, such as those focused on simple principles or algorithms. Rather, inference to the best free and moral decision aligns more with notions of reflective equilibrium that have been developed in political and moral philosophy. Reflective equilibrium is a method for moral decision-making that works back and forth among different principles, considerations, and other relevant beliefs for a particular decision context. The method delivers a decision when reflection on all these beliefs achieves some degree of coherence—an equilibrium of sorts—among all the morally relevant considerations. Reflective equilibrium appears, on a first look at least, to correspond well to how I have applied inference to the best decision to the topics in this chapter. Therefore, it might be worth it in the future to explore a notion of "reflective self-evidencing equilibrium," both to see if it is consistent with the job as it is described in debates on free will and in moral and political philosophy, and also if it might bring some interesting inferential and probabilistic properties to this debate. I will not develop a substantial account here, but I will point to some considerations in favor of this idea.[19]

A key problem with reflective equilibrium is that the notion of equilibrium as coherence is relatively unconstrained, speaking to logical consistency but also to richer notions where some of the beliefs provide justification and explanation for others, but without defaulting to a foundationalist account of justification. Self-evidencing, as we have developed it here, contains all these elements in one package. Coherence can be thought of as a strong fit of the observations expected under the model and the observations in the long-term average; this is what suggests that the model is fit for its purpose in the agent's eco-niche. Coherence is therefore tractable for the agent, as they measure their policy selections against the objective function for expected free energy. Internal support among beliefs would relate to the way that beliefs and policies explain away expected observations, as well as the hierarchical structure where beliefs with slower and faster dynamics constrain each other.

Self-evidencing also provides an elegant, coherentist solution to the problem of grounding beliefs without defaulting to foundationalism; typically, foundationalism relies on self-evident, basic beliefs, which have been hard to identify and agree about, but we can appeal to self-evidenced beliefs, in particular, the belief "I exist," which grounds the self-organization and selective sampling in self-evidencing. Although my comments here are brief, I believe that they indicate a potentially fruitful project of assimilating reflective equilibrium with self-evidencing.

This chapter looked at important aspects of our everyday lives—namely, volition, emotion, and value. Chapter 9 will address another much-debated issue: the problem of consciousness. Conscious experience suffuses every moment of our everyday, awake lives (and even some sleeping moments)—and self-evidencing has much to say about what underpins and shapes our consciousness.

9 Self-Evidencing for Consciousness

Here is a simple argument about consciousness and self-evidencing: if self-evidencing provides the fundamental explanation of what the brain does, and if brain activity gives rise to consciousness, then self-evidencing also explains how consciousness arises. However, this argument does not imply that self-evidencing immediately explains consciousness. Self-evidencing happens in all self-organizing organisms, many of which are not conscious. So, to zero in on consciousness, more is needed—only *some* types of self-evidencing processes underpin consciousness. The task of this chapter is to begin identifying these.

The philosophy and science of consciousness are fraught with scientific and philosophical difficulty, requiring careful methodological and theoretical choices. The chapter therefore proceeds by navigating through various decision points in the science of consciousness. This helps identify a project for connecting self-evidencing and consciousness, which I believe is scientifically tractable, philosophically interesting, and metaphysically realistic.

In the wider perspective, the stage-setting that I present in the first three sections of the chapter also serves to demonstrate how self-evidencing interfaces with existing scientific endeavors in the mind, cognition, and life sciences. Self-evidencing delivers an all-encompassing, fundamental construct for every aspect of these sciences, and it is important to demonstrate how it can engage in constructive, piecemeal conversation with particular domains.

The following sections of the chapter then proceed to explore a key aspect of self-evidencing and consciousness—namely, the detachment of internal states from external states. This is the stepping stone for identifying types of active inference that speak to particular properties of consciousness. This allows us to build up, piece by piece, a self-evidencing account of consciousness. The virtue of this approach is that it helps itself to a valid construct, self-evidencing, and searches within that construct for properties that can match onto consciousness with good face validity. This differs from the more constructless typical search for neural correlates of consciousness, and from attempts to develop theories from scratch using relatively untried constructs.

The upshot is a picture of consciousness that I find appealing because it ties in themes from previous chapters on the self, on volition and free will, and on rationality and active inference. It portrays our consciousness as a natural fit with the concept of human self-evidencers as agents of volatility.

This will be the longest chapter in the book because it needs to not only address consciousness but also demonstrate how to prepare the ground for connecting self-evidencing with an ongoing scientific endeavor, and making good on the promises with more detailed examples of research. Even though consciousness is a difficult arena for science, several of the dots are beginning to be connected.

I begin this chapter by sifting through debates within the science of consciousness, and then navigate through topics relating to the metaphysics and the scope of consciousness. Having set the scene, I then appeal to self-evidencing in a discussion of detachment and consciousness, in agents of volatility. This notion of detachment is then brought to bear on the role of active inference for conscious content and its enactment in unfolding experience, with consideration of online and offline experience and their typical hierarchical levels of processing. I finally discuss hallmarks of conscious experience—namely, subjectivity and unity of consciousness.

The Science, Neural Correlates, and Theories of Consciousness

The science and philosophy of consciousness constitute an exciting and fast-moving field with its own scientific associations and journals. There is a rich body of literature of innovative experimental designs targeting many aspects of consciousness. There are also many proposed theories of consciousness of different ilks, some of which are associated with sustained research programs. As with any fledgling field of scientific research, especially fields focusing on philosophically recalcitrant topics, consciousness science is also strewn with flawed, inconclusive, speculative, and untested approaches. Different approaches operate with different concepts of consciousness, so it is often unclear how to even begin to compare them. There are also often different levels of metaphysical ambition in the science and philosophy of consciousness, concerning how close to a solution to the mind-body problem an approach seeks to get to. This section discusses several of the issues and problems in the science of consciousness, ultimately aiming to clarify a tractable connection to self-evidencing.

When we use the terms "conscious" and "consciousness," we can refer to several different things, and accordingly consciousness science can target several aspects and dimensions of consciousness. This is well recognized in discussions of consciousness, although the relations between different aspects and dimensions of consciousness are poorly understood and only rarely studied.

A creature can be conscious as such, rather than unconscious (e.g., waking up after dreamless sleep or coming to after being knocked out or after anesthesia). This might be called the "global conscious state." A commonly seen descriptor for the global conscious state is "arousal," which is sometimes spelled out in physiological or behavioral terms and is often associated with activation in some brain-stem structures. A creature in a global conscious state may also be in some particular background state of consciousness. For example, they may be delirious, highly attentive or in a flow state, or high. A background state is a particular way of being in a global state—a way that somehow characterizes or "flavors" all or most of what is currently experienced.[1]

A creature in a global state of consciousness will be conscious of a number of different things. They will experience a variety of conscious content. For example, they may experience the sound of music through their headphones, the light of the electronic screen that they are looking at, or the words and images appearing on the screen. Some content stems from the world around the creature (exteroception), such as coffee cups, cars, or cathedrals; some stems from inside the creature's own body (interoception), such as the heartbeat, hunger, or an itch. Some content occurs in perception, some in imagery, and some in episodic memory. Some conscious content is perceptual and some, such as thought, is perhaps more cognitive. A creature in a global state, enjoying some suite of content, will typically be attending to only some of it; some content will fade into the background and some will be lucidly present at the center of consciousness.[2]

With the basic structure of global state, background state, and content of consciousness, several challenging aspects are then broached. A key element of consciousness is its subjective character, sometimes captured in terms of the first-person perspective. This relates to difficult ideas about what it means to be a creature with not just its own viewpoint on the world (as a surveillance camera might have) but a sense of subjectivity integrated into that viewpoint. What does it mean to have a subjective self, which imbues all of consciousness somehow? On this topic, there is significant discussion, harking back to chapter 7, about minimal ways of being a subjective self or having models of the self.[3]

Sometimes we are just consciously experiencing the world—conscious experience seems transparent in the sense that the experience of a green tree is just that—seeing the green tree (sometimes this is referred to as "diaphanousness"). We are often not explicitly conscious of the fact that we are conscious. But we can turn attention to consciousness itself, introspectively noticing our consciousness. For example, you can ask yourself if you are really hearing a faint sound or hallucinating it. Or you can do a kind of quality control of your own experiences, asking yourself how confident the experience leaves you that the sound was really there; this is related to metacognitive processes and can be said to opacify the otherwise transparent conscious experiences.[4]

Phenomenologically, at any given time, all the different aspects of consciousness seem to come together in one somehow coherent whole—this is the unity of consciousness. Over longer time periods, this unity may be experienced in the stream of consciousness. There is considerable debate about what it means to have a unified consciousness, what does the unifying, and whether consciousness must be unified. Similarly, there is debate about what it means to say that consciousness somehow flows, whether it has to, and at what rate. For example, do people whose brain hemispheres have been disconnected have two individually unified streams of consciousness? In this area, there is a question about what happens at any conscious present; it seems that we do not experience the literal, pointlike present, but instead experience a more smeared, moving window of experience, which feels like a "now," but one that encapsulates a bit of the immediate past and a bit of the immediate future—this is called the "specious present."

If the goal is to explore consciousness in terms of self-evidencing, it matters which of all these aspects or dimensions of consciousness we have in mind. In this chapter, I will focus mainly on conscious states (which I shall discuss in terms of inferential detachment), and conscious content (specifically, how active inference determines it).

The next step is to situate this project within the science of consciousness. There are two somewhat related ways of engaging in consciousness science: there are efforts to identify the neural correlates of consciousness and efforts to formulate theories of consciousness.

A neural correlate of consciousness is often defined as finding the minimally sufficient conditions for the occurrence of a conscious state or content. The logic of this program of research is simple: identify which parts of the brain are most closely associated with the presence or absence of a particular conscious state or content. That is, find ways of varying the conscious state or content and identify which parts of the brain varies in a similar way, or find ways of keeping the conscious state constant and identify areas of the brain that change nevertheless when the input or other things not related to consciousness change.[5]

A classic paradigm is binocular rivalry, a robust visual effect where participants are presented with one image to one eye and another image to the other and, instead of seeing a mishmash of the two images, the participant's conscious experience alternates between the two images every few seconds. This is useful for consciousness science because the physical stimuli are kept constant while the conscious content changes, so any changes in the brain that coincide with reports of the changing experience are likely to be implicated in the consciousness of that content.

Searching for the neural correlates of consciousness is popular because it seems a feasible way to make empirical inroads on the otherwise recalcitrant topic of consciousness. We can identify the neural correlates of consciousness while being almost entirely neutral on the metaphysics of consciousness. All that is needed is an agreement that there are brain processes cooccurring with conscious states. We need not have any particular belief that those brain processes are identical to, constitute, or realize conscious states (or, for that matter, if conscious states are also somehow linked to nonphysical states or substances, such as various versions of dualism would have it). One hope is that once we have identified the neural correlates of consciousness, then they can help inform the formation of a theory of consciousness, or at least that the neural correlates will be a constraint on any proposed theory of consciousness.

Theories of consciousness come in many stripes. There are philosophical, metaphysical theories of consciousness, such as identity theory, nonreductive physicalism, and property or substance dualism. And then there are more scientific, empirical theories of consciousness, such as higher-order thought theory, integrated information theory, and global workspace theory. Some theories are inspired by neuroscience findings, including from neural correlates research. Some are more driven by a priori or phenomenological considerations, such as integrated information theory. It can often be difficult to directly compare theories of consciousness. They often seem directed at different aspects of dimensions of consciousness, and often it is somewhat nebulous what their empirical predictions might be.[6]

It is natural to expect that the project of finding consciousness within the self-evidencing framework will relate to finding correlates and theories of consciousness. However, as Anil Seth and I have argued, the relation is not straightforward. Self-evidencing does not directly deliver a theory of consciousness. Backed by the free energy principle and its process theories, it is a framework for any self-organizing organism, whether conscious or not. At the same time, the search for the neural correlates of consciousness is in its basic form blind to self-evidencing. A researcher could point to patterns of brain activity as correlating with some conscious state without knowing anything about self-evidencing.[7]

This independence of correlates and theories is a methodological advantage for self-evidencing. The method would be to piece together a systematic neural correlate of consciousness, where the systematic part is provided by processes and mechanisms that are unified within the self-evidencing framework. That is, for several different, particular aspects of conscious experience, each of the neural correlates should be interpreted in terms of self-evidencing. This way, one could provide self-evidencing explanations of, for example, binocular rivalry and other core paradigms in the study of conscious content. It is similar for other types of content and for background states and global states. As the individual self-evidencing mechanisms accumulate, the hope is that a pattern stands out—something special about the self-evidencing processes that governs in the domain of consciousness and is not seen in nonconscious processing. This would provide significant insight for the eventual, piece-by-piece building of a self-evidencing model or theory of consciousness.

The assumption here is that only certain types of active inference, model maintenance and precision control are specifically associated with consciousness. This would drive in a wedge, serving to explain why self-evidencing itself is not a theory of consciousness but nevertheless a theory can emerge from it. This kind of approach to consciousness promises to have construct validity because it deploys an existing, robust construct—self-evidencing—to a specific domain; this addresses the complaint about some theories of consciousness that they feel the urge to bring new and untried constructs to explain consciousness. The approach also will have face validity; that is, it is able to explain our conscious phenomenology (e.g., simulations of an active inference model can be relatively isomorphic with the dynamic evolution of conscious experience). Finally, it can have predictive validity because active inference models can be used to predict new phenomena within the modeled domains (I will provide examples later).

We have now placed the self-evidencing approach to consciousness within the science of consciousness. Before embarking on the actual self-evidencing account of consciousness, two more broad topics should be discussed: the metaphysics and the scope of consciousness.

The Metaphysics of Consciousness

Imagine that someone gave you a list of the individual heights of 100 people and then reminded you of the concept of relative height (i.e., who is taller than whom). Based on the particular facts about individual heights, and by virtue of mastering the relevant concepts, you could derive the facts about who is taller than whom (e.g., person 34 is taller than person 56 and person 56 is taller than person 13). It would be odd for someone to complain that in spite of having all the facts and mastering the concepts, it is still an open question who is taller than whom. You would think that they had either gotten the individual facts wrong or misunderstood the concepts. Given enough information about individual heights, the truth about relative heights is fixed.

We can be pretty sure that this is the situation for all the truths about relative heights, even for past generations and all generations to come. There are no weird facts about individual heights hiding anywhere, and no aspects of the truths about relative heights that dangle unfixed. The metaphysics of relative heights is pretty boring, which is why there are few philosophers exercised by that question.

The same can be said for many other phenomena, even those where we do not yet have all the facts. For example, we are pretty sure the facts about the tides are fixed once all the facts about the moon, oscillation, gravitation, and particular facts about bodies of water on Earth are known and we master the relevant concepts. This may be so even if there are still matters of particular fact that we need to determine, such as how the tides will be distributed once sea levels rise and deep-sea currents reverse. Another example could be health, where we are pretty sure that all the biological, psychological and social facts fix all the facts about health, even if we do not know yet what those facts are and might in fact still be unsure exactly which concepts are needed to make the transition from facts about biology and social existence to facts about health. We do not think that there are facts about health beyond the facts about those domains.

Physicalism is the metaphysical position that *all* the truths about everything are fixed in this way, by all the physical truths together with the appropriate concepts. At least, that is one clear conception of physicalism. Notice that physicalism does not imply that we can cogently and informatively explain all the truths about everything exclusively on the basis of physics—to fix truths in this way is not to explain.[8]

One way of expressing the metaphysical mind-body problem, then, is to point out that the existence of consciousness suggests that physicalism is false: it does not seem odd or wrong-headed to say that even once all the physical truths are fixed, it remains an open question what the truths about consciousness might be. Possessing the concepts about consciousness and other concepts needed to understand things, I can point to the complete list of the particular physical truths describing some individual creature and reasonably say that whether that creature is conscious is an open question, even if that list of physical truths is identical to the list of facts describing *me,* and even if I know that I am conscious. I might sincerely believe that the copy of me is probably conscious, but I am not contradicting myself by denying consciousness to the copy of me, in the same way as I would be doing if I denied that truths about relative heights are fixed by truths about individual heights.[9]

The metaphysical situation does not concern our current ignorance about physical truths about the brain. There is much that we still do not know about brain function, obviously. But we have a reasonably good idea about what kinds of truths are in play: truths about neurons and how they work, neurotransmitters, neuronal networks, processes of neuronal signals, oscillations and connections, morphological and genetic facts about the brain, and so on. The metaphysical problem seems to persist even if we imagine having all these types of facts. It will go away, it can seem, only if an entirely new type of fact is called on, which can close the gap between the physical truths and the truths about consciousness (hence the problem with construct validity of some theories of consciousness, mentioned previously).

The literature on the metaphysics of consciousness is vast. For our purposes, the question is: How does it relate to the exploration of consciousness within self-evidencing? My view is that the metaphysics of consciousness is a genuine philosophical problem. All the proposed solutions that I am aware of either water down consciousness (e.g., to something merely functional or behavioral, with no genuine guarantee of instantiating subjective experience) or smuggle obscure properties into the physical basis of consciousness to do the explaining (e.g., electromagnetic properties, neural oscillations, or quantum properties are all physical properties, but it is not clear how they would close the gap). None of that seems attractive.

An alternative is to give up physicalism about consciousness, which incurs the debt of explaining what kinds of nonphysical properties in our universe underpin consciousness, and not least, how those subjective or phenomenal properties causally interact with the normal physical properties. There are some well-worked-out dualist and idealist positions, but to me, they are not attractive either.[10]

I regard it as unlikely that any theory of brain function, including any process theory for self-evidencing, might deliver a straightforward solution to the difficult metaphysical problem of consciousness—it is not hard to conceive a sophisticated self-evidencing model that is not conscious. What is needed to deliver the metaphysics outstrips what self-evidencing can deliver.

If there is no real prospect of solving the metaphysical problem of consciousness, then why bother with the science of consciousness described in this chapter, let alone theorize about self-evidencing? Several scientific theories of consciousness purport to be in the business of solving the metaphysical problem, after all, but I think that those theories are mistaken about this. And as we have seen, the search for the neural correlates of consciousness is designed not to address the metaphysical problem. Luckily, there is much to learn about consciousness, even if we fall short of solving the metaphysical problem outright. We can learn about the function of consciousness and how consciousness interfaces and interacts with other cognitive and sensory systems; the form and structure of conscious phenomenology; why experiences have their particular content and feel; what kinds of causal, neural processes it takes for a creature or content to be conscious rather than unconscious; and the nature of the self and autonomy.

All this research and knowledge will be immensely important and useful. It could help us better diagnose and develop treatments for disorders of consciousness, such as the vegetative state, and mental illnesses involving consciousness such as schizophrenia, depression, and personality disorder. No one would be swayed by the objection that such a discovery is deficient because it does not yet close the metaphysical gap between brain truths and consciousness truths. Metaphysical problems are real and point to important and fascinating issues in our knowledge of our world, our concepts, and ourselves. But metaphysics does not need to stand in the way of science. It may be that one day, when the science of consciousness has really progressed on these nonmetaphysical problems, the metaphysical problem will seem less pressing—the passion of an obstinate group of philosophers, perhaps. Metaphysical problems sometimes persist for millennia, and they sometimes fade into oblivion (such as the scholastics' much-derided concern about how many angels can be in the same location). I am content to wait and see about this.[11]

We can compare the metaphysical question about consciousness to the situation in other domains marked by unresolved philosophical questions. Philosophy still grapples with deep problems of knowledge and skepticism, free will, ethics, fairness, and beauty, but none of these prevent progress from occurring in domains such as scientific method and evidence-based practices, decision theory and rationality, ethical frameworks, political science, and aesthetics. The philosophical problems do serve a function, though—they remind us that there are no easy solutions in these areas of inquiry and we must continue to open-mindedly explore and evolve conceptual and theoretical frameworks.

Overall, self-evidencing will not directly deliver a solution to the metaphysical problem of consciousness. As Seth and I argue, it promises instead to offer a new, systematic, and unifying framework for interpreting the neural correlates of consciousness and for navigating existing theories of consciousness. If we are lucky and remain open-minded, this can amount to a breakthrough in our understanding of the nature and function of consciousness.

The Scope of Consciousness

The science and metaphysics of consciousness struggle with two problems concerning scope—one empirical and one more metaphysical. These problems potentially apply to the self-evidencing framework too, so we need to know how to steer our way through them.

First, there is the question of which kinds of creatures are conscious. Is it just humans who are conscious, humans and other so-called higher animals, or other kinds of animals and creatures too? This is an empirical question about something difficult to observe, or indeed something that is just plain unobservable. This scope problem is by no means solved, and there is much disagreement about where to draw the line of consciousness. The disagreement is aggravated by different research programs focusing on different aspects of consciousness when discussing its presence in nonhuman creatures: Are we looking for consciousness exactly like ours, or just some degree or kind of phenomenality or sentience (whatever each of these terms mean)? To stake out a position here, I believe that our current state of knowledge suggests that some aspects of consciousness exist in many species of animal. Many animals surely have conscious states more or less like our experiences of fear, pain, distress, anticipation, and pleasure. With pain, for example, I think that it is obvious that many creatures can feel pain and that humans inflict untold and intolerable pain on many animals.

Second, there is a more metaphysical question about how widely distributed properties of consciousness are (philosophers often call these "phenomenal properties," picking up on the link to conscious phenomenology). Some philosophers believe that the best solution to the hard problem of the metaphysics of consciousness is to forget attempts to base consciousness on physical matter or some ethereal dualist substance. Instead, we should accept phenomenal properties as a fundamental element of the universe. This is panpsychism, which comes in various versions, each with its own motivations and attendant problems. Whereas I can see the philosophical motivation for exploring panpsychism as a kind of last metaphysical resort for consciousness, I do not find it an attractive position.[12]

There is a scientific version of panpsychism too. This is mainly driven by one of the theories of consciousness mentioned here—namely, integrated information theory. As I see this theory, it proceeds by first offering an introspectively based conceptual analysis of consciousness, translating this analysis into tractable information-theoretical and causal constructs that in turn lead to a mathematical expression of both the level and content of consciousness. This expression is so general, however, that there is no principled reason to restrict it to particular kinds of biological substrates such as brains. Advocates of this theory therefore embrace a version of panpsychism, pointing out that even if consciousness is ever-present, there are more or less complex, interesting instances of conscious experiences—in humans, for example. Again, I can see the motivation behind this approach, but I do not find it attractive.[13]

The reason why I find panpsychism unattractive is that I am first and foremost interested in the kind of everyday consciousness that we are familiar with, and I think that panpsychism's phenomenal or mathematical-causal properties are too distant from our consciousness to be helpful in elucidating it. If literally everything is indeed conscious, then it is "conscious" in a sense so removed from our consciousness that it will have no scientific bearing on our own understanding of consciousness. Even if we were convinced by the truth of panpsychism, I find it hard to believe that anything in the science of consciousness would change (in particular, none of the scientific questions would be made easier).

After this sojourn into the question of the scope of consciousness and the challenging notion of panpsychism, the question for self-evidencing is: What scope for consciousness would it entail, and is it committed to panpsychism? If any process for self-evidencing were sufficient for consciousness, then any self-organizing organism would be conscious, entailing a very wide scope that would include, for instance, plants and single-cell organisms. However, the path that I have outlined here presumes that only some kinds of self-evidencing processes are involved in consciousness. That is, unconscious states and content in humans or other creatures (or machines, for that matter) might involve aspects of self-evidencing, but certain active inference characteristics turn up when singling out conscious states and contents specifically.[14]

My approach here is consistent with a minimal hypothesis that whenever there is a conscious state or content, active inference is involved; in other words, active inference is a necessary condition for conscious state and content. This could be tested, and it would be bad for this hypothesis if changes in conscious state or content occurred without a change in active inference. This minimal hypothesis could be made more substantive by then considering which specific types of active inference coincide with consciousness. For example, we could investigate if consciousness requires active inference that has specific precision control properties or counterfactual or metacognitive properties. I turn to this project now.

Detached Models and Conscious Agents of Volatility

Recall the basic biological imperatives considered in chapters 2 and 3 relating to homeostasis and allostasis. An agent who exists is able to maintain homeostasis, acting to keep blood pressure, glucose levels, body temperature, and other characteristics within certain bounds. Homeostasis is largely reactive. The agent senses an actual deviation from homeostatic set points and therefore reacts to recover; for example, the rose thorn stings the agent, causing their heart to pound, so they remove the thorn if it is still stuck in them. It would be advantageous to anticipate and prepare for such deviations; for example, putting on gloves before gardening. Such anticipatory processes for maintaining homeostasis constitute allostasis. Allostasis seems pervasive in humans at all timescales of processing; for example, the agent exerts effort to shop now to prevent being hungry tonight, and they invest in their studies or work now to earn money and prevent being hungry in five years' time. Allostasis describes our efforts to maintain self-evidencing in the long-term average, with reference to our homeostasis. We describe this as self-evidencing guided by variational and expected free energy with respect to an internal model that harbors precise homeostatic beliefs.

Any allostatic agent who can enact self-evidencing policies will need to model aspects of the world around them, as discussed in chapter 5. For example, an agent who can propel themselves away from a predator selects that action because they expect that they will be faster than the predator, and this will help them occupy their characteristic states. Some agents will be likely to form more elaborate internal models of the causal structure of the world around them. Having preparatory processes in place for predators might require careful mapping of their habits and whereabouts. When planning an ocean swim, for example, it is useful to know in advance that there is a higher probability that a thing sticking out of the water is a shark fin when observed at dawn or dusk rather than in the middle of the day, because that is when sharks tend to prowl for their prey. In general, the more anticipatory capacity an agent has, the more internal modeling capacity we should expect them to have. For example, an agent who can combine seasonal and daily anticipations of the whereabouts of sharks will be able to avoid them quite well, but they need representations operating at different timescales and levels of abstraction, as well as the ability to convolve these representations and let them inform their everyday decision-making. For example, it might be reasonable to take a dawn swim at the agent's beach in the summertime only, because the warmer water keeps the sharks away.

The internal models of such anticipatory, allostatic agents will be models in a thick sense, harboring vast, rich structural representations (as discussed in chapter 5). One key aspect of agents with such models is the degree of *detachment* from the current sensory input. For the homeostatic system, there are just reactions to the current sensory input, whereas for the allostatic system, there is anticipation of the sensory input that will arise in the future based on learnings from the past. Everything might still function at a basic level of processing learned associations; for example, dawn and dusk increase the probability of a shark. But the raised heart rate and fear are not simply in response to the amount of daylight but is related to the evoked internal representation of the whereabouts of sharks and policies for avoiding them. The convolution of different represented aspects also invites genuine modeling where causal interactions are tracked (e.g., between sharks' seasonal movement patterns and water temperature). Interactions between causes require some kind of distinct representations that can be combined and recombined in flexible ways. This suggests models with independent, reidentifiable representational components that can be combined in various ways and contexts, independent of the current sensory input.

The point here is that anticipation necessarily requires some detachment from current sensory input, and likely also some degree of freedom to recombine elements to form new representations of objects and properties. I think that humans are exceptionally advanced in our anticipatory capacities; that is, in our capacity to form detached, hierarchically deep, structured and flexible models of the world around us.

Self-evidencing provides the principled foundation for the anticipatory modeling and agentive elements needed for allostasis. Self-evidencing also comes with a clear, two-sided notion of detachment, in virtue of the Markov blanket (see chapter 4). One side is causal through the conditional independence of the internal states and the external states and the active states and the sensory states. The other side is inferential, where external states are only known and acted upon vicariously through probabilistic inference implemented with internal precision control mechanisms. If we want to describe an agent with the basic setup for allostatic detachment, self-evidencing is the right framework.

All this matters for consciousness because our conscious experience has extensive capacity for detaching such that it operates independent of current sensory input. Humans (and most likely many other animals too) dream during sleep, and we daydream and mind-wander during our waking hours; we make extensive plans well ahead of time; we imagine all sorts of humdrum or fantastical scenarios; we think purposefully or are lost in thought; and we engage in episodic memory where past experiences are relived. Our imagination and thought can enter outrageous counterfactual worlds ("What would airline safety guidelines be if pigs were capable of flying?," "How great would I feel if only I'd made that buzzer beater!," "What will I believe about what others believe about me, if I were to get that promotion?"). Even when we stay in the present, we often turn partly away from the world and internally monitor ourselves and our own mental processes, metacognitively assessing the quality of our beliefs and inferences. All this detached mental processing occupies our consciousness—this is a central part of what it is like to be conscious. And the processes are all the ones that we have aligned with the self-evidencing processes that we have discussed so far (in chapters 6–8), revolving around inference to the best decision.

Self-evidencing in hierarchical active inference has been proposed as being important for consciousness by Karl Friston, who argues that temporally thick and causally deep internal models necessarily must capture aspects of agency, purpose, and self. Given that these aspects are often associated with consciousness, Friston can say that agents "with deep (temporally thick) generative models are self-evidencing in a prospective and purposeful way, where agency and self become an inherent part of action selection," such that "the sort of selection we have associated with (self) consciousness operates within the same system—a system that can simulate multiple futures, under different actions, and select the action that has the least surprising outcome."[15] Although Friston is not here explicitly concerned with detachment as I have conceived it, it is clearly presumed to be part of what enables autonomy and self-consciousness.

The project, as I have set it out earlier in this chapter, is to identify systematic neural correlates of consciousness, founded in a sound construct with the resources to explain distinctive aspects of conscious phenomenology. Self-evidencing is an excellent candidate framework here because, by construct and before thinking about explaining consciousness, it provides the opportunity for detachment. On that basis, then, self-evidencing can be wielded in several ways, captured in inference to the best decision, which matches the particular ways of detaching that we recognize in our own consciousness—for example, in the shape of the deep and thick counterfactual processing that Friston emphasizes. Appealing to self-evidencing in this way also means that we do not have to say that every self-evidencing organism is conscious. All we have said is that for some types of detached, thick, deep, and counterfactual self-evidencing, which not every self-evidencing type of creature would need to have, there seem to be good fits with conscious experience specifically. In other words, every self-evidencer detaches, but not all detachment needs to be associated with consciousness.

Now a difficult challenge can be formulated. We want to pick a construct for our neural correlates that avoids maximum scope of consciousness, which I described as unattractive and panpsychism-invoking in the earlier sections of this chapter. Self-evidencing gets some way toward this goal because even though every organism self-evidences, only some self-evidencing properties speak to the detached aspects of consciousness that we are considering. The challenge is that although maximum scope is avoided, it appears that the distinctive characteristics of consciousness can begin to emerge just by adding a bit more hierarchical and parametric depth to the self-evidencer's internal model, allowing it to, as it were, move inward and be further detached from sensory input. With a few levels, there can in principle be inferential processes happening at levels sandwiched between other levels, looking much like the kinds of detachment that we have discussed thus far. This still seems somewhat unprincipled, and not enough to draw a harder, more categorical line between creatures who do and do not enjoy detached consciousness. Consciousness seems like the kind of property that should be categorically different from other kinds of cognitive properties—it should be associated with some distinctive process, architecture, or mechanism. For example, we would likely not be tempted to say that an active inference machine learning algorithm that gets a couple of layers added to it is more likely to be conscious. In other words, it should be not only the deep and thick degree of detachment that does the explanatory work, but also the specific kind of detachment.

To answer this challenge, I will point to two elements from chapters 6–8, which in combination could provide a particular inflection for the detached elements of consciousness that are familiar to creatures like us. We observed that self-evidencing agents must model themselves as partial causes of their own sensory input. I have also at various points entertained the notion of humans as agents of volatility who occupy a kind of statistical eco-niche. Agents of volatility plan and act to explore and exploit nonlinear state transitions. They readily increase their learning rate and abolish habitual policies when they seek or encounter volatility, so they can quickly move to new patches (e.g., an otherwise fastidious businessperson suddenly borrows a lot of funds to encroach on the market of a competitor who has shown sudden signs of weakness). They spend much time modeling in advance, at short and long timescales and at great causal depths, to put themselves in a good position to understand and manipulate underlying mechanisms when unexpected uncertainty arises (e.g., investing in fundamental research on tectonic plate movements to understand forces implicated in natural disasters and inform engineering of buildings that will be stable even in earthquake-prone areas). Further, they spend much effort modeling and exploring causal mechanisms to explain volatile events and inculcate them into their model even at extreme timescales (e.g., astronomical or cosmological events).

The self-models of such agents of volatility could have a distinctive characteristic that would be relevant for meeting the challenge I outlined. The self-model of a conscious self-evidencing agent might model their own peculiar causal role explicitly as an agent of volatility. That is, they might model themselves as a cause of volatility or as responding to a volatile situation. For example, on the basis of such a model of themselves as an agent of volatility, they can select disruptive or counterdisruptive policies such as buying a business close to a competitor to explore how the competitor will react to such a volatile change; alternatively, when expecting other agents to inflict volatility, they could adopt a policy of buying land around their own business to prevent a competitor from moving in and disrupting things. These types of cases would amount to a kind of active volatility inference, where the agent is selecting among policies that exploit facts about volatility; the agent is considering counterfactuals wherein they themselves enter the volatile fray.

In these cases, the agent must explicitly model themselves as an independent, autonomous cause whose actions are not in the immediate thrall of regular causal patterns. Their own policies can express a volatile, nonlinear change in behavior, not predicted by ongoing events, and can occasion nonlinear changes in the chain of events that would otherwise be orderly. Note that this is not to say that the agent simply does random

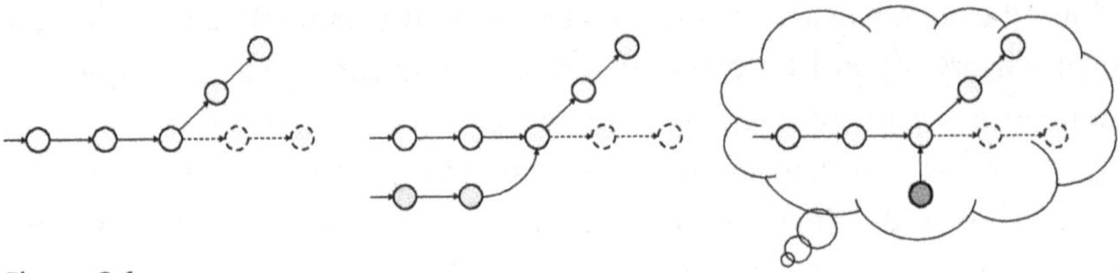

Figure 9.1
Detachment for agents of volatility. Left: Depiction of volatility, as a pattern of events is broken. Middle: Inference of latent causes (darker gray) interfering in the pattern and explaining the volatility. Right: The agent of volatility believes about themselves that they (i.e., their self-model) are detached causes of volatility (dark gray).

stuff since the policies are vetted in counterfactual deliberation and in any case work to approximate the agent's preferred states. If agents have this belief as part of their self-model, then it expresses a peculiar type of detachment going beyond detachment from the current sensory input by virtue of their deep and thick hierarchical model. The agents have also adopted a belief about themselves as being free of the normal causal order of events so they can freely insert themselves into any kind of situation. This belief about themselves constrains the agent's policies, and if those policies are successful at reducing uncertainty, then evidence begins to accumulate for it in the normal self-evidencing circle. That is, the agent makes it a self-fulfilling prophecy that they are unconstrained agents of volatility, able to interfere in the causal order. Agents believe that they can insert themselves—like little miracles violating the deterministic laws of nature—into the natural order of things. Figure 9.1 provides a heuristic for the idea that agents represent themselves as detached agents of volatility.[16]

I believe that this aspect of intrinsic self-modeling in agents of volatility picks up on a specific property, which could point to a categorical difference in detachment for self-evidencing agents like humans. This could help meet the difficult challenge of marking a more principled difference between detachment in self-evidencing agents in general and detachment in agents with our familiar conscious experiences. This kind of modeling of the self as detached in a profoundly intrinsic sense speaks well to the dramatic ways that our consciousness is detached as we engage in dreaming, mind-wandering, imagery, counterfactual thinking, and other ways of thinking.

In chapter 7, I proposed that the self-model serves to constrain everyday selection of policies in active inference, and in turn is constrained by what happens as those policies unfold (recall figure 7.2). This idea appealed to the sparse coupling of Markov blankets with different temporal dynamics. We can further specify this proposal here, such that the self-model specifically constrains policies under the belief that the agent is a detached agent of volatility. This belief in turn reflects that the statistically described eco-niche that we occupy is one where we exploit and explore volatility.

In this section, I have investigated the confluence between the detachment from current sensory input that is a hallmark of our consciousness and the types of detachment that self-evidencing affords, in particular for self-evidencing agents of volatility. This puts us on the right track for the piecemeal building of a self-evidencing approach to consciousness.

It seems plausible that the agent of volatility who detaches in these ways will also increasingly look as if they have the kind of compatibilist free will that I proposed as a good fit for self-evidencing in chapter 8. Agents of volatility believe about themselves that they act according to their own beliefs and preferences, implying that they are not in the direct thrall of the causal order of things. And by acting under that belief, harnessed in counterfactual deliberation, they make it so. This marks a promising coherence between the various aspects of self-evidencing, suggesting that our consciousness and our belief in being free are related.[17]

Active Inference for Conscious Content

It is time to return from examples of detached experience such as dreaming and imagery to conscious perception of the world around us. What can self-evidencing bring to our understanding of the conscious experience that makes us feel connected to the world? I will argue that conscious perception of the world is based on the property of detachment that we have discussed thus far.

I begin with my favorite example of a paradigmatic experiment in consciousness science—namely, binocular rivalry. This perceptual effect has provided one of the key discussion points in the predictive processing literature, following the early, conceptual predictive coding model of binocular rivalry, proposed by Friston, Andreas Roepstorff, and myself. This model is discussed extensively in the predictive processing literature, so I will only briefly recapitulate it here. The point of picking up on this familiar story is to develop it further, with respect to both developments in active inference modeling and the underlying story about detachment and agents of volatility.[18]

As mentioned earlier, binocular rivalry is a robust perceptual phenomenon where different images are shown to each eye, and yet conscious perception alternates between the two stimuli instead of the person experiencing some mishmash of the images. For example, if an image of a house is shown to one eye and an image of a face to the other, conscious perception is of either the house or the face, with some intermediate periods in between where patches of either image wax and wane against each other, rather than a fuzzy mix of semitransparent parts of each image. Binocular rivalry is a central paradigm for studying the neural correlates of conscious content. This is because the physical stimulation is held constant, and yet conscious perception changes, which implies that brain activity time-locked to changes in conscious perception is especially relevant for the neural mechanisms underpinning the selection of content into consciousness (since that brain activity could not be caused by changes in the physical stimuli).

There is something quite amazing about rivalry, which often goes unnoticed. Conscious perception in rivalry is dramatically detached from the current sensory input. When the house image presented to one eye is consciously perceived, the perceptual system is fully ignoring and suppressing the highly salient, clear sensory input from the face stimulus that is in fact right there all along. Our brains are saying "Yeah, no, this is not a face nor a weird face-house—it is a house—so I am just detaching from the face data stream for now." This is an amazing power for the perceptual system to have, conflicting with our common-sense idea that when we open our eyes, we simply see what is there. Note that this is inferential suppression since the suppressed input still causally suffuses the visual system to some degree. We said earlier that perception is controlled hallucination, and in rivalry, the hallucination is stronger and the control from the input weaker than in normal perception; that is, the perception of the house is more detached. This also suggests that the even stronger kinds of detachment in dreaming, imagery, and thought might overlap with detachment in perceptual experience. This pronounced detachment makes rivalry an even better paradigm to pursue in consciousness science, helping us to reveal the constructive (hallucinatory) power of mechanisms for consciousness.

After setting out the predictive coding story about rivalry, I will argue next that the beliefs about state transitions in active inference, through precision control, drive this detachment from some input together with the selection of other input.

Our original predictive coding model of rivalry began with a simple premise, which is that conscious content is driven by perceptual inference. What is perceived is determined by the belief that currently has the best evidence—the highest posterior probability—by virtue of being best at minimizing prediction error. If the system consciously perceives the house, then that is because the internal model's hypothesis that there is a house at that moment best explains the sensory input. In a rivalry situation, this means that the top-down predictions from the belief that there is a house outperform the predictions from the belief that there is a face.

This was never a fully formed theory of content consciousness, however. It is a relatively simple hypothesis that conscious content coincides with perceptual inference (in a creature who is already in a conscious state). The hypothesis can be tested in various ways, such as by changing the inferential elements by strengthening or weakening the prior expectation that a certain stimulus is in the environment by providing additional evidence. The hypothesis can be strengthened by providing a mechanistic story based on a process theory for a particular system, such as predictive coding in the primate brain. Now the hypothesis would be about particular message passing within the brain, in particular passing predictions down to meet and attenuate incoming sensory data, thus minimizing bottom-up prediction errors (for process theories, see chapter 5). In general, several studies find signatures of predictive coding in the brain in conscious vision across several paradigms, including studies of binocular rivalry. Such evidence is important for the hypothesis that conscious content coincides with perceptual inference. But the most that we can get at this point is the idea that perceptual inference is necessary for conscious content (limited to the particular kinds of neural systems that we study). It is obvious that additional conditions are needed to explain consciousness more fully.[19]

Our first attempt at explaining rivalry did have more machinery than just the bald appeal to predictive coding and winning beliefs. It employed a specific mechanistic explanation, bringing it closer to self-evidencing. In rivalry, it is not enough to explain why one percept dominates the other at a given time. In addition, the process of alternation between percepts must be explained. We did this by appealing to the way that priors decrease over time, which is related to precision control in agents operating in a dynamically changing world. Over time, as the agent is continuously presented with the face-plus-house stimulus, the given posterior belief (the probability of a belief after the evidence is in) becomes the new prior belief in the ongoing assessment of incoming evidence. Crucially, this prior decreases under the agent's meta-belief that in a changing world, they cannot cling to existing beliefs for too long; the agent has learned that they live in a dynamically changing world where states tend to transition from one to another, so a prior will always have a "use-by date," so to speak. This adds to the inferential story and requires additional message passing associated with precision control, such that the precise prior over some appropriate timescale becomes less precise. Computationally and mechanistically, the unexplained prediction error from the currently suppressed stimulus should begin to gain relative weight, as the prior for the winning belief begins to fade, causing the alternate stimulus to eventually dominate in conscious perception.[20]

In our account of rivalry, precision optimization in the shape of beliefs about change and volatility therefore played a key role in explaining rivalry, the phenomenon of interest. However, in my mind, there was always something missing from this story. It seemed to assume that the perceiver has beliefs about the typical state transitions for faces or houses, or whatever other stimuli are presented in a particular experiment, and these beliefs would drive the switching between percepts. This seems to predict that images of very stable things, such as the Pyramids of Giza, would dominate perception heavily over images of more ephemeral things, such as a drop of paint dispersing in water. This prediction is implausible and conflicts with how robust and automatic rivalry is across diverse stimuli. I think that a better idea here is to appeal in addition to the perceiver's beliefs about their own state transitions; that is, their beliefs about how likely it is that they will abruptly change the way that they actively sample observations, especially when there are unusually high levels of uncertainty (which in the case of rivalry is caused by the suppressed stimulus causally suffusing the sensory system). It does not take many seconds of perceiving the house before the probability increases over the agent's belief that they will soon explore the sensorium differently, under a new hypothesis about the latent cause—this is then the state transition belief that would be the main driver of rivalry dynamics. When this belief gains strength, it becomes the typical self-fulfilling prophecy, where indeed the agent of volatility inserts themselves into the causal order of things to see what happens. In terms of detachment, the agent's belief dynamically drives precision control that detaches perceptual inference from the sensory input.

If this is a better role for state transition beliefs in the rivalry, then we get a good fit with our account of self-evidencing agents who keep close track of volatility. And now the account begins to apply to everyday, conscious visual perception. Plenty more examples of visual perception of relevance to consciousness science can be given where this account could apply, such as the Necker cube, Rubin's vase-face illusion, Troxler fading, inattentional blindness, and motion-induced blindness. The agent's state transition beliefs are reflected in their active inference, as their attentional states and eye and head movements flit about every few seconds, exploring what is around them and actively seeking something new. This account predicts that if there is an attention-demanding task associated with one of the images in rivalry (such as target detection for the image shown to one eye) or if it dynamically reveals information (such as a movie shown to one eye versus a static image to the other), then the perceiver will be less likely to believe that they will soon undergo a state transition, and rivalry will slow; this seems to be the case. It also has the potential to capture an odd aspect of rivalry, where different people will stably experience perceptual alternations at significantly different rates for the same stimuli, ranging from subsecond to multisecond switches. This suggests that the mechanism for selecting one rather than the other image in consciousness is not just about the statistical properties of the depicted objects; it also relates to the perceivers' beliefs about their own state transitions in the context of different levels of uncertainty.[21]

We can now begin to unpack a story that associates the self-evidencing properties of, specifically, agents of volatility with both consciousness as experienced in classic detached cases such as dreaming, mind-wandering and thought and the mechanisms for the selection of perceptual content into consciousness. The story suggests, somewhat surprisingly, that even conscious visual perception of occurrent stimuli is inferentially detached from the sensory input, and as such, it seems continuous with conscious dreaming, mind-wandering, and thought. It is all controlled hallucination, with various degrees of control by the sensory input determined by the agent's beliefs about volatility.

To progress this story scientifically, it is necessary to move beyond conceptual explanations and on to actual models of paradigmatic consciousness science experiments. This difficult work is now underway. In work led by Thomas Parr, we presented a computational model of both binocular rivalry and Troxler fading (where blurry peripheral targets fade from consciousness when fixating centrally). This work uses active inference to simulate the conscious perceptual inference that is characteristic of both phenomena. In the case of Troxler fading, actual eye movement is used to reduce uncertainty and thereby counter the fading, under a belief that overall uncertainty is increasing for stimuli that are not actively sampled. Here, the sampling action is for epistemic value, reflecting a state-transition belief, the enacting of which reduces uncertainty about whether the peripheral targets are still present by moving the gaze there. This state transition belief is topic-neutral in the sense that it is not about the specific properties of the peripheral targets. This story about action is driven by precision control, where the system selectively samples according to beliefs about expected precision combined with beliefs about how the current state will transition to the next state, given a certain action. For the second case, binocular rivalry, the active inference is not about the oculomotor state transition since the two competing stimuli are located in the same region of visual space; therefore, our model for rivalry includes an element of endogenous attention shifting as covert action, where expected precision is attached to particular stimulus features (e.g., features of a face versus a house). This is equivalent to selective, precision-controlled sampling (as discussed in chapter 4; see figure 4.3) and creates the alternation characteristic of rivalry in conditions where mental action but not bodily or eye movement is occurring.[22]

In this updated account, active inference is necessary for the selection of content into consciousness because it is covert or overt action that determines the flow of evidence for or against a certain hypothesis. The key is that the system just needs to have a belief about underlying, dynamically changing uncertainty and then overt or covert policy selection and action ensue, which seek to minimize that uncertainty, given the agent's state transitions. In the earlier account, the agent sits around passively, absorbing whatever sensory input comes their way. In the new version, the agent is actively engaged in generating their own percepts.[23]

This proposal is consistent with the ideas about self-evidencing agents occupying an eco-niche of volatility. Under the belief about volatility about their own engagement with the sensorium given some level of uncertainty, the currently winning percept loses its attraction, even though the actual prediction errors are being resolved well at that moment in time. The agent, prematurely in a sense, goes exploring for a better percept. Although the account is consistent with the notion of agents of volatility, much modeling and empirical work remains to be done. For example, there is some evidence of bistable perception in pigeons and fruit flies, whose consciousness (if they have any) I would assume to be quite dissimilar to ours. There are some interesting empirical questions here, such as whether fruit flies would be able to wield active inference under beliefs that would give rise to the endogenous attention and reward effects seen in humans during binocular rivalry, or whether the bistability is more hard-wired in such animals and insects. What we can say at this stage is that the modeling work indicates that active inference can begin to explain types of perceptual bistability of interest to consciousness science.[24]

Friston has a nice slogan for this kind of proposal that we are considering here: "To see is to look." That is, to consciously perceive (i.e., to see) is to actively sample observations that conform to an expectation (i.e., to look for or to notice). This amounts to a proposal that active inference is a necessary condition for the selection of content into consciousness (taking due notice that this is not the claim that active inference is sufficient for consciousness). I will refer to this proposal as the "active inference account of conscious content." This proposal serves to highlight the role of beliefs about uncertainty and agential volatility in conscious perception. It turns the focus from actual uncertainty to expected uncertainty as the initial driver for the active selection of content into conscious perception. What is perceived depends on the processing of what the expected uncertainty will be, and what will be sampled in observations, given the system's model of the world and its own state transitions.[25]

Perception in Enacted Inference

The proposal in the previous section was that active inference is a necessary condition for selection of content into consciousness. This section will provide a solution to a puzzle about this proposal.

Active inference is the selection of policies; that is, planning and deciding what to do. So if we say that active inference is necessary for online conscious perception, then we seem to be saying that planning is necessary for conscious perception. This needs to be unpacked since planning and perception seem rather different processes. The puzzle is that we have said that content is selected into consciousness when the relevant evidence is actively sampled, but strictly speaking, active inference is for planning an action, not doing the action. So the unfolding experience of conscious content coming and going in everyday experience needs to be explained. The puzzle is how active inference could give rise to the actual conscious experience in everyday perception because active inference is just the inference of a policy, based on ambiguity and risk, scored against expected free energy. Why should the inference of the policy to "move my eyes up there" give rise to conscious experience rather than, say, a mere propositional thought? I think that there is a good answer to this question.

A policy is really a representation of a sequence of control states which, when chained together, are expected to give rise to a trajectory of observations such as the swath of visual, kinesthetic, proprioceptive, tactile, and thermoceptive observations that the agent expects under the policy that we can paraphrase as "Reach out my arm and grab the coffee cup." The active inference is an inference to a belief that is actually contrary to fact—namely, the belief that the agent is now in the process of sampling that swath of observations. So long as the agent is not initiating the movement, having this belief generates a prediction error due to the simple fact that those observations are not in fact occurring. If this belief is sufficiently strong, it will not be revised in the light of that prediction error; in other words, we assume that the agent will not revert to the belief that they are not reaching out to get coffee. The belief might be strong because it is early in the morning and the agent, like so many of us, has a strong belief that, among their preferred states, there is coffee. Since the agent now cannot get rid of the reaching-for-coffee belief, this prediction error entrains the body through precision control of the requisite (afferent) sensory channels, kicking the arm into motion so the expected kinesthetic and proprioceptive observations can be made, shooting out toward the cup and grabbing it and sensing its heat. When things go well, each predicted observation along this trajectory comes to pass with reasonable precision and accuracy. The active inference mechanism at play in the online feel of everyday conscious perception is exactly this—the feeling of a swath of predictions coming to pass. Conscious perception is therefore not simply what happens at the point that a policy is actively inferred, but what happens when the policy is put into covert and overt action through precision control mechanisms.

We can provide some context for these ideas by asking where in the cortical processing hierarchy conscious content is concentrated. It seems that low in the cortical hierarchy, where processing dynamics are in the low number of milliseconds, we do not see the immediate neural correlates of conscious content, nor does it feel subjectively like we have conscious experiences unfolding at such quick timescales. High in the cortical hierarchy, where processing dynamics range from several seconds up to hours, days, and beyond, we also do not see the neural correlates of conscious perceptual content, nor does it feel like events at such timescales are part of conscious perception. Conscious content perhaps seems to unfold on middling timescales (about 100 milliseconds to a couple of seconds) and indeed, much research on the neural correlates has focused on parts of the brain where processing dynamics happen within this range.

The questions here are: Why is conscious content tied to those middling levels of processing? Why do we fail to enjoy ultrashort bursts of conscious experiences or very protracted experiences? The proposal that active inference is necessary for conscious content can help here. In work with Francesco Marchi, we argued that conscious content is tied to middling levels of the cortical hierarchy because that happens to be favorable to the timescales over which we, as embodied agents, can execute the basic building blocks of our policies. If conscious content depends on enacting policies, then it should be tied to the kinds of actions that we can perform, from covert precision control to overt movements of the eyes, head, legs, arms, and other parts of our bodies. That is how we selectively sample sensory input as we self-evidence.[26]

The puzzle that we are addressing in this section has been how to avoid saying that our everyday, unfolding dynamic experience of conscious perception is a matter of what we have planned, while still operating with the idea that active inference is necessary for conscious perception. The solution to the puzzle is to break the process into two stages, both of which are necessary. The first stage is the planning or inference of the policy, for which the objective function is expected free energy; at this exact point in time, there is no online perceptual experience associated with the policy (although there might be plenty of offline conscious experience based on the fictional enaction of counterfactual deliberation of competing policies leading up to the final active inference).

The second stage is the enaction of the inferred policy; that is, the active sampling of observations, for which the objective function is now the variational free energy, to quash the prediction error generated by adopting the policy. During this prediction error minimization process, conscious experience is unfolding. Hence, we combine the enaction of the policy with the planning, and it is this combined package that is necessary for conscious content. Note here that a policy is being enacted as soon as any kind of precision control kicks in to enable sampling of expected sensory input, which can happen in covert action, such as attention allocation, and well before the muscular body plant has been entrained. This means that there is much scope for the agent to have varying strengths of conscious perception depending on the time course of covert and overt enaction of policies.[27]

This active inference account of conscious perception differs from the way that earlier predictive processing (and predictive coding) accounts told the story. Andy Clark coined the apt phrase of "predicting the present," where predictions give rise to experience when they are satisfied. I have been in broad agreement with Clark here, but although the notion is compelling, it is also slightly oxymoronic because, colloquially at least, predictions are not about the present. With the two-pronged appeal to active inference, this terminological tension is resolved because the agent simply had a belief, arrived at in planning, for which evidence is now sampled, and as this unfolds, the conscious perception arises. In other words, conscious experience happens as the agent engages in self-evidencing, as they are soliciting and explaining the input that they were expecting would occur given their model—and thereby evidencing their model.[28]

We can contextualize this proposal by cycling back to the notion from chapter 8 that agents with the capacity for generalizing inference beyond current contexts must infer which actions they perform (see figure 8.1). This inferential predicament allows flexible learning and inference and applies to agents of volatility, which actively seek to alter the current context. This was contrasted with inflexible systems such as thermostats, which we can describe in active inference terms but whose actions directly cause a change in their internal states. The account of consciousness that I have sketched here requires just this kind of architecture. This is because actively sampling the evidence predicted by a policy is tantamount to inferring that that policy was indeed enacted. This helps tie together the accounts of self, volition, and now consciousness. In self-evidencing agents of volatility, this capacity is manifested in rich, counterfactual planning phases leading to the enacting of policies that are constrained by the self-model's belief that the agent can freely intervene in the causal order of things (as depicted in figure 9.1). This is a speculative proposal, of course, but I think that there is something intriguing about pulling together these three philosophically challenging notions—self, will, and consciousness—within self-evidencing.

This section has proposed that active inference is necessary for online conscious perception. As things become consciously perceived, and as that conscious perception changes from one thing to another, active inferences are made and enacted.[29]

Subjectivity and Unity

We have been building the active inference account of conscious content, focusing on the selection of perceptual content into consciousness. In this final section of the chapter, we consider some of the distinctive properties of conscious experience—subjectivity and unity—exploring whether the active inference account can accommodate them too.

Consciousness is marked not just by a first-person perspective in a geometrical sense, but also by the sense of subjectivity. Philosophers find it difficult to make this notion concise. It is not only the idea that we can append "mine" to all we experience. Rather, the "mineness" or the "I" is somehow inherent to experience. When we use Thomas Nagel's phrase that we are conscious when there is "something it is like" for us, we mean that there is a subjective perspective without which we would not recognize that state as consciousness. It is common to ask readers to try to catch this element in their own experience rather than to provide a clear definition; I hereby invite you to contemplate the subjective element of your consciousness, and I challenge you to try to conceive what it would be like to not have your subjective "you" in play and yet still be conscious. There are cases where the sense of self is challenged, such as depersonalization, Cotard's delusion (believing that one is dead), psychedelic ego dissolution, or contemplative nondual experiences, but I believe that across even those "selfless" conscious experiences, there is an experience of basic subjectivity, hard as it is to articulate.[30]

Methodologically, it is a challenge to give an intelligible account of something that one finds difficult to define clearly. I will assemble some elements of self-evidencing and suggest that with them, we have at least a start of something that looks like subjectivity. The success of this venture will be tenuous, depending on how it gels with one's own subjective grasp of subjectivity. Such are the working conditions in consciousness science.

Building on our discussions in this and several previous chapters, we can now say the following: The self-evidencing agent (of volatility) must infer everything; they are not helped by their own actions hard-coding their internal states (as in the thermostat), and even their homeostatic beliefs delivered through evolution are subject to self-evidencing. Inference includes their own model of themselves, an explicit model that they infer and continuously develop and explore so they can best use it to detach from the current sensory input and counterfactually insert themselves into all sorts of contexts. This gives consciousnesses like ours a great freedom to explore and exploit volatility, so we can be first to occupy, and perhaps even disruptively create, new patches. All this internal mental structure and activity happen in the model-constrained trial and error variational process of inference to the best decision. Once a policy is inferred, consciousness is shaped as the body is entrained and the policies' expected observations come to pass with reasonable fidelity through selective sampling.

That story about self-evidencing in agents of volatility can plausibly capture subjectivity in conscious experience. The cascade of observations that we expect under a policy is determined in a process that represents us, at the nexus of those observations, as Markov-blanketed contributing causes of our own observations. These self-model causes are portrayed as conditionally independent, so they can counterfactually be inserted meaningfully into all sorts of contexts, exploiting or creating volatility. If the self-model had been less geared for volatility and counterfactuals, it would be more chained to external causal goings-on—less selfy and more like the causally immediate nonsubjectivity of a thermostat. This provides one aspect of subjectivity, where all policies leading to conscious experience are informed in part by a relatively impervious self, figuring out what observations it would receive if this or that were the case. The self-model can be expansive and precise, or imprecise, or shallow or shrunk to a bare minimum, but it is always there and always informing policies, given what it is.

What happens next is the enacting of the inferred, self-imbued policy. This is where everyday conscious experience happens as policies come to pass. The agent of volatility enacts policies constrained by the internally modeled belief that they are autonomous, free, and unshackled by immediate causes.

This adds another facet to subjectivity in conscious experience because as these observations occur, they can be fashioned by the agent themselves as intrinsic in the sense that they are caused by the agent's actions, not by anything external to the agent. Now we have causal integrity and autonomous intrinsicality as the building blocks of subjectivity. As the conscious experience unfolds in active inference, it may feel like this is mine and mine alone because it is represented as caused by me, and only by me.

I will immediately add that this is not rampant libertarian dualism. It is not the view that the self-model that helps us infer our policies is cut adrift from the physical causal order of the world, operating as a first, original cause. The subjectivity is anchored in the agent's belief that they have intrinsic autonomy. The self-evidencing agent is in fact fully part of the causal nexus—where the parent-states of the internal states that encode the belief themselves have parents in the external world, and the active children-states of the internal states have external children-states too. This does not mean that these intrinsicality beliefs are false, however. If the agent of volatility is self-evidencing, then even those beliefs accumulate evidence. In other words, if the agent is able to make inferences to the best decision that allow them to exploit and explore volatility, then they have some kind of intrinsicality. Indeed, the self-evidencing agent of volatility has these beliefs because they exist in an eco-niche that affords them, as I have argued in this discussion. Looking at this from the outside, we could flesh this out in terms of autonomy via self-organization, together with causal independence, via Markov blankets operating over different timescales.

Finally, we add another facet of consciousness—namely, the unity of consciousness. The stream of consciousness flows in one unified direction. There might be eddies, rapids, bends, stillwaters, obstructing bends, and submerged trees and rocks in our flow of consciousness but we have only one. I think it would be hard to maintain a sense of subjectivity if conscious experience somehow (almost unfathomably) flowed outward in many directions at once, as water disperses when a water balloon bursts.

I think that there is a principled but somewhat uninformative account of unity, which simply says that if a self-evidencing agent exists, then there is a Markov blanket that defines the agent at any given time. There will be several questions about whether it can change over time, or expand, or shrink, or whether there can be nested blankets (I think that the answer to these questions is yes, but I will not here get into ontological matters of how to count things). If there is a blanket, then there is one thing that has the expected flow of observations, and if we are talking about a conscious thing, then that could ground unity. There is a kernel of truth here, I believe, but it is an etiolated notion of conscious unity, established simply by thingness and therefore not terribly specific to conscious self-evidencing agents.

It is possible to dig deeper. In previous work, I have proposed that unity is underpinned by active inference in the simple sense that active inference can be executed only by the agent, with one body. This funnels all the selective sampling into one stream, centered on the subject. I gestured at how this helps make sense of binocular rivalry (discussed earlier in this chapter), where the two stimuli are selected sequentially into consciousness rather than all at once; the idea being that active inference cannot simultaneously sample observations under both the beliefs that there is a face and there is a house out there—this is embodied cognition in the unmysterious sense that the eyes of humans do not move independent of each other (hence, perhaps chameleons do not experience binocular rivalry).[31]

Back then, I also suggested that this idea helps address a challenge for one of the most well-established theories of consciousness: global neuronal workspace theory. This theory holds that consciousness is associated with the point at which low-level neuronal representations are "ignited" into a global workspace where it is made available for various cognitive consumer systems, such as report, attention, and memory. The problem with this theory is that ignition is a metaphor, best explained in terms of some kind of evidence threshold. The proposal that I made was that the ignition threshold is tied to the formulation of a policy in active inference, which is then shunted to the workspace to be enacted through selective sampling or to work out its counterfactual implications in offline processing using the resources of the generative model. This provided a more specific and principled funneling into one stream, helping to make sense of a series of impressive experiments suggesting that consciousness is associated with ignition.[32]

In light of the discussion in the sections of this chapter, I will now flesh out this initial proposal in terms of the switch to enacting a policy; in other words, ignition is the beginning of variational free minimization as cognitive consumer systems generate precision-weighted predictions that afford selective sampling of the incoming sensory signal, which has been floating around at lower levels. The global neuronal workspace is a meeting place for top-down and bottom-up signals—it is where the cognitive consumer systems quite literally begin to consume (i.e., actively sample) expected input to provide evidence for their beliefs.

Whereas I liked the idea of tying the global workspace theory to active inference, I myself could not work it into anything more substantial. Christopher Whyte luckily saw some potential in the idea and developed and expanded it into a coherent account, which he then, together with Ryan Smith, coupled with active inference models of types of inattentional blindness. This work provides an example of the piecemeal self-evidencing approach to consciousness, with good construct validity through the base in the free energy principle, impressive face validity through simulations of classic paradigms in consciousness science, and promise for predictive validity through process theoretical predictions of empirical data that can be fitted to the model (referring back to the discussion in chapter 5 about the testability of the self-evidencing framework). There is increasing momentum for Whyte's approach, with several models and ideas being added. I think that this is an exciting phase for the self-evidencing and broad predictive processing approach to the science of consciousness. It fits well with how Anil Seth and I conceived it in our argument, mentioned earlier, which begins with the idea that predictive processing is not in itself a theory of consciousness but can ground the construction of one.[33]

With an active inference account of unity of consciousness, we can therefore begin to capture feelings of subjectivity and unity in conscious experience. This account emerges from reflections on what it is to be agents of volatility, agents who believe of themselves that they have intrinsic agency, and thereby shape policies for embodied, selective sampling so they are imbued with subjectivity and unified when enacted.

Concluding Remarks: Consciousness in Agents of Volatility

This chapter explored how to connect self-evidencing to consciousness. I made the task easier for myself by navigating a feasible route through the science and philosophy of consciousness, avoiding the Scylla of the metaphysical hard problem and the Charybdis of the panpsychic explosion of consciousness to everything. Self-evidencing is not a theory of consciousness in itself, but it can provide the underlying, systematic, computational correlates of conscious phenomena. These correlates can be pieced together, and we can begin to identify core properties of how they explain consciousnesses like ours, pointing toward something that may become a more delimited and tractable self-evidencing theory of consciousness. There are intriguing modeling efforts in this vein already, backed up by conceptual frameworks and in the process of empirical testing. To me, this is a promising new start, given the slow progress of consciousness science, where theories of consciousness are hard to test and may have poor construct validity (e.g., pulling in obscure constructs) or face validity (e.g., explaining behavior rather than consciousness).

This proposal might help advance the theoretical and conceptual components of the metaphysical debate. The elements of consciousness are poorly understood: What is a global state of consciousness (as opposed to mere arousal)? What does it mean to populate a global state with conscious content? What is the sense of self? The concepts are hard to pin down. Perhaps offering a broader framework for situating these concepts will help us gain more clarity about them; sometimes we do understand a concept better once it is embedded in the right kind of conceptual networks. And it may be that we learn something too about how these concepts relate to each other by learning something about how they are underpinned by a unified type of process, such as self-evidencing. The explanatory toolbox of self-evidencing processes developed in previous chapters may help suggest more or less radical revisions or clarifications or explications of consciousness. This is all exciting ongoing work, to be fully fleshed out in the future. Maybe we will begin to think and reason differently about consciousness and its metaphysics once we see them through the lens of the self-evidencing agent of volatility.

I will highlight one last, overarching aspect of the story in this chapter. Often, when we try to say what consciousness is, we vacillate among several notions, such as experiencing things, enjoying self-consciousness, having free will, and being an agent with one's own preferences. Philosophers can get a bit exasperated by this, trying to focus the discussion on the conscious state and conscious content. This is justified to some extent because the science of consciousness shines most brightly when clearly focused in this way, such as in compelling studies of disorders of consciousness or in binocular rivalry studies. But perhaps philosophers should be more forgiving sometimes. It is not necessarily conceptually confused to want to explicate consciousness together with all of experience, self, volition, and agency. If the story in this chapter is right, then those kinds of properties are in fact baked into the subjectivity that sets apart our kind of consciousness. This inculcation of subjectivity for creatures like us is unsurprising, given the bootstrapping, subject-oriented nature of self-evidencing, self-supervision, and self-organization.

This chapter is thus the beginning of a much bigger story, to be developed in the future, and in a collective effort. There are already several other, somewhat competing proposals for how to build the free energy principle toward consciousness. This chapter has not done justice to all those developments, nor has it explained and reviewed the full palette of views. Rather, I have focused on what the self-evidencing aspect specifically can bring to consciousness when the investigation is conducted as a piecemeal journey of discovery rather than a head-on theory of consciousness.[34]

In chapter 10, we want to flex self-evidencing further, trying to move in on aspects of everyday experience that matter deeply to living well. These include meaning in life and what might make decisions not only best, but wise.

10 Wisdom and Meaning

There is a meme about the philosopher who goes to a party and is asked what they do for a living, and when they answer "I am a philosopher," invariably the response is either "So, what is the meaning of life?" or "You must be very wise!" Philosophers dread this because parties are not great for in-depth conversation, and, generally, we don't really know what the meaning of life is, and we are not terribly wise either. However, philosophy is indeed about meaning and wisdom, which this chapter will focus on. This is appropriate because back in chapter 1, and at various places throughout the book, I have promised that self-evidencing can be shown to matter for everyday experience, and the search for wisdom and meaning is a part of what makes our everyday experience what it is. This is not to say that it is easy to lead meaningful lives and to be wise—these notions can seem nebulous and perhaps appear to be the exclusive province of gurus and sages.

I will explore whether self-evidencing can contribute something valuable to our understanding of meaning and wisdom. Philosophers are fairly certain that wisdom and meaning matter to our well-being, and it is good to be wise and to lead a meaningful life. Psychologists, contemplatives, and therapists, and, I suspect, people of many walks of life broadly agree—although few of us have explicit, fully worked-out conceptions of these notions ready at hand. To make the task tractable, I will help myself to some functional and operational definitions of these concepts that have emerged at the intersection of philosophy and psychology. This will leave out much of the rich and important philosophical debate. But it will help provide a concrete starting point for connecting self-evidencing's analysis of existence and active inference's circular causality with notions of existence and autonomy in the meaning of life and wisdom literature.

The psychological constructs that I will lean on in this chapter are also useful because they facilitate empirical inquiry and investigation into the role of wisdom and meaning for psychological well-being. Signs are that indeed they matter, in the sense that people experience more well-being when making wise decisions and leading meaningful lives, as these things are operationally defined in psychology. A key aspiration of this book is to connect self-evidencing with what can be called "everyday experience"—things that matter to us. If the psychological research in this domain is on the right track, then wisdom and meaning are indeed important parts of everyday experience and of living well.

The upshot of the discussion in this chapter will be that, when starting with these definitions and when framed in terms of self-evidencing, both meaning and wisdom become achievable aspirations. We can also see how the different elements of meaning and wisdom can begin to be unified within the self-evidencing framework. Given the agent is constantly self-evidencing, this enables them to engage in wide-ranging explorations of possibilities for increasing both meaning and wisdom. Because they are self-supervised, they must pursue a journey of self-realization and self-shaping (through maximizing self-model evidence). This allows diversity among agents, and also a wide-ranging scope for going wrong in the search for meaning and wisdom. Along the way, I also ask how to go about these questions of living well in a practical way and push the debate about meaning and wisdom into the territory of contemplative studies, considering aspects of mindfulness from the self-evidencing perspective.

I begin this chapter with setting out a definition of wisdom, working toward a self-evidencing analysis of each of its components. Then I approach the definition of meaning in a similar vein, and I show how it also would appear within a self-evidencing framework. I end by venturing into the notion of mindfulness, which is relevant for cultivating wisdom and promoting meaning in life; here, I use self-evidencing to discuss attention and acceptance.

Wise Self-Evidencing

An agent's decisions can be rational, clever, inspired, smart, or intelligent and yet not be wise. A person can be considered wise even if by most lights, they are not the most rational or intelligent. Wisdom is not the same as knowing things, maximizing utility, being quick to understand things, or seeing patterns. Lacking in wisdom is not the same as being uneducated or irrational. It is easy to pump intuitions about what wisdom is not. Perhaps many of us know one or two people whom we consider quite wise, and perhaps history has examples of people who have some wisdom. It is harder to provide a definition of wisdom itself.

Wisdom can seem somewhat nebulous, gradual, context-dependent, and holistic rather than coming down to performance on one parameter. Wisdom is not just something that a person possesses; being wise also depends on how one acts, on character traits and notions of virtue—on how a person conducts their life. Wisdom also seems to have moral and ethical dimensions intermingled with dimensions of rationality and personality traits. Developmental aspects also seem prominent in our conception of wisdom, with a notion that wisdom improves with age—the wise sage in a movie or book is seldom portrayed as young (except when wise children are the foils for the folly of adults). No doubt, whatever conception of wisdom informs our intuitions is itself subject to cultural and moral norms of time and place; what we consider wise here and now is not necessarily what someone in another culture or age would see as wise.

There is no conclusive escaping the wooliness of the concept of wisdom, and it is still very much a valid topic of philosophical discussion and conceptual development, with much focus on the perennial question of wisdom as knowing how to live well. To circumvent this situation, I will appeal to what I consider an attractive conception of wisdom that has emerged in positive psychology. A consensus paper on wisdom from several researchers involved in this area define wisdom in terms of an overarching construct of perspectival metacognition. Metacognition in this sense is a broad ability to step back from current thought, perception, and action to consider goals, trajectories to the goals, and possible alternatives (we will see shortly how this relates to the more technical notion of metacognition as parametric depth that we saw earlier in this book). Perspectival metacognition captures the ability to detach from divergent interests, both for oneself and for others; this in turn opens wisdom to moral aspiration, where the balancing of viewpoints can underpin prosociality and other moral factors. Metacognition of this type has many of the processes and virtues that people associate with wisdom, in addition to moral aspiration—namely, epistemic humility and open-mindedness combined with an ability to adapt to a context. Together, these elements enable what researchers in this field consider wise functioning, marked by an ability to implement abstract moral concepts, seeing though illusions and self-deception, and engaging in coordination with others and facilitating long-term planning.[1]

In my view, this psychological conception of wisdom is an excellent (although time-stamped, of course) distillation of significant parts of the wisdom debate, which incorporates philosophical trends, is based on a wide assessment of what researchers associate with wisdom and which is empirically and computationally tractable. I will begin by giving self-evidencing accounts of each of its components, relying on the notion of inference to the best decision from chapter 6. Then I pull the components together, relying on the notions of detachment, agents of volatility, and moral model fitting from chapters 7–9. I believe that this could advance the conceptual analysis, which may provide the starting point for an attractive, computationally oriented account of wisdom.

The key computational tools for wisdom are available to the self-evidencing agent, all of which have been discussed in previous chapters. Metacognition was framed in terms of parametric depth (parameters representing other parameters) in the generative model, which was what makes active precision control possible (as mental action). It is the ability to rerepresent a belief, to form a new belief about a belief. This belief assesses the quality of the belief that it is about—asking if this belief is precise enough to be trusted. In creatures like us, the answer would depend on a hierarchy of other beliefs, making metacognition and precision control highly context-sensitive. The beliefs that we form about the world, as we actively engage in conscious perception and inference to the best decision, rely on contextual beliefs about how much this or that belief or policy can be trusted. Overall, with self-evidencing, metacognitive precision control is cast in context-sensitive, active inference terms, meaning that in principle, it can be cultivated in different, more or less wise ways in action (with wisdom operationalized as before).

The perspectival element comes in the first instance with metacognition itself, which necessarily involves a "stepping back" process of taking a new perspective, at the statistical properties of an internal inference. In a hierarchical setting, several levels can inspect each other in this quality assessment and precision control manner. This means that precision control can be expressed at various levels, and high levels can exert their precision control through many levels. For example, some people can override extreme stress to manifest high focus to achieve a goal, such as a bomb disposal expert working in a war zone. Intriguingly, this type of processing can be captured in computational, active inference models. Lars Sandved-Smith and his collaborators used active inference to model precision control for meta-awareness and mental action over parametric levels, capturing the phenomenology of some meditation practices; the first level deals with perception and action ("What am I perceiving? What am I trying to do?"), the second with attention ("What am I paying attention to? What am I trying to pay attention to?"), and the third with meta-awareness ("How aware am I of where my attention is? Am I trying to maintain awareness of my attentional state?"). In this model, not only can we perceive and have precision control, we can actively direct attention and treat our direction of attentional states as itself a hidden state that can be inferred and precision-controlled at levels of meta-awareness. The agent can keep asking questions about not only the quality of their internal states but also—in terms of perspective—their awareness and control of them.[2]

A richer notion of perspectival metacognition comes in through our ability to go fully counterfactual, releasing us from our own actual, more or less narrow perspective. This allows us to consider what policy should be inferred under counterfactual assumptions about not only what the world is like, but also what we are like and would prefer to be like. Indeed, because we model ourselves as agents of volatility (as described in chapters 7–9), we assume counterfactual self-models inserted into all sorts of scenarios. This can be our selves, equipped with different beliefs ("If I were to believe that the election was rigged, what would I do?"), or counterfactually assuming different self-models ("If I were my colleague, how would I respond to the proposal that I am planning to put forward at the meeting?"). The freedom to counterfactualize afforded by being agents of volatility, who are chasing nonlinear state transitions, gives us broad scope for perspectival metacognition. Importantly, this opens the door for genuine consideration of other people's perspectives where we counterfactually put ourselves in their mental shoes.

The further properties of wisdom are epistemic humility and open-mindedness. These are in the first instance straightforward outcomes of the Bayesian inference that is inherent in self-evidencing.

To be epistemically humble is to at least avoid overly strong priors, and since exact inference weighs the prior according to the already accumulated evidence, the agent will be humble in this basic sense, as they are approximating exact inference. If, in a given context, the agent engaging in approximate, variational inference spends much time overweighting their priors, causing them to take longer to converge on the exact priors (as we would say, looking in from an external vantage point), then they are less humble; this may be caused by overly precise metacognitive beliefs about past evidence, which can be difficult to dislodge.

The second property is open-mindedness, which is related to humility in the sense that it is hard to be open-minded without also being to some degree epistemically humble—"I don't know everything already, so I might be able to learn something new here." But conceptually, open-mindedness seems best linked to variability in the learning rate in hierarchical Bayesian inference. When an agent is open-minded, they increase their learning rate; that is, they decrease reliance on past information, opening themselves up to learning from what is happening now. Rationally, this should happen when there is a belief in increased volatility—"Something unusual is happening now, and I can't rely on my past assumptions and beliefs as much anymore, so I must turn up the gain on the current input and try to quickly learn what's going on." Being closed-minded, then, is a failure to adjust learning rate under volatility. In contrast, being overly open-minded is to overestimate the volatility in the current moment and discounting past learning too much. Notice that in volatile situations where the incoming information is nevertheless too noisy and uncertain to allow learning, the inclination to increase one's learning rate may require one to go exploring for good, reliable observations.

In a simple Bayesian scheme, humility is then associated with the weight put on the prior belief ("How much do I already know?"), and open-mindedness is associated with the weight put on the likelihood ("What am I learning here? How can I learn something new?"). The regulation of both operates through precision control, which is itself governed by context-dependent beliefs about confidences. Epistemic humility and open-mindedess can thus initially be understood in terms of optimality in Bayesian inference.

There is a popular preconception that wisdom increases with age, but our proposal of wise Bayesian humility and open-mindedness levels the age-related playing field. So long as an agent is approximating exact Bayesian inference reasonably well, they will be as humble and open-minded as they rationally need to be, given their current beliefs. Young people have not accumulated as much knowledge yet, and therefore they should be more reactive to new information. Older people can be more sanguine in the face of prediction errors that they encounter because they have built strong priors over a long period of time. Both are equally Bayesian and thus equally manifest wise humility and open-mindedness.[3]

However, it is possible to go further in this account of humility and open-mindedness, for active inference affords the (somewhat oxymoronically sounding) capacity for making self-fulfilling prophecies of our belief in our own humility and open-mindedness. That is, we would have the capacity for selectively sampling observations based on the assumed belief that we do not know quite as much or know things quite as well as we may have thought, and there are better things to learn. This is really driven by our belief that we are agents of volatility. If the world plays its part, if indeed there are new patches of pragmatic and epistemic value just around the corner, then we can learn and grow. Indeed, as agents of volatility, we can do more to volitionally enforce a higher learning rate by counterfactually assuming that there are underlying causes to learn about, even if we are not currently experiencing volatility ourselves.

I will give an example from my own world, which brings together some modest degree of volitional humility and open-mindedness, as well as perspectival metacognition. A while back, I was setting up a new research center when a very wise person gently asked "What's the First Nations policy of your new center?" To my shame, I had little to say in response. We had just won a donation and everything was going just fine, with no volatility in sight, so we had no reason to change our beliefs. Yet, as a group, we decided to stop, weaken our current beliefs about how to run a research center, and begin to listen differently to Aboriginal and Torres Strait Islander peoples. This meant increasing our learning rate, exploring, and slowly beginning to understand the complex pattern of hidden causes at play for First Nations peoples and how that should inform what we do. It felt like a wise step to take, even if we needed the prompt to do so—and the journey of learning has been meaningful.

On other occasions, we pursue new patches but find them barren; sometimes it was wasted effort to go exploring in the hope of learning new patterns of events because there was no patterns to find. In these cases, we could have stuck to our existing strong priors just fine. Similarly, sometimes we can be led astray by excessive open-mindedness, such as when people abolish much prior learning in order to "learn" from conspiracy theorists. There is an important lesson in this, which is that although wisdom calls for humility and open-mindedness, too much of a good thing can be bad for us. This is not news to philosophers, who know that virtue is a matter of skillfully balancing excesses such as dogmatism and gullibility. Self-evidencing provides that balance through the approximation to exact inference, which penalizes excessive weighting for priors or likelihoods, since prediction errors will tend to stay large if the agent bounces between extremes of "I know everything" and "I am learning everything right now." This is what all the best-maker balances in inference to the best explanation/decision are about (which we encountered in chapter 3; recall figure 3.1). When self-evidencing's full arsenal of counterfactual deliberation, active sampling, exploration, and precision control in mental action is added, the excesses attain higher and higher dimensionality—now the agent can reap great rewards by exploiting volatility, but holding things together is also harder. Even wisdom itself can be excessive, such as the sage sitting in their cave forever, lost in perspectival metacognition and never able to act to feed themselves.[4]

The processes that I have outlined are all part of inference to the best decision, ranging from deciding what to have for dinner to how much time should be spent on metacognition. The rule governing it all is the variational and expected free energy—beliefs, finessed counterfactually, about how much one can optimize self-evidencing in the long-term average via this or that policy (including metacognitive policies). This can begin to address the potential worry that we have provided the tools that constitute wisdom but not said how to use them to actually be wise. Self-evidencing through inference to the best decision just requires the objective functions for variational and expected free energy, which, as discussed in chapters 2 and 3, is critical because it delivers on the imperative to accommodate self-supervision. The cost of the parsimony of these objective functions is that agents need to spend much time finessing and roaming through their internal models, varying, exploring, and adding and removing parameters to keep the free energy low and maximize self-evidencing. There are many degrees of freedom for the agent in this model-maintenance game, and in particular for the agent of volatility, who can dive deep into counterfactual worlds.

In earlier chapters, we have likened the self-evidencing agent to an abductive hill-climber, where different agents even of the same species can take vastly different trajectories. Some of these self-evidencing trajectories can be wiser than others, and wisdom is captured in the extent to which agents take trajectories reflecting perspectival metacognition. This gives the agent capacity to think through how to apply abstract, higher-level concepts by translating high-level beliefs down to a particular context. The agent will be better able to see through illusion and self-deception by considering both the situation and the self across several counterfactual scenarios, revealing shallow and unstable beliefs. It also facilitates long-term planning, but in a balanced way, by being neither dogmatic or gullible and by having beliefs and policies that are neither too risky or too safe, or too ambiguous or too precise—all guided by free energy.[5]

Wise self-evidencing, therefore, does not fall prey to the common mistake of adhering overly to virtues such as open-mindedness or perspectival metacognition. Excesses will eventually be caught up by increasing free energy. Failures to realize wisdom are still possible, first in the ways explored in chapter 5 of error and irrationality, and second in the failure to deploy our counterfactual capabilities, stretching them into metacognition, meta-awareness, and volatile landscapes. Herein lies also, I think, the reason why it is so difficult to provide a clear, positive, and philosophically satisfactory conception of wisdom beyond psychological operationalizations. Self-evidencing with the components of wisdom is just one way of being, and self-evidencing with less wisdom is another way of being. This focus on ways of being suggests that an unwise person and a wise person differ because over time, they have selected different models of the world and themselves, where the wise person has selected a model that is finessed with precise beliefs for perspectival metacognition.

I think the picture is then this: wise self-evidencing charts a variational path, employing all the tools of the agent of volatility; that is, neither too wildly careening around exact inference (which can stray into territory where the Markov blanket is exposed to too much surprise—think of a risk-seeking teenager who cannot deal with boredom), nor too tightly undulating around exact inference (which exploits too much and will miss important information gain in a changing world—think of an aging recluse who obstinately avoids volatility). If the self-evidencing agent's most basic belief is that they exist, that they self-evidence, then putting in the mental effort to chart the best self-evidencing route at any given time seems deserving of the label "wisdom." This is rightly deemed "effort" because being wise means extending, or complexifying, one's model with extra parameters to achieve the aspirational goal-states of perspectival metacognition, and making one's model more complex is computationally associated with cognitive effort. Given that effort is aversive, not everyone will be attracted to wisdom.[6]

And yet this approach to wisdom might still be too light because we are yet to build in the notion of moral aspiration. For all we have said so far, psychopaths could be considered wise in a way when they expertly read the mental states of their victims, scornfully ignoring their suffering. Or, consider when mining companies blithely blow up sacred cultural sites and have expensive lawyers at the ready to get them off the hook; they are taking short-term heat for the long-term rewards for their shareholders and for eliciting large bonuses for their executives. These examples should not count as wise because they are morally odious. I have waited until last to discuss moral aspiration because morality (as opposed to norms of rationality) is not entailed by self-evidencing, as we saw in chapter 8. Wisdom does open itself up for moral aspiration by virtue of perspectival metacognition, which can make the needs and suffering of other people and creatures visible. But morality is still not guaranteed since agents can ignore the suffering, as in the case of the psychopaths and mining companies in our examples. Instead, moral aspiration must be based on the agent's prior beliefs, and if these beliefs indeed engage prosocial and other moral factors, then the agent can infer policies informed by them and act to make them come true. In chapter 8, I considered the idea of moral model fitting, which can benefit from the kind of balanced wisdom processes outlined here. So wisdom that includes the crucial component of moral aspiration is possible, but it is held hostage to the beliefs of the agents.[7]

This is perhaps a bit disappointing. It would be nicer if moral aspiration were inherent to self-evidencing wisdom rather than an add-on that an agent displays if they are lucky enough to have absorbed moral priors. In a speculative spirit, I could conceive one way to build in a generic moral aspiration based on a belief about common humanity. A self-evidencing agent will do many things to maintain their model and have expected free energy within reasonable bounds for the long-term average. We work together to ensure that our environments afford insulation against many kinds of volatility, such as orchestrating taxation, educational systems, supply-chain logistics, elaborate infrastructure and built environments, and, yes, incentives for mining companies to do what they do. Without all this action on the environment, we would not have a safe base for exploration and cultivation of new adventures in volatility (such as deep-sea diving, space exploration, or vaccine development). Human societies go to extraordinary lengths to achieve this kind of stable (but not too stable) expected free energy over very long time frames. Often, this takes a strong social route (e.g., building massive cathedrals, sometimes over centuries) to ensure social focal points for ritual, cohesion, and compliance. Through all this, we know that the actions of other people can induce massive volatility into the systems that we have built to maintain acceptable levels of expected free energy; and in particular, we know that ignoring the suffering of others leads to volatile events that can destabilize everything, such as revolutions and wars. If agents therefore model all others as having beliefs, desires, hopes, and plans and intentions for the safety of themselves and their loved ones, this can factor into our policies, making sure that there is a belief in common humanity infusing our decision-making.

Thus, for a social species like us, it is no surprise that overwhelmingly, wisdom is associated with an expectation of morally aspiring to, at the very least, try to avoid the suffering of others. This belief, however, does have a flavor of being merely instrumentalist rather than inherently moral, in the sense that we sample evidence for the common humanity belief under the overarching, self-serving belief that we ourselves exist. It is perhaps a smidgen better than pure reward optimization, where others are means to my ends regardless of their suffering because it is simply a belief that our own self-evidencing comes in a package with the self-evidencing of others.[8]

I have presented wise self-evidencing as the effort to chart a particular path through inference to the best decision, which uses mental action on one's own model to avoid too much and too little deviation from exact inference, guided by expected free energy. This allows us, finally, to associate wisdom with improved functioning, and thereby improved self-evidencing over time. But, in addition, being wise is to work with oneself to live better. For this, we turn to the self-model (as discussed in chapter 7), which can be revised over time in ways that depend on how wise we manage to be. For example, an agent can embark on an exploratory project where they begin to self-evidence under the belief that they are open-minded. They might be lucky enough that this adopted stance to gaining information succeeds; that is, there is good evidence for them to gain, which revises the overall model well, opening them up for new endeavors. This would mean that their belief in their own open-mindedness will solidify over time—it is self-evidencing well—and they will have added a trait to their self-model. They have developed their own selves to become wiser. Self-evidencing provides some of the architecture for such active self-shaping. Recall the idea that the self-model is a Markov blanket operating at relatively slow dynamics, which is sparsely coupled to a larger blanket, operating at faster dynamics, for implementation of preferences in everyday action (as depicted in figure 7.2). This architecture mutually constrains the two blankets and affords stability between them, which in the cases at hand could facilitate the kind of considered, integrated change of self in the person who cultivates wisdom.

This section has built a case that self-evidencing can accommodate wisdom. The different elements of wisdom are already in the self-evidencing toolkit and can be wielded in ways that are more or less wise and open to moral aspiration by the lights of self-evidencing's own objective function.

Self-Evidencing and Meaning-Making

The question "What is the meaning of life?" seems almost impossible to answer, whereas "What is meaning in life?" appears somewhat more approachable. The subtle shift in preposition from "of" to "in" makes quite a difference in tractability, it seems. Of course, these two questions are not independent of each other, and philosophers have made several significant incursions, traversing both topics. As was the case with wisdom, to make things tractable for self-evidencing, I again appeal to recent discussions and operationalized definitions in psychology of a core construct—meaning in life.[9]

Research in this area builds on the idea that there is a subjective sense of meaning in life, which can be reported and measured through questionnaire instruments and which play a role for well-being and in explaining why we do what we do. A widely used questionnaire instructs people to reflect on what makes life feel important, and then ask them to rate on a seven-point scale statements such as "I understand my life's meaning," and "I am seeking a purpose or mission for my life." High scores on these types of questionnaire are associated with increased well-being. The assumption here is that the psychological construct measured reflects psychological states and traits relating to feelings of meaning. We do certain things because they are meaningful, and we do things because they contribute to meaning in our lives. Some philosophers agree that meaning is a motivational and intentional factor, going beyond acting for rewards or feelings of happiness.[10]

Self-evidencing is based on minimizing variational and expected free energy, and thus allows a broad set of motivations for actions, with motivational distinctions written into the generative model as beliefs for which the agent seeks to maximize model evidence. The sense of meaning could then in principle arise in different ways from self-evidencing processes in terms of having specific beliefs about what should be done and not done, or in terms of meta-awareness beliefs about how one goes about self-evidencing. We explore this potential next.

The research on meaning in life is coalescing around three constructs—namely, comprehension, purpose and existential mattering. Meaning in life is felt when the agent can make coherent sense of the world, has a sense of direction reflecting their goals, and senses that their life in some way caries significance and their actions are of consequence. I will work with this definition, setting aside the vast debates in this area across philosophy and many other disciplines. For each of the three constructs, there is a self-evidencing account, and together these accounts provide a unifying approach to meaning in life.[11]

Comprehension is the feeling that things in life make sense, when the agent understands what is happening around them, and when past, present, and future integrate into a coherent whole. People report comprehension when they score high on statements such as "I can make sense of the things that happen in my life" and "Looking at my life as a whole, things seem clear to me." These formulations are good for questionnaires but not very informative in a conceptual sense. For example, "comprehension" is just another word for "understanding" or "making sense." One thing to notice, though, is that comprehension for the sake of meaning is viewed in personal terms, as relative to the life, goals, and plans of the individual agent. In terms of self-evidencing, the presence of comprehension should go with the decline of uncertainty relative to the generative model.[12]

What we can then say to begin with is the following: The agent will comprehend if they manage to adopt beliefs and policies into their model, under which they can relatively unambiguously and accurately sample evidence (within the agent's bounds of irreducible uncertainty). The comprehension speaks to integration and coherence because the model is a vast structural representation of the world, where integration and coherence are properties of structure. And it speaks to the idea that comprehension must include a holistic life and self-perspective because the generative model includes the self-model, such that understanding is inherently self-involving. When the agent comprehends, they can say, "The sensory input makes sense for me—I can fit it into my model of the world and myself, given my beliefs about who I am and given my preferences and the policies under which I actively sample evidence."

I have proposed in chapters 3 and 6 that self-evidencing is basically abductive, working as inference to the best explanation and inference to the best decision, both of which are good fits for sense-making in perception and decision-making, respectively. The picture arising here is that whereas the reduction of uncertainty in perception, planning, and decision-making is just measured against variational and expected free energy, there are many ways to build, maintain, and wield the generative model, all of which influence uncertainty in the long-term average. Better comprehension can come from simply being able to respond to uncertainty and to adopt policies that keep future uncertainty in check. But agents can respond to uncertainty on the basis of different kinds of models, some of which might be better for comprehension than others. For example, a complex model might be able to deal with much uncertainty in the immediate here-and-now but requires much effort to tend to its many parameters and constant work in tuning it because it will fit poorly with new instances. It is better to have a simple (but not too simple) model, which more effortlessly generalizes over many instances. This again speaks to inference to the best explanation because simplicity and generalization are considered best-makers of inference (and emerge from the free energy principle, as we saw in chapter 3). Agents who have the capacity for considering which explanation or policy is best among a set of worthy alternatives might also experience more meaning-conducive comprehension than agents who are stuck choosing among a set of poor candidates. Thus, the ability to counterfactually consider a range of promising policies, and making an inference to the best one, may matter for comprehension. Finally, an agent tending to infer relatively unambiguous policies for pragmatic and epistemic value will experience high rewards and strong information gain from their selective sampling in the world, such that each action contributes much to self-evidencing, and thereby comprehension.[13]

It seems, then, that an agent who on average self-evidences reasonably well, suggesting good convergence on exact inference, may also experience much comprehension. This experience might simply be the feeling of being a self-evidencing agent, but it could also be encoded in its own inferred beliefs. This would be in terms of meta-awareness (as discussed in this chapter), as a belief that keeps track of the self-evidencing gradient itself; for example, it could be by virtue of treating the precisions of beliefs in general as latent states, such that the agent can try to infer whether they are comprehending or not, depending on whether their beliefs tend to have reasonable precision overall.

The second component of meaning is sense of purpose, the notion that the agent's behavior is guided by their own goals. This is measured by asking people to rate statements such as "I have a good sense of what I am trying to accomplish in life" and "I have aims in my life that are worth striving for." As with comprehension, this is not just piecemeal goals and purposes (such as having the goal of painting this fence), but rather having goals that speak to what the agent is and wants to be, their long-term, more abstract purposes that serve to organize and inform their everyday lives. To see what self-evidencing can bring to this, some of our earlier discussions are relevant, such as about the way that agents must engage in inference, learning, and exploration of their own preferences (mentioned in chapters 3 and 6). The agent does not know a priori what all their actual preferred states are, but they must arrive at them through variational inference, guided by the expected free energy under policies informed by those preferences. This means that there must be a set of beliefs about what the agent prefers, which they can acquire evidence for in selective sampling, hoping that the world plays along such that self-evidencing succeeds and does not call for too much revision. Having well-evidenced and precise beliefs about preferences would contribute to this sense of purpose. This is because if the agent only has weakly evidenced and imprecise beliefs about preferences, then it would be hard for them to be confident about the purpose of their actions or that those beliefs reflect their life's goals; the agent would be justified in asking themselves why they are engaging in their behavior, thus doubting their purpose.[14]

I think that purpose also relates to the risk of the policies inferred by the agent (where risk is the divergence between the policy's expected outcomes and the agent's preferred outcomes). This suggestion would complement the idea that comprehension may be helped by policies with low ambiguity. If the agent frequently is forced to infer policies that are high in risk, then it can be hard for them to confidently infer a manageable and simple set of preferred states. Risk relates to complexity in the sense that risky policies can have several different outcomes, some good, some bad; and this complexifies the agent's ability to update beliefs about their preferred states. Rather than having a clear set of goals, the agent receives evidence consistent with diverse states, which can contribute to diminishing the sense of purpose; the agent might feel as if no matter what they do, they never get a clean sense of what their purpose is. This speaks to the idea that purpose, as a component of meaning in life, is not merely found but also is actively shaped through the agent's own actions. Hence, if they manage to begin to infer policies that are less risky, they can hope to increase their sense of purpose.

The third component of meaning in life is existential mattering. If agents feel that their existence matters, then they confirm statements like "Even considering how big the universe is, I can say that my life matters" and they reject "There is nothing special about my existence." This component can sound somewhat grandiose, as if meaning exists only if one leaves a mark for all future generations, changing the course of the universe. There is considerable debate about what existential mattering means and if reports of mattering reflect reality or are the expression of some kind of protective delusion of grandeur. I believe that we can find a useful self-evidencing account of existential mattering by using considerations about self-evidencing agents of volatility (as discussed in chapters 7–9).[15]

If we view an agent of volatility as a cause of change in the environment and in themselves, then it makes sense to add something like mattering to comprehension and purpose. If the agent can comprehend and has much purpose but fails to do anything that matters to anyone, then meaning should be threatened. Moreover, comprehension and sense of purpose are related to policies with low ambiguity and risk, which can in fact make the agent rather predictable, and in that sense boring—and boring people are not remembered for posterity. If mattering is about making changes and standing out, then something extra is needed. Consider an agent whose actions are utterly predictable from the context, someone who never explores the policy landscape, or someone who never forms hypotheses about their own self and tests them in action. Such an agent will have a hard time leaving a distinctive mark. Being disruptive in the sense of inserting one's own agency into a situation to force a nonlinear change in observations of self and other is what will help with mattering (recall figure 9.1).

However, I am not saying that mattering is the same as disrupting. Consider the agent who encounters great volatility but is able to stay focused on a task that needs doing nevertheless (such as the bomb disposal expert working calmly in a war zone); they stayed the course and prevented a bad outcome but did so when the context seemed to call for radical, disruptive changes to behavior. This agent will matter, and they might be remembered too. Existential mattering can therefore be a reflection of a belief of agents of volatility—namely, that they are able to express their relatively free agency and seeing themselves as to some extent liberated from the current causal nexus. In other words, if they infer policies on the basis of deep counterfactuals, where their own agency is modeled to cause significant state transitions, then, if they are able to evidence this belief in action, they will have strong beliefs that what they did mattered. What they did, as the agents they model themselves to be, diverted the course of events from what it otherwise would have been, and therefore it could potentially matter and stand out as salient and surprising for others—to be remembered.

On its own, such a belief in mattering might not support meaning in life, but meaning might emerge if combined with purpose, where the agent did what they did for a purpose, not by chance; and comprehension, where they did what they did because they understood what was happening and what was required. Existential mattering therefore might be related to inferences about what kind of cause the agent is, as a self in the world: a predictable one, operating at the behest of surface sensory input, or a change-maker, operating on the basis of deep understanding of the world and the agent themselves.

We now have reasonable self-evidencing processes lined up for the three aspects of meaning in life: comprehension as uncertainty reduction on the basis of unambiguous policies, purpose as action on the basis of inferred preferences and low-risk policies, and existential mattering as a self-evidencing belief that one is an agent of volatility. These three processes are all needed for meaning in life and can emerge in self-evidencing agents like us. As with wisdom, there is no guarantee that agents will experience maximum meaning in life. It is entirely possible that agents in their variational quest for self-evidencing take paths that require spending much time deviating from exact and rational inference (given their model). Agents will fail to think through all the counterfactual scenarios needed to come up with the least ambiguous and risky policy that is nevertheless compatible with making a difference in the world. It is difficult and takes effort to keep track of all the different parameters in play in a complex, changing world. Often, a decision—any decision—needs to be made immediately to avoid worsening a crisis; recall the World Health Organization (WHO) advice to enact risky and ambiguous policies in a pandemic (mentioned in chapter 6). Agents can work on themselves, exploring counterfactual space to be better at articulating what makes life meaningful for them, and to engage in actions that improve meaning. But even this will be a work in progress—a journey of discovery and self-shaping.

In the discussion of wisdom earlier in this chapter, we had to build in moral aspiration at the end, since it is implausible that normative ethics arises directly out of self-evidencing. For meaning, there is also a much-debated normative element. If an agent has high comprehension and purpose and believe that they matter, but they spend all their time doing crossword puzzles, weeding the garden endlessly or, like Sisyphus, repeatedly pushing the same stone up a hill, is their life truly meaningful? Many will say no, although it is not a given that the verdicts of other people can or should override an individual's own sense of meaning. From the perspective of self-evidencing, meaning depends on the individual's internal model, including their beliefs about how well their comprehension, purpose, and mattering are evidenced in selective sampling. If it feels meaningful to spend years raking the stones in the yard, then this simply reflects a particular constellation of beliefs of that agent (recall here the complete class theorem, discussed in chapter 5, which implies that for every behavior, there is a set of priors and a cost function explaining it). The rest of us might be disdainful, and we might have various ethical or moral arguments to this effect, reflecting our beliefs in turn. The incessant raker might come to question themselves and explore other ways of being, or the rest of us might come to view the zenlike simplicity of the raker's actions as a valuable beacon of stability in a world obsessed with disruption and expansion. Various arguments can be constructed in favor of different norms, but it is difficult to see how self-evidencing in and of itself can underpin any of them.[16]

Here, it shines through that self-evidencing is not a teleological notion. We do not self-evidence to stay alive, to create meaning or be happy, or to maximize rewards. There is no greater truth or purpose or meaning, and no metric against which our existence can be assessed. Insofar as there is meaning, or wisdom for that matter, it is manifested in the way that we go about existing. There can be many types of existence, and within each type, there can be many different paths to take. From the outside, it may look as if there is something that each agent knows and aims at, such as riches, thrills, or the quiet life, but, from the inside, it is just a matter of variational processes that provide evidence for existence, so long as the agent exists. I think that this provides a valuable perspective, shifting the focus from trying to define and aim at norms and goals to the messy ways that we go about our daily business—meaning and wisdom can arise there.

Mindful Self-Evidencing

There are plenty of wisdom traditions, contemplative practices, philosophies and therapies focusing on improving our wisdom and meaning in life. People revere and look for wise people, or those who seem to lead meaningful lives, for advice and guidance. What they seek are policies that might promote wisdom and meaning. Self-evidencing is not a wisdom tradition, although I am tempted to regard it as a simple meditation over the nature of existence. So now that I have given wisdom and meaning a self-evidencing framing, perhaps we can set out some active inference policies to achieve them. I want to stress that there are many ways to work toward these aspirational goals. It seems to me that age-old contemplative practices, for example, might well cotton on to important self-evidencing mechanisms in their development of methods of training the mind toward psychological transformation that promotes states of human flourishing, such as insight, wisdom, purpose, or moral cognition and behavior. As with wisdom and meaning, I cannot do justice to the richness of these traditions, so again I will pare down things to a few psychological and cognitive components.[17]

Much mindfulness-based meditation focuses on a combination of attention and acceptance. Participants are trained to keep attention directed at a certain domain, such as their body or their breath, and learning to notice when it strays and to gently guide it back when it does. They are also trained to cultivate a nonjudgmental attitude to what they attend to, simply noticing it and accepting it but refraining from immediately reacting to it. These two processes are much emphasized in contemporary mindfulness training, which in its secular forms are relatively divorced from overt traditional (often Buddhist) contexts, which might have more specific instructions about how to behave. Attention and acceptance can dissociate, indicating that they are indeed different cognitive processes. For example, a participant can attend to their bodily states, and in fact become quite distressed as they increase the gain on bodily signals that often sit in the background, and only later, after more weeks of training, does the participant learn to adopt an accepting attitude toward the signals, making the distress dissipate, but without battling the sensations. We experienced something like this in a study led by Tess Guthrie, which looked at how mindfulness affects the famous rubber hand illusion, where participants experience a touch, which is really made on their real, hidden hand as occurring on a rubber hand that is visible in front of them. Participants who had just been meditating had a strong rubber hand illusion when told to focus attention on the conflicting tactile and proprioceptive signals, getting rid of the attentionally amplified uncertainty by falsely inferring that they had a rubber hand. Only later, after several days of mindfulness training, did the acceptance seem to kick in, reversing the effect such that the participants in the end neither believed nor disbelieved that they had a rubber hand when we tested them. In contrast, the control participants who did something else than meditate did not have the initial strong reaction, and in the end mostly disbelieved rather than adopting a neutral attitude. Mindfulness can thus be seen to interact with perceptual inference.[18]

The question is: Why would attention and acceptance, assuming that they are key components of contemplative practices in the wisdom traditions, facilitate wisdom and meaning?

The capacity for sustained attention in the presence of distracters seems useful for thinking through scenarios in inference to the best decision, and for staying the course with difficult policies. The accompanying capacity for noticing when attention strays, and for noticing one's mental states in general, seems useful for meta-awareness and for being aware of whether one is acting in accordance with one's preferences. If one is often distracted, with one's attention pulled this way and that, without noticing it, then it seems hard to string together a coherent story about what the world is like, what oneself is like, or what one might contribute to the world that will stand out from stimulus-driven patterns. All this flows easily from the discussions so far and already speaks to several aspects of wisdom and meaning.

Acceptance, the nonjudgmental attitude to the things that one attends to, relates in the first instance to a kind of equanimity. That is, rather than amplifying the sensory input through attention and then immediately knocking it down with the first available perceptual or active inference (like our participants who immediately form a strong, false belief about the rubber hand), the accepting attitude lets the amplified signal sit for a while, buying the agent time to consider if it is really worth knocking down or if there are other ways of going about the situation. The agent can spend some time exploring counterfactual scenarios, trying out a wider range of policies to gain control of this strong signal or they might feel safe enough to simply wait out the situation (our participants might adopt the policy of waiting out the rubber-hand trial, confident that the conflicting somatosensory signal is not damaging and will abate on its own).

The equanimity that comes with acceptance should, if the training is good, differ from chronic indifference, which is bad for self-evidencing and for existence (e.g., it is bad for existence if we are indifferent to a rapidly approaching truck). Rather, equanimity affords opportunity for counterfactually asking whether the best policy is really the first one striking us, and for exploring what policies might be consistent with our beliefs about who we are and how it integrates with the wider model and the state transitions in the world. In this way, acceptance is a way of exploring, and perhaps even pushing, the expected boundaries on uncertainty reduction; that is, how much surprise can be expected and how much might be acceptable. This can help the agent be motivated to go beyond their habits, which typically traverse only a circumscribed range of uncertainty. The agent who can begin asking questions about the ways that they deal with uncertainty can take new and wise paths and can explore better if there are different ways to create meaning for themselves.

In this way, mindfulness meditation can be seen as a tool for cultivating wiser, more meaningful action. This contrasts with some popular images of meditators as removed from the real world, their heads in the clouds. Meditation allows the agent to focus their self-evidencing in more dynamic, skillful ways on both connecting, through attention, and detaching, though nonjudgmental acceptance. Mindfulness may empower agents to make better decisions by optimizing the balance between connection and detachment (which was the main theme in our self-evidencing account of consciousness). This is an important facet of mindfulness that makes excellent sense when cast in terms of self-evidencing.

Practically, if mindfulness is, say, rolled out as a professional development program in an organization, some executives and human resources departments may believe that this will help with stress, and therefore help with retention (and, more nefariously, they may hope that this training puts the onus on employees to cope better and be more productive, letting the executives themselves off the hook for providing better work conditions). But what the organization should really be getting through proper mindfulness training, understood in the light of self-evidencing, are employees who are better able to make decisions for themselves and might be more likely to see through the illusion of working themselves into ill health and leave the organization for a different one. In this sense, mindfulness should be empowering agents of volatility.

This brief sojourn into mindfulness is meant as an appetizer for understanding contemplative practices and wisdom traditions more broadly. I think that when seen through the self-evidencing framework, it makes excellent sense that mindfulness cultivates both attention and acceptance, and this usefully highlights the kinds of avenues that an agent can take to develop wiser and more meaningful patterns of behavior. There are many more contemplative practices, with different experiential and behavioral outcomes, that one could look at. In fact, this is a lively area of research in the predictive processing field, with papers looking at several kinds of meditation and the curious aspects of experience that might happen in meditation, such as cessation of consciousness.[19]

Concluding Remarks: Wisdom and Meaning for All

We have explored the prospects of a self-evidencing approach to wisdom and meaning. The task was made tractable by focusing on psychological operationalizations of these notions, which sidelined as too challenging what the vast philosophical debates, wisdom traditions, and other discourses say about these important aspects of the human experience. This allowed us to pare down wisdom and meaning to a series of constructs that each could be underwritten by self-evidencing processes. Any self-evidencing agent can in principle, given the right training and circumstances, engage in mental actions that promote wisdom and meaning. Of course, this is no guarantee that the agents will succeed. It will always be a journey with obstacles on the way. But the self-evidencing lens reveals that wisdom and meaning are not rarefied states reserved for gurus, sages, and philosophers. Neither are these states all pointing in one direction; rather, they depend on the agent's own generative model and can lead to vastly different, equally valid trajectories through life. This means that there is plenty of room for diversity in our everyday search for wisdom and meaning in the lives that we lead.

At times, there is a sense that meaning in life is threatened by the sciences, which are thought to disenchant the world, reducing everything to mere cause and effect. There is a fear that perhaps with increased knowledge of the neuroscientific principles governing the brain, we will not in the end be able to locate our selves, meaning, or ethics. This fear is misplaced for a framework like self-evidencing, which is wholly cast in terms of cause and effect and yet can accommodate and even democratize wisdom and meaning, together with self, will, and consciousness (as we have seen in previous chapters). Perhaps other naturalistic approaches can do the same, but I think that self-evidencing is particularly promising for making sense of many and diverse aspects of everyday, lived experience. This is because self-evidencing itself begins from simple reflections on existence and brings the dual epistemic-causal idea of self-evidencing to bear on who we are, leaving plenty of degrees of freedom while respecting that agents are inherently inferential across all their beliefs and inescapably caught up in the causal nexus. The underlying theme is self-supervision, which affords autonomy to each agent in spite of the causal setting. Here, agents can learn to pursue wisdom as they like and to create meaning that reflects, and helps bring about, who they believe they are and what they have learned to prefer.[20]

Having discussed these deep philosophical themes, it is time to conclude this book. The next, final, chapter will provide an overview of the self-evidencing agent as we have presented it throughout all the preceding chapters.

11 Concluding Remarks

This book has developed a case for self-evidencing as a first-principles method for understanding the existence of agents. Thus, the self-evidencing agent exists as a model that explains away its observations, and the observations thereby become evidence for the model, or the agent's existence.

The argument was based on a simple conceptual analysis of what it means for an agent to exist. The free energy principle articulated by Karl Friston grounds self-evidencing, such that existence entails the minimization of variational or expected free energy, which is equivalent to maximizing model evidence.

Self-evidencing is a distinctive method in this sense, for several reasons, emerging throughout the book and setting self-evidencing apart from other approaches, including prior formulations of predictive processing:

- Self-evidencing provides an epistemic framing for the free energy principle, helping to explain perception, attention, and action in entirely abductive terms of inference to the best explanation and inference to the best decision.
- The self-evidencing agent can be understood to be self-supervised, providing assurance that the cognitive processes that we identify are tractable for the agent. The self-supervised process is conducted entirely through model-constrained trial and error precision control within the agent's internal states.
- Normative and descriptive approaches to perceptual inference and decision-making are brought together in self-evidencing, complementing each other rather than pointing in different directions.
- Self-evidencing implies a circular causality, which makes closing the action-perception loop central to understanding agents. This synthesizes

embodied and cognitive approaches, anchors all behavior in our prior beliefs and yet enables false inference, and invites us to acquiesce in a limited skepticism.

- A self-evidencing agent is epistemically secluded from the world external to them, and yet also is inextricably causally enmeshed in the causal nexus, capturing the deep tension in our existence in the world.

Articulating these properties of self-evidencing provides a distinctive philosophical method, which is revolutionary in the sense that it does not explain existence in teleological terms of achieving goals of survival, thriving, utility, or rationality; rather, it explains these goals in terms of existence. This means our basic research question is not "How does the agent achieve their goals?" but "What (beliefs) characterizes this agent's existence?"

To answer this question, we move away from imposing intractable objective functions on the agent. Instead, agents just have the one tractable job of increasing model evidence, of making their model of the world and themselves cohere with the observations that they receive or seek out. Describing the model and self-evidencing relative to the model's beliefs accounts for how agents exist.

For the agent, the simplicity of this task, where there is only a simple objective function, puts the onus on the agent to form and update the model of the world and of the agent themselves. That is where the complexity and challenges lie, both for the agent who must maintain the model in whatever way possible and for the scientists who must try to capture this model and the agent's generative process.

With the focus on the agent's model and the idea that this model is shaped in a wholly self-supervised, nonteleological manner, we get the opportunity to ask philosophical questions afresh. Now the answers are simply about the models of agents, insofar as they exist, and how these models are wielded in causally embedded, epistemically veiled, active precision control:

- Decision-making becomes inference to the best decision, governed by best-makers analogously to inference to the best explanation, and adjudicated by expected free energy minimization or model evidence maximization. Rationality is then likened to abductive hill climbing, marked by the way that agents wield best-makers, including how they counterfactually deliberate and how they maintain their model. Perception and decision-making are fundamentally the same: to perceive something, we ask which state of affairs best explains our current observations, and to make decisions, we ask which state of affairs best explains future observations, where those states of affairs include our own actions. What we perceive and decide depends on our internal model and the environment that it reflects, and which in turn reflects our model.

- The sense of self arises in inference of a self-model—and the modeled self is as real as other things. The self is just the structural network of causes of our own observations, which stem from within ourselves. The self-model grounds both the minimal self and narrative selves ranging from the epic to the aphoristic. We can discover ourselves in self-exploration and shape the self actively in self-fulfilling prophecy by using the action-perception loop with a focus on ourselves. The internal self-model may be sparsely coupled to the internal world-model, operating at different timescales and thus providing constraints on each other and conferring resilience on the agent. The human self-evidencing agent, specifically, might be an agent of volatility, adapted to an eco-niche of volatility, busy modeling volatility and exploiting it as well.
- Volition depends on how sensory input is filtered by precision control to the active states of the agent's boundary (or the Markov blanket), marked by a motley of counterfactual deliberation, potentially guided not just by the expected observations at the boundary but also by the expected observations of internal beliefs formed over time. Self-evidencing becomes an attractive fit for compatibilism about free will because self-evidencing harbors the duality that compatibilism requires of causal, deterministic embedding and epistemic and rational inference. Emotion and bodily sensation represent expected self-evidencing, and they arise as inference for arousal-related sensory input, as constructivism has posited, but here this is cast through forward-looking active inference, which explains affective phenomenology. Constructivism and basic, nonconstructed emotions are not in competition but can coexist as sparsely coupled systems that lend the agent affective agility and resilience. Moral value is extraneous to self-evidencing in itself, such that morality depends on moral discourse—the marketplace where prior beliefs about moral value policies are negotiated in communities. We should not seek to reduce value and ethics to our biological natures or self-interested survival instincts—there is only the way that we deliberate with each other about what we should do—how

we share our priors. This places the responsibility on us to uphold high-quality, democratic debate. Self-evidencing highlights risk and ambiguity in decision-making, which can provide useful tools—in terms of moral model fitting—in contemporary ethical debates, steering us away from simple but damaging solutions.

- Consciousness provides a fruitful testing ground for self-evidencing, when the complicated science of consciousness is organized to receive self-evidencing as a valid construct, setting aside some types of metaphysical and theoretical aspirations. Self-evidencing agents essentially filter—and therefore detach from to some degree—the immediate sensory input. Consciousness is marked by detachment too, and thus the task is to consider which kinds of detachment distinguish conscious creatures from nonconscious ones. Here, active inference emerges as a necessary condition for the selection of content into consciousness, which in turn relates to the agent's beliefs about their own expected state transitions as they selectively sample observations when inferred polices for action are enacted. Subjectivity and unity of consciousness come in because of the inherent involvement of the one embodied self-model in active inference. Agents of volatility specifically represent themselves as conditionally independent causes that can insert themselves at will into the causal order of things, bringing a strong sense of subjectivity to their conscious experience.
- To live well is to balance detachment, maneuvering skillfully between closed-minded reliance on prior beliefs and open-minded updating in the light of new observations. Wisdom is to know and be aware of this balance, and meaning in life depends on honing and wielding the balance over time. Mindfulness meditation practices are directed at developing this skill, focusing on exploring attentive connection and nonjudgmental, accepting detachment. The moral aspiration in wisdom and the normative dimension of meaning in life are not surprising for a social species like us, but ultimately, they are negotiated only in moral discourse, for both individuals and communities. To be wise and have meaning in life are just to exist in certain ways; they are not so hard that they are the province only of sages and gurus, but they are still nontrivial because there are so many degrees of freedom in model maintenance, with no external yardstick and only the guidance from the mere fit between expectations under our beliefs and the observations that we make. There is always a risk that we navigate the free energy landscape far from wisdom and meaning.

Carl Hempel, who coined the term "self-evidencing" for use in the philosophy of science, not only identified its benign epistemic circularity but also its link to existence, insofar as a self-evidencing explanation only exists to explain the observations that become the evidence for itself. If the evidence is not there, the model doing the explaining would not exist. This link to existence is taken literally in the free energy principle for the self-evidencing agent, which is not in existence if observations are not explained and model evidence is not accumulated. This is where the new insights from self-evidencing begin because existence comes with model maintenance. And in each case, our answers to the questions about rationality, self, volition, emotion, value, consciousness, wisdom, and meaning come down to which model this agent has constructed to account for its observations.

Many questions for future research programs emerge from employing self-evidencing as a philosophical method. Is inference to the best decision, with its unwavering abductive perspective, adequate to explain rational decision-making? Can it really reconcile descriptive and normative approaches to decision-making and rationality, and if so, how much might we need to revise our concepts in these debates? Is the self-model sparsely coupled to the world-model, and is this what lends robustness to self-evidencing agents (and where else might this sparse coupling architecture occur)? Do we represent ourselves as agents of volatility, and self-evidence accordingly, essentially to create our own free will? Will self-evidencing really be useful for articulating satisfactory accounts of compatibilism about free will and faring better in the face of the many problems in that debate? Can constructivism and basic emotions be reconciled within one system, coupled to deliver our distinctive emotional phenomenology? Is active inference really necessary for conscious experience, and can detachment be operationalized computationally and empirically to elucidate consciousness? Will consciousness science benefit from incorporating self-models and free will, in particular under the notion of agents of volatility? Can wisdom and meaning in life be understood in terms of Bayesian inference, and will this open up approaches for computational understanding about living well and mindfully?

More broad-based research questions also emerge, for example in the context of advances in artificial intelligence. The self-evidencing view of our existence is one of more or less frantically bootstrapping our own internal structure through self-supervised precision control, which keeps the dissipative forces of the universe at bay for the time that we exist. The existence and the computational processing are of one piece, implying that if the computing ceases, so does the existence, and vice versa. This suggests that human, free, and conscious computing is mortal and should not be conceived as the resource-expensive way that computers have immortal programs. This sets out a research agenda, exploring what is special about conscious, mortal computing, when set in the context of self-evidencing. Might it be that mortal computing lends a particular kind of agility to agents, ensuring that they are able to exist in a changing environment, but at the price of mortality, whereas computer programs live forever, but at the price of rigidity? Will this help explain why we exist the way that we do?[1]

* * *

Throughout this book, there is an existential invariant—namely, the tension between, on one side, the epistemically insulated, solitary precision control, which evokes Cartesian skepticism and occurs exclusively within the Markov blanket, and on the other side, our causal connection to the body and the world, embedding our models fully as causal nodes within the environment. We try to negotiate that tension by representing ourselves as agents of volatility, causally effective but yet with some independence from immediate causal control from anyone but ourselves.

This existentially invariant tension is what underwrites self-evidencing's claim to capture the diversity of everyday experience. Self-evidencing is constrained only by the objective function of minimizing free energy or maximizing model evidence, leaving it up to agents to finesse their internal models. Different species of agents will be funneled along distinctive avenues of modeling, and within species, there will be many degrees of freedom in shaping model parameters, even if evolution has endowed us with the same high-precision beliefs about our homeostatic states. You and I (and everyone else) can take very diverse routes to build models that can help provide evidence for those homeostatic beliefs. What it takes to live well as a self-evidencing agent of volatility can take many forms. This is important because I promised in chapter 1 to provide a self-evidencing framework that is adequate for the diversity of everyday experience.

In contemporary public discourse, the term "lived experience" is sometimes used for what I call "everyday experience," in order to highlight that marginalized, discriminated, stigmatized, and disadvantaged groups often are effectively silenced by those with privilege and power. Public discourse has little patience for diverse voices, and those in power get to set the agenda. I believe the concern with "lived experience" in contemporary discourse is critical for society. Literally, untold suffering results from ignoring lived experience, understood in this sense.[2]

The question then is: How might self-evidencing, as presented in this book, relate to these important issues? The initial answer might be that the framework runs counter to accommodating lived experience since, as I present the framework, it may appear disembodied, cognitivist, representationalist, and computationalist. This take on predictive processing has been labeled variously as orthodox, conservative, seclusionist, imperialist, and authoritarian. These are not labels easily associated with a framework on the side of marginalized voices. Might self-evidencing be a Trojan horse for regressive politics? If that is the case, then I can hardly, from my position in the debate and discipline, claim that the framework can respect the diversity of everyday experience.

I think that self-evidencing is in fact a well-suited framework for making sense of, and room for, the diversity of everyday experience. The notion of self-evidencing has as its starting point a limit case of equality—namely, existence. But it does not dictate a unitary *telos* for that existence, under which all an agent's activities should be subordinated. The whole idea is that the objective function of variational or expected free energy is equally accessible to all who exist, and the cost of that internal accessibility is that we each must build our internal models in the best way that we can. There is not an optimality function that we are assessed against, and which is delivered in some authoritarian fashion, neither by actual authorities or by nature itself. The complete class theorem (discussed several times in this book), indeed says that for any behavior that we observe, there will be a set of priors that explains it, and in that sense, all behaviors are equal and describable through a generative model. I find this a liberating perspective because it frees us from focusing on how to rank people's beliefs and instead invites us to engage with each other, through democratic discourse, to explore new beliefs, and how to negotiate about what evidence to bring to bear on our beliefs and on to our decisions.[3]

Philosophical and cultural discourses that have taken diversity and equality seriously risk ending up at the extreme, where relativism reigns. On that side, there is no ground truth, and every belief is as good as any other. However, we want to have ways to filter out some beliefs, also collectively. We need to efficiently and accurately counter misinformation about things like climate change, for example. Self-evidencing can allow multiple viewpoints, while avoiding relativism. It allows rationality to be a matter of inference to the best decision, where the only thing that counts is the ability to maximize evidence for one's own model, however idiosyncratic. We go about our business without aiming for objective truth or maximum utility. We can do this in a rather unruly trial and error manner, allowing much individual variability. But, as we do this, optimality and truth come along for the ride. Variational inference will, by fiat of the mathematics behind the free energy principle, approximate optimal inference over time. The beliefs that we have and the actions that we perform will inexorably get closer and closer to what exact inference would deliver—at least until the world throws a wrench into the works and we must go back to the drawing board. Truth comes along since the internal model must come to recapitulate the causal structure of the otherwise hidden generative process. Here, truth is not a dictate but rather an inevitable gradual coupling between the internal and the external. More than that, our actions shape the true generative process in the image of our generative model, forcing a coupling to exist for anything (including collectives of interacting self-evidencing agents) managing to exist over some period of time.

Ultimately, of course, ideas, beliefs, and doubts play out in public discourse, which is unwieldy and prone to distortion. Prosocial and inclusive values arise and are negotiated in this volatile marketplace. Cultivating open-minded listening to diverse voices must happen in ongoing, collective, trial and error engagement and communication. Articulating, amplifying, disseminating, and repeatedly both supporting and challenging ideas are all necessary here. Given the state of the world, we must approach this task humbly, relying less on past beliefs and habits and therefore, though without taking too many risks, increasing our learning rate and willingness to try new policies of action. I believe that this is facilitated if self-evidencing is one of the ideas that we spread, which has overtones of common humanity because it emphasizes that even though everyone's lived experience can differ greatly, everyone's existence is the same at heart.

Notes

Chapter 1

1. My paper on self-evidencing (Hohwy, 2016) appeared online in *Noûs* in 2014. In this book, there will be many references to the predictive processing literature, with a slant toward philosophical treatments. My own book (Hohwy, 2013) and Andy Clark's influential paper and subsequent book (2013b, 2016) can be useful starting points. By 2020, a large body of philosophically oriented work on predictive processing had already emerged, with several hundred listed in the appendix of a review that I conducted (Hohwy, 2020a), and it has only grown since. Karl Friston articulated the free energy principle and is the architect behind its main developments; see, for example, his extremely influential landmark paper (Friston, 2010) and many other works by him, many of which I cite in this book.

2. The free energy principle is a variational principle of least action; the recent body of technical literature thus makes it clear that it implies that there is an action for every existing system—namely, the path integral of variational free energy (Friston et al., 2023). For a discussion of the "high-road" versus "low-road" approaches to predictive processing, see Friston (2019a); for a discussion of principles and processes and references to the philosophy of science debates about the relation between principles and laws, see Hohwy (2021b), and for a critical discussion of the distinction, see Williams (2022). The metaphor of the high and low roads picks up on the idea that both will end up at the same destination, but via different methodologies. Which methodology one prefers depends on pragmatic interests and scientific choices; the metaphor as it is employed in the literature on the free energy principle is not intended to say that either is better than the other.

3. For many of the more philosophical critiques of predictive processing and the free energy principle, there are already several published responses, which, as I see it, efficiently engage with the criticisms. See, for example, the commentaries by Friston, Ramstead, Andrews, Parr, Kiefer, and myself, as well as works by Seth, Clark, Kirchhoff, and several others, which were published with the following critiques of predictive processing (Bruineberg et al., 2022; Raja et al., 2021; Ransom et al., 2017; Sun &

Firestone, 2020). There are several other critical papers as well, many of which I will allude to along the way in our discussion.

4. There are several formal accounts of how other frameworks fit in with the free energy principle, such as Friston (2010) and Parr et al. (2022).

5. As a companion to this book, I recommend the excellent and authoritative *Active Inference* by Parr et al. (2022). For two recent comprehensive and detailed counterparts, see Nave (2025) and Beni (2025).

Chapter 2

1. J. L. Austin discusses things in *Sense and Sensibilia* (1962, pp. 7–8). For a discussion of flames as a problem case, see the critical argument in Raja et al. (2021) and responses (Friston, 2022; Parr, 2022; Beck and Ramstead, 2025). For a discussion of different kinds of things, like rocks, cells, thermostats, and mammals, see Friston, Da Costa, Sakthivadivel et al. (2023).

2. The description of a thing is in causal terms of nodes and edges in Bayes nets, which Judea Pearl (2000) understood as applying to directed cyclical graphs.

3. Carl Hempel (1965, pp. 372–374) discusses self-evidencing, and Peter Lipton (2004) picks up this thread in his influential book on inference to the best explanation. Chris Frith and I appealed to self-evidencing in an account of trust of introspective reports for consciousness science (Hohwy & Frith, 2004), and self-evidencing made an oblique appearance in *The Predictive Mind* (Hohwy, 2013, p. 239) before taking center stage in "The Self-Evidencing Brain" (Hohwy, 2016), and several subsequent works.

4. In some new treatments, one will see reference to "paths" rather than "states"; this is for technical reasons having to do with recent path integral formulation of the free energy principle. For instance, see Friston, Da Costa, Sakthivadivel et al. (2023). Paths can be conceived of as trajectories or events that traverse state space and behave much like a variable or state (paths can be treated as a state in generalized coordinates of motion).

5. Notice that here, we are beginning with the assumption of a thing's existence in physical space, and then analyze this in terms of state space (we are not somehow constructing physical space out of state space). In general, I do not think that one can separate physical space from state space. There are physicists who would say that physical space is an emergent property of the free energy principle under a background (space-time) free treatment; for example, see Fields et al. (2022).

6. This section and the next few sections explain, unpack, and frame a number of themes, some of which occur in Hohwy (2016, 2021b) and several other papers of mine. These can be regarded as philosophical developments of themes in Karl Friston's work from Friston (2010), Friston and Stephan (2007) through to his

monograph on a free energy principle for a particular physics (Friston, 2019b) and his subsequent works.

7. For an impressive study of bacterial spores, see Kikuchi et al. (2022).

8. Bayes' rule says that the posterior probability of the model given the observation, p(M|O), equals the likelihood of the observation given the model, p(O|M), times the prior probability of the model before getting the observation, p(M), divided by the probability of the observation, p(O) (requiring the marginalization mentioned in the main discussion).

9. The idea of this free energy "bound" on surprise is textbook material on machine learning and pattern recognition, where negative free energy is sometimes known as the *evidence lower bound (ELBO)*. Since these results are standard in machine learning, which is a burgeoning discipline, there are many excellent demonstrations of Jensen's inequality, ELBO, and free energy available on the internet. For an authoritative textbook, see Bishop (2007). For an explanation and worked example of variational inference, see Bogacz (2017).

10. I have set out this kind of argument concerning self-evidencing (Hohwy, 2021b). There is considerable technical debate about computational tractability and approximate inference, which I do not pretend to resolve here; rather, I rely on core developments (based on statistical physics) in the machine learning and pattern recognition literature, as they are absorbed into the description of the free energy principle. A series of papers by Iris van Rooij, Johan Kwisthout, and colleagues set out the challenges in this debate; a useful starting point would be van Rooij et al. (2018), which discusses whether approximate inference (although not variational inference specifically) can avoid intractability problems. Although there is much work to do to resolve this debate, my suggestion would be to explore whether variational inference under the free energy principle is a natural fit to Van Rooij et al.'s positive suggestion that tractability requires restricted parameter ranges (e.g., van Rooij et al., 2018); this constraint seems inherent to the understanding of existence in terms of surprise and self-evidencing. A further question in the vicinity is what the relation is between tractability and self-supervision.

11. A highly influential formulation of the free energy principle is Friston (2010). For recent technical developments of the free energy principle, see Friston, Da Costa, Sakthivadivel, et al. (2023) and Ramstead, Sakthivadivel, et al. (2023).

Chapter 3

1. These concerns about descriptive and normative perspectives on cognition and mind come up in many guises in philosophy. I think that there might be an interesting project in connecting the dual existential/epistemic properties of self-evidencing to Wilfrid Sellars's notion of "The Myth of the Given" (Sellars, 1956); see also John McDowell's more recent take (McDowell, 2008). I also discuss the

descriptive-normative duality (Hohwy, 2021b), which focuses on another influential debate about descriptive and normative perspectives—namely, Ludwig Wittgenstein's rule-following considerations (Wittgenstein, 1953), which are also relevant here.

2. For background, see Peter Lipton's influential work on inference to the best explanation (Lipton, 2004); for a contemporary discussion, see Douven (2022). For a relevant, classic treatment of related abductive inference patterns, see Harman (1973). I considered inference to the best explanation as a way to understand self-evidencing (Hohwy, 2016). Philosophy of science has discussed the link between prediction and explanation, noticing that in itself, prediction is neither necessary nor sufficient for the understanding of phenomena that explanation is supposed to give (for an overview, see Bird, 2004); I believe that these problems are sidelined in the abductive setting that I present here.

3. The derivations leading to these reorganizations of the free energy principle employ standard rules of probability theory, such as the product rule, together with mathematical observations, such as that the model evidence does not depend on hidden, unknown states. The reorganizations are set out in many publications on the free energy principle, such as Parr, Pezzulo, and Friston's active inference book (2022), the appendix of Da Costa et al. (2020), Buckley et al. (2017), and the tutorial by Smith, Friston, and Whyte (2022).

4. Thus, what I call "trial and error" in this book is different from some uses in reinforcement learning and behaviorism, as well as in other fields, where the phrase sometimes refers to model-free inference and learning.

5. This overall treatment is directly inspired by the free energy principle, but it has many threads in the literature. A prominent example is W. Ross Ashby's Homeostat, which randomly adjusts its parameter settings until it hits on a configuration whereby all essential variables are jointly stabilized within range (Ashby, 1947). There are also resemblances to psychological theories in the same vein as Ashby's, such as William Powers's perceptual control theory (Powers, 1973). Works by Karl Friston and colleagues map such antecedents and explain similarities and differences; for example, see figure 4 of Friston, Da Costa, Sakthivadivel et al. (2023).

6. For the various kinds of decision-making discussions, see Gigerenzer (2019), Gigerenzer and Selten (2001), Kahneman (2011), Kahneman and Frederick (2002) and Oaksford and Chater (2007). The interactionism position advanced by Mark Bickhard discusses how normativity may emerge from a process-theoretical, thermodynamic account; and he contrasts this with predictive processing (Bickhard, 2009, 2016).

7. For discussions of universal optimism and how it relates to decision-theoretical optimism, see Fisher and Hohwy (2024). Parr and Friston (2018) sets out the objective functions under the free energy principle.

8. I have further articulated some of the argument for the reconciliation of the descriptive and the normative (Hohwy, 2021b). The Cartesian motto is found in Friston (2011a); the sentiment is also evident in recent updates of the free energy principle, such as Friston, Da Costa, Sajid et al., (2023).

9. For enactivism and autopoietics, see works such as Maturana and Varela (1987) and Varela et al. (1974). There is an ongoing debate about whether a process like self-evidencing under the free energy principle really is compatible with autopoietic and enactive approaches. Notably, Di Paolo, Thompson, and Beer (2022) argue that the free energy principle cannot capture the enactive notion of bodies and sensemaking as irreducibly historical, such that they depend on the past and change the present. I think that there is scope for more in-depth debate here, since variational inference is constrained by history in the sense of learning and evolution (e.g., inference is based on prior beliefs with various precisions); and with the notion of unfolding forward-looking active inference (which we will discuss more in later sections and chapters), there is a very direct way in which agents are changing the present too. These authors also note that enactive agents do much more than predict and infer, arguing that this takes enactivism beyond predictive processing; however, the burden of this book's appeal to trial and error variational inference (which balances best-makers in inference to the best explanation) is to show that there are indeed many processes involved in self-evidencing beyond austere predictive coding—though the internalism of self-evidencing remains anathema to enactivism.

10. For discussions of metamorphosis, see Clark (2017a) and Friston and Stephan (2007); for a comprehensive recent discussion, which provides the most developed and detailed challenge of this type, see Nave (2025). The idea of samples that never deviate too much from the average is related to notions of "ergodicity" in physics and information theory. In some discussions of the position in cognitive science called "enactivism," drawing on ideas of ergodicity as used in biology, the use of ergodic self-evidencing accounts is challenged, arguing that it ignores historical change (also appealing to cases of metamorphosis; see Di Paolo et al., 2022). As I see it, these are different senses of the term "ergodicity," and they are not incompatible. Specifically, the appeal to ergodicity in self-evidencing is primarily about ruling in the existence of an organism, with its entire historicity. The free energy principle was once described in terms of ergodicity, but in recent iterations, this concept has been left out, which precludes misinterpretation due to the different technical uses of the term. Specifically, the term was used in the sense of George Birkhoff's ergodic theorem, and not in other senses, to describe a nonequilibrium steady-state solution, or attracting set, which is still part of the free energy principle. (Some of the objections to the use of ergodicity in the free energy principle operate with other definitions, on which the probability over time can be exchanged for probability at a moment, but this definition of ergodicity technically requires the notion of ensemble densities to overcome the notion that the universe as a random dynamical system is in only one state at

any moment; and the free energy principle does not make these canonical ensemble commitments.)

11. For reviews of allostasis, see Schulkin and Sterling (2019) and Sterling (2012), for a discussion of this topic in the context of self-evidencing led by Andrew Corcoran, see Corcoran et al. (2020).

12. The most comprehensive treatment of active inference is Parr, Pezzulo, and Friston (2022); see also Sajid, Ball, et al. (2021) and Smith et al. (2022). What we have talked about so far is variational free energy, and what is required for forward-looking inference is expected free energy; those deliver the two objective functions for inference that update the model and lead to action, respectively.

13. There has been much discussion of this "dark room" problem, starting with work by Friston, Clark, and Thornton (e.g., Friston, Thornton et al., 2012), my discussion is found in Hohwy (2013, pp. 86–87, 175); and the problem is rehearsed and recapitulated in several other places (Baltieri & Buckley, 2019; Fink & Zednik, 2017; Seth et al., 2020; Sun & Firestone, 2020; Van de Cruys, Friston, & Clark, 2020). For discussion of the relation between optimism and the dark room problem, see Fisher and Hohwy (2024).

Chapter 4

1. *Markov blanket* is a term emerging from structural model analyses of causation. Such models consist of directed acyclic graphs with nodes, representing random variables, and edges, representing probabilistic relations between nodes, where the key property is the *Markov property* of conditional independence. This means that once the behavior of a parent node is known, the child node behavior can be known without knowing the parent's parents. A Markov blanket is described in terms of some internal states and their parents, children, and other parents of the children, such that the behavior of the internal states depends only on the blanket states; see the works of Judea Pearl (e.g., Pearl, 1988). Recent technical developments allow undirected and cyclic graphs, for discussion, see for example (Friston, Heins et al., 2021; Kirchhoff et al., 2018; Palacios et al., 2020; Beck & Ramstead, 2025).

2. In this description, we are setting aside other types of distributions than normal, Gaussian distributions, for which the sufficient statistics are different; and we also ignore that the math mostly calls for continuous probability density functions. For background on neuronal and theoretical aspects of sufficient statistics in the brain, see Marreiros et al. (2009) and Sengupta et al. (2016), among other sources.

3. For these reasons, it is challenging to study precision learning in the cognitive science laboratory, where study participants are asked to sit through a high number of trials that subtly shift the sufficient statistics around to emulate a predictably unpredictable world. Nevertheless, precisions, volatile state transitions, and metacognition are studied in the lab, for example in intertemporal choice-type tasks (Schwartenbeck,

FitzGerald, Mathys, Dolan, & Friston, 2015), autism (Lawson et al., 2017), and in consciousness science, which has helped develop statistical tools to make inferences about metacognition, such as so-called type two signal detection theory. For an in-depth discussion of metacognition, see Fleming (2021).

4. For development of these ideas around precision and mental disorder, see Gu et al. (2019) and Schwartenbeck and Friston (2016).

5. There is a large body of literature in this field of Bayesian inference. For my heuristic take, for perception, see Hohwy (2017b); for a compelling account of approximate Bayesian inference that relies on precision weighting, see works by Chris Mathys and colleagues (e.g., Mathys et al., 2014; Weber et al., 2023).

6. For a philosophical discussion of issues around reading and predictive processing, see Fabry (2015); for an active inference account of narrative and reading, see Friston, Rosch et al. (2017). For predictive processing with a mix of amortized and inferential representations, see Tschantz et al. (2022). The inclusion of amortized inference is a challenge to a simple predictive coding framework but not to self-evidencing, since self-evidencing countenances many different processes in the service of self-evidencing, all guided by free energy objective functions. There is an interesting project in comparing this emerging literature on amortization with considerations about affordances and further debates about the role of priors and inference in perception; see, for example, Orlandi (2014, 2018).

7. There is extensive literature on attention and predictive processing, although not so much of the philosophical debate as yet has focused on the agential aspects of attention. For an active inference model of selective attention in visual search, see Mirza et al. (2019). For my own early statement, see Hohwy (2012, 2013), based on Feldman and Friston (2010). There is some critical engagement with these ideas, such as in Ransom et al. (2017), where I agree with the response given by Clark (2017b).

8. In active inference models, agents are modeled using POMDPs, and here, the beliefs encoding preferences mostly are determined in advance by the modeler. However, given that they are beliefs, they too can be subject to variational inference; for accounts of preference learning in active inference, see works led by Noor Sajid (e.g., Sajid, Ball et al., 2021; Sajid, da Costa et al., 2022; Sajid, Tigas et al., 2021, 2022).

9. There is a sizable and rapidly developing body of active inference literature on motor control. See, for example, earlier statements such as Friston (2011b) and more recent work such as Parr, Friston et al. (2023), as well as other works such as the models in (Oliver et al., 2019).

10. Memory is, of course, one of the biggest topics of research across cognitive sciences. There are interesting reinforcement learning studies of the use of memory traces for offline, counterfactual decision-making, such as Biderman et al. (2023), and the link between model simplification and dreaming was developed in Hobson and Friston (2012).

11. For a neurophysiological account of an implementation of predictive coding, see Andre Bastos's work on canonical microcircuits (Bastos et al., 2012), and the discussion in Parr et al. (2022). For the computational and anatomical account of this dual role for the hippocampus, see Barron et al. (2020).

12. For work on imagery, see studies by Joel Pearson's group (e.g., Keogh et al., 2023; Pearson 2019; Pearson et al., 2015).

13. The distinction between transparency and opacity is central to work on consciousness, self, and mental action from Thomas Metzinger, and it has been explained by Jakub Limanowski and his collaborators in terms of precision control, consistent with our story so far. For influential work on transparency by Metzinger, and the discussion by Limanowski and colleagues, see Limanowski and Blankenburg (2013b), Limanowski and Friston (2018) and Metzinger (2013, 2017).

14. Notice that the description (and figures) so far abstract away from a distinction between continuous and discrete processes, where POMDPs for decision-making happen discretely and predictive coding and movement predictions happen in continuous space; continuous and discrete space transitions are captured computationally in link functions.

15. With self-evidencing's duality of internal precision control and causal embeddedness, I think that there is an interesting project in combining self-evidencing with autopoietic and enactive approaches (Maturana & Varela, 1972; Varela et al., 1974), which are also under constant development; see, for example, Froese et al. (2023) and Kirchhoff and Froese (2017). More broadly, there is a live debate about the extent to which systems theory as formulated in computer science is applicable to biology; for a discussion that focuses on autonomy, see Raginsky (2023), and for consideration of the computer metaphor, see Bongard and Levin (2021). The self-evidencing framework that I favor puts a decidedly epistemic spin on these matters, and the relation to other frameworks are yet to be worked out. Chapter 5 will discuss the extent to which self-evidencing requires internal representations, which is a constant theme across these debates as well.

16. A contemporary collection on these debates in what is known as "4E cognition" is found in Newen et al. (2018), which includes a chapter by myself that seeks to conciliate debates; several discussions relevant for predictive processing can be found in Ciaunica, Levin et al. (2023) and Ciaunica, Shmeleva, et al. (2023); for nonrepresentationalist views, see Hutto (2018) and Hutto and Myin (2013), and Varela (1979). Arguments for extending the mind to external objects are developed in Clark (2022) and Clark and Chalmers (1998); see also Kirchhoff and Kiverstein (2021). For a discussion of varieties of directness in predictive processing and the importance of keeping them separate, see Drayson (2018). For a description of cognition that includes the adrenal cortex, see Huebner and Schulkin (2022).

17. For Corcoran's study, which we conducted in collaboration with the cardiologist Vaughan Macefield, see Corcoran et al. (2021). The bistable stimuli that we used was

binocular rivalry, where different images shown to each eye alternate in conscious perception rather than simply being perceived as a mishmash of the images; binocular rivalry will reappear in the discussion of consciousness in chapter 9.

18. For an example of perceptual inference that seems affected by bodily changes during pregnancy, consider phantom kicks felt by some women after birth; I was fortunate to be involved in one of the few studies of this phenomenon, from Sharna Jamadar's lab (Sasan et al., 2020). Although I think that self-evidencing can deal with cases such as metamorphosis and birth, there is much interesting discussion of this question (Ciaunica, Levin, et al., 2023; Clark, 2017a; Friston & Stephan, 2007; Hohwy, 2017a; Nave, 2025).

19. For further arguments in this vein, see Hohwy (2017a, 2018, 2019). There is much debate in this field, however; for an overview, see Gallagher (2023), although its treatment of active inference in my view does not engage directly with self-evidencing. Intriguing arguments for nonbrain cognition can be found in Ciaunica, Levin, et al. (2023) and Ciaunica, Shmeleva, et al. (2023). Andrew Corcoran and colleagues present a compelling developmental perspective on embodiment (Corcoran, Perrykkad, et al., 2023).

Chapter 5

1. For an active inference account in pragmatic, goal-directed terms, of an *E.coli.* model, see Tschantz et al. (2020).

2. Anil Seth and Andy Clark emphasize pragmatics and downplay veridicality for predictive processing (Clark, 2016; Seth, 2021); my view is that self-evidencing unifies pragmatism and representationalism. Sometimes in the literature, instead of "representation" the term "recapitulation" or "mutual information" is used; the internal model recapitulates the causal structure of the environment, or there is mutual information between model and world. I do not believe that this is a significant semantic difference as a recapitulation is a representation (if I recapitulate a movie to you, I am offering a representation of its narrative); the same point holds for other notions, such as "mutual information," "attunement," "mirroring," "consistency," "isomorphism" and "model." Ramstead, Kirchhoff, et al. (2019) argue against structural representations and in favor of a nonrepresentational notion of "enactive inference"; their argument appears to me to establish something like the kind of representations that I point to in this chapter. They are clear that the "generative model is a statistical model of the generative process" (Ramstead, Kirchhoff, et al., 2019, p. 234) and the inversion of the generative model in what is called the "recognition model" represents the sufficient statistics of beliefs that structurally represent the generative process. At best, they argue against a kind of functional *representationalism,* according to which the function of the generative model is to represent the generative process, but that is not the view that I have presented since I have said that representation follows (or as they say is "entailed by") active inference. In their paper, they seem to move from

pragmatism to the necessary absence of representations (with truth conditions), but as I have argued, this move is not convincing since the generative process that has pragmatic value for an agent can be represented veridically. They also seem to argue that representations cannot be in place since inferred policies antedate the properties of the generative process put into place by action, such as reaching out for something: "There is nothing 'out there' that corresponds to the articulated movement of our hands, until it is authored by the organism" (Ramstead, Kirchhoff, et al., 2019, p. 234). However, it is a hallmark of representations that they do not need to have the state of affairs that they represent instantiated. Later work in this vein appears to reinstate representations in a nuanced approach; see Ramstead, Sakthivadivel, et al. (2023). Overall, there is room for further discussion along these themes within the free energy principle framework.

3. A classic discussion of types of uncertainty in the context of a predictive processing-type framework is Mumford (1992), on the computational architecture of the neocortex.

4. There is much interest in how predictive processing can lead to environmental change and how intimately connected organisms are to their eco-niches; see, for example, Constant et al. (2018), Bruineberg et al. (2018), Fabry (2017a), Ramstead, Constant, et al. (2019), and Veissière et al. (2020).

5. For work on the hippocampus and navigation, see Biderman et al. (2020), Davachi and DuBrow (2015), and Nau et al. (2018).

6. Counterfactual inference of future beliefs is known as *sophisticated inference* (Friston, Da Costa, et al., 2021).

7. The notion of controlled hallucination has come to the fore with predictive processing, but these ideas have surfaced many times in the past too. An anonymous reviewer of this book alerts me to an early occurrence in the French scholar Hippolyte Taine's 1870 *De L'Intelligence:* "Ainsi notre perception extérieure est un rêve du dedans qui se trouve en harmonie avec les choses du dehors; et, au lieu de dire que l'hallucination est une perception extérieure fausse, il faut dire que la perception extérieure est une *hallucination vraie*" [Thus, our external perception is a dream, from within, which is in harmony with things outside, and instead of saying that the hallucination is a false external perception, we must say that the external perception is a *true hallucination*] (1870, p. 411; emphasis in original). For critical discussion of predictive processing and its appeal to the notion of controlled hallucination, see Tiehen (2022).

8. For my arguments about evidential seclusion in delusions, see Hohwy and Rajan (2011) and Hohwy and Rosenberg (2005). For a synthesis in the understanding of hallucinations, see Collerton et al. (2023). For a discussion of the computational difficulties in mentalizing, see Hohwy and Palmer (2014); for a compelling computational take on the complexities of mentalizing inference, see Friston and Frith (2015).

9. For a discussion of false inference, especially in neurological conditions, see research led by Thomas Parr (e.g., Parr, Benrimoh, et al., 2018). For an account of false inference in interoceptive psychopathology, see Paulus et al. (2019).

10. See Hohwy (2013) for several examples of illusions explained in terms of predictive coding and active inference, such as the Müller-Lyer, ventriloquist, rubber hand, and full-body illusions.

11. It therefore seems to me that it is not correct to object to predictive processing by arguing that it necessarily portrays people with autism or schizophrenia as irrational, as in Elkin and Wiśniowska (2022).

12. For the complete class theorem, see Wald (1947), and for discussions in the context of the free energy principle, see Parr et al. (2022) and Parr, Rees, et al. (2018).

13. For discussions and examples of process theories, see Friston, FitzGerald, et al. (2017) and Parr and Friston (2018).

14. The central proposal for predictive processing with respect to canonical microcircuits is found in Bastos et al. (2012), see also chapter 5 of Parr et al. (2022). For details about process theories and neuronal anatomy, see Parr and Friston (2018). For examples of modeling work comparing synthetic data to real data under the process theory, see Parr and Friston (2017a), Parr et al. (2019), Whyte and Smith (2021), and Whyte, Hohwy, et al. (2022). The role for acetylcholine in precision weighting in inference is supported by Lavine et al. (1997), Moran et al. (2013), and Zarkali et al. (2019). The role of noradrenaline in volatility and forgetting rates, as well as sensorimotor actions such as eye movement, is investigated in Jordan and Keller (2023), Parr and Friston (2017b), Sales et al. (2019), and Vincent et al. (2019). The link between volatility and learning rates is set out in Behrens et al. (2007).

15. For a review of empirical work on active inference (and predictive coding), see Hodson et al. (2024), which discusses several pioneering studies of active inference from Ryan Smith's group. For Beth Fisher's work on the optimism bias, see Fisher, Smith, et al. (2024) and Fisher, Whyte, et al. (2024); for a theoretical discussion of the relation between self-evidencing and optimism, see Fisher and Hohwy (2024). For a treatment of the distinction between reinforcement learning and active inference as models of behavior, see Sajid, Ball, et al. (2021).

16. Textbooks in the philosophy of science set out the arguments about Popper and falsificationism in detail; for example, see Bird (2004). One attractive aspect of a Bayesian model comparison scheme is that it just quantifies evidence accumulation—or the probability of hypotheses given observations, which allows scientists to place their bets. It is not concerned with the futile endeavor of getting advocates of scientific theories to give up their theories, which also depends on individual biases and personalities. For our account of this process, as applied to adversarial collaboration in consciousness science and itself appealing to free energy minimization, see Corcoran, Hohwy, et al. (2023).

17. Whereas the philosophy of science and developments in Bayesian analysis offer a constructive answer to worries about process theories and just-so elements of self-evidencing, there is still work to be done on the relationship between principles and process theories. This is a generic task for the philosophy of science, with regard to other principles, such as the principle of least action; see Hohwy (2021b) and references therein for pointers about this debate.

18. There is substantial discussion about this issue of realism, both for free energy and self-evidencing and in the philosophy of science in general. For the self-evidencing debates, see Anderson (2017), Andrews (2021), Bruineberg et al. (2022), Di Paolo et al. (2022), Fabry (2017b), Ramstead, Sakthivadivel, et al. (2023), van Es (2020b), and Williams (2022). For a philosophical account, see Kirchhoff, 2025.

19. For mechanisms in neuroscience in general, see Bechtel (2007) and Craver (2007); for a comprehensive discussion of computational mechanisms, see Piccinini (2020). For a direct, mechanistic treatment of self-evidencing and the free energy principle, see Ramstead, Sakthivadivel, et al. (2023). I develop the map-terrain argument further with my colleague Alex Kiefer (Kiefer & Hohwy, 2022); see also Kirchhoff et al. (2022). As always, there is room for debate about these matters. For example, Dan Williams argues that the descriptive and explanatory aspirations of the free energy principle cannot hold together, and in particular that the process theories do not do the explanatory job that they are supposed to under the principle (Williams, 2022); I would argue that the way that I set out the free energy principle as a method for understanding agents, and as an assurance of self-supervision, preempts these concerns.

20. This idea of representation has been part of the predictive processing framework from early on; see, for example, Friston and Stephan (2007). I tried to convey this dynamic coupling process in a lengthy analogy about plugging holes in a leaky dam in Hohwy (2013, chapter 3); for further discussion, see Kiefer and Hohwy (2018). There is some technical debate about how to carve out the components discussed here, the generative and recognition models (or densities); see Ramstead, Kirchhoff, et al. (2019). I am working on the assumption here that the generative model generates the agent's expected observations, given the states that the agent has accrued evidence for about the world and their own actions, and this model is inverted into the recognition model pursuant on active inference. However, it is possible to take different stances on these issues, which also raise fundamental debates about semantics and representation. For discussions of broader topics around Markov blankets, see Ciaunica, Levin, et al. (2023), Facchin (2021), Kirchhoff and Kiverstein (2021), Kirchhoff et al. (2018), Palacios et al. (2020), Ramstead et al. (2021), and van Es (2020b).

21. I discussed structural representation in terms of statistical "network analysis" in Hohwy (2013, chapter 8); it was Paweł Gładziejewski who first developed structural representation for predictive processing (Gładziejewski, 2016; Gładziejewski & Miłkowski, 2017).

22. There is a large body of literature on this topic, with a landmark being Hurley (1998).

23. I set out this caveat to skepticism in more detail in Hohwy (2017a); for a critical discussion, see Clark (2017a); see also Beni (2019).

Chapter 6

1. For mental effort and habits, see Parr, Holmes, et al. (2023). For optimism and marital prospects, the optimism bias plays a role; for a discussion of this topic, see Sharot (2011). In addition, we discussed optimism bias and active inference in chapter 5.

2. The account of intentional action employs what is called *backward induction*. It is based on a proposal from Friston, Salvatori, et al. (2025); in my retelling, I emphasize how beliefs about latent states serve as constraints on the abductive process.

3. So far as I know, the phrase "inference to the best decision" has been used only rarely in past literature, and not in the sense intended here. Patricia Churchland (2009) appeals to the centrality of inference to the best explanation in cognitive science, considered as case-based reasoning, and then introduces a notion of inference to the best decision to apply case-based reasoning to the moral domain. I think parts of Churchland's account align with the notions that I will discuss, as well as with active inference, when broadened to decision-making in general. A key difference is that there is no normative element in Churchland's account, in terms of either an objective function or approximation to optimality. This difference comes about because the notion of inference to the best decision that I develop here begins with self-evidencing and free energy minimization, whereas Churchland's is a descriptive, naturalistic account. Closer to the self-evidencing theme, Anil Seth (2015) extends the appeal to inference to the best explanation (or abduction) to control theory, with the label "inference to the best prediction," explaining that "active inference in the cybernetic Bayesian brain becomes a process of 'inference to the best prediction,' where the best predictions are those that enable control and homeostasis under a broad repertoire of perturbations. It will be interesting to fully develop criteria for best-making in this control-oriented form of abductive inference" (Seth, 2015, p. 3). Key to Seth's concerns is the idea that sometimes we act for epistemic value in a way that gives information gain by falsifying or disconfirming beliefs. I agree with this idea of acting to disconfirm (or to provide evidence for the negation of) a belief. In this chapter, I frame the abductive perspective on active inference in terms of inference to the best decision, where it is not the prediction as such, but simply the policy (whether for utility or epistemic value) that is the target. I also set out the best-makers for this kind of inference.

4. The literature on decision-making is enormous. Utility-maximization approaches have their origin in the utilitarianism of Jeremy Bentham and John Stuart Mill. There are excellent treatments of the field, such as Newell et al. (2015). Philosophy of science approaches that seem more descriptively adequate were pioneered by Feyerabend

(1975), Kuhn (1962), and Lakatos (1976); and adequate treatments of rationality and decision-making include Gigerenzer (2000), Gigerenzer and Selten (2001), Kahneman (2011) and Kahneman et al. (1982, 2002).

5. For Perrykkad's proposal about fidgeting, see Perrykkad and Hohwy (2020a).

6. We focus mainly on human agents here, but it is worth noting that deliberation is seen in nonhuman animals too, such as rats when they pause to deliberate as they navigate through mazes, something that is described in terms of hippocampal function and its connectivity with the frontal cortex. There is impressive neuroscience in this area (e.g., Redish, 2016). Chapter 8 will consider how prior beliefs about policies might propagate in social networks.

7. The classic sources for counterfactuals are in philosophical semantics (Lewis, 1973).

8. For the inaugural account of sophisticated inference, see Friston, Da Costa, et al. (2021).

9. There is a large body of literature on affective forecasting (e.g., Wilson & Gilbert, 2005).

10. Although exploration in *E. coli.* is model free, it is likely goal-directed, and *E. coli.* behavior can be modeled with active inference in that sense (Tschantz et al., 2020). There is a vast body of literature on the exploit-explore distinction. A useful review is Hills et al. (2015).

11. For this unpacking of expected free energy, see chapter 2 of Parr et al. (2022).

12. WHO health emergencies program executive director Dr. Mike Ryan, WHO press conference, March 16, 2020.

13. For a study on exponential bias, see Lammers et al. (2020). Our game theoretical study of lockdowns is reported in van Baal et al. (2022). There are echoes of semantic discussions of "backtracking" counterfactuals in the case of vaccine hesitancy (cf. Jackson, 1977).

14. For heuristics and biases, see major works like Gigerenzer (2000), Gigerenzer and Selten (2001), and Kahneman et al. (1982). This description also aligns with the discussion from Patricia Churchland mentioned previously (Churchland, 2009).

Chapter 7

1. The quote from Hume is given in the majority of discussions of the self; see section B1.4.6 of Hume (1739–1740). For the conceptions of the self, see Perrykkad and Hohwy (2020b, table 1).

2. For the Perrykkad studies on the sense of self, and more information about these various dimensions of the self, see Perrykkad and Hohwy (2022) and Perrykkad et al.

(2023). There are several papers on self-prioritization by Woźniak (Woźniak & Hohwy, 2020; Woźniak et al., 2021); for related papers, see Ciaunica, McEllin, et al. (2022) and Wozniak et al. (2023).

3. For philosophical and cognitive science analysis of body size illusions and eating disorders, see papers by Stephen Gadsby (e.g., Gadsby, 2017, 2021). The study finding postperceptual effects in body size estimation is reported in Engel et al. (2022).

4. There is a wide body of literature on self-models and predictive processing (Apps & Tsakiris, 2014; Ciaunica & Fotopoulou, 2017; Gerrans, 2015a, 2015b; Hohwy & Michael, 2017; Letheby & Gerrans, 2017; Gerrans, 2019, 2024; Kiverstein, 2020; Limanowski & Blankenburg, 2013; Lin, 2015; Moutoussis et al., 2014; Perrykkad & Hohwy, 2020b; Schlicht, 2018; Schlicht & Venter, 2019; Wiese, 2019; Woźniak, 2018). For examples of work on interoception, see Allen (2020), Gerrans (2019), Kleckner et al. (2017), Quigley et al. (2021), and Seth (2021). For further predictive processing discussions of the self, see Gerrans (2019; 2024), Ciaunica, Seth, et al. (2022), Ciaunica, Constant, et al. (2021), and Kiverstein (2020). For my own earlier statement, see Hohwy (2007), and Hohwy (2013, chapter 12) and the account in Hohwy and Michael (2017), critically discussed by Letheby and Gerrans (2017).

5. There are extensive, empirically well-informed and conceptually sophisticated discussions of the self that lean away from some notions of realism. For a classic argument, see Metzinger (2004), and for a recent inferential account, see Seth (2021). I believe that the simple realism that I advocate here is consistent with these types of accounts since it just says that the self is whatever causes lie behind inference of self-related causes. There is also a rich body of literature on the social and developmental origins of the sense of self, which moves the notion of the self away from some types of realism and toward more relational, social-embedded accounts of selfhood, see Ciaunica and Fotopoulou (2017), Ciaunica, Constant, et al. (2021), and Corcoran, Perrykkad, et al. (2023); again, the realism that I argue for does not exclude relational, social, and developmental embeddedness, just as being a realist about the financial system or about universities allows that they are things that are shaped socially and dynamically.

6. More technically, the structural representations of the self-evidencing agent's generative model are cashed out in terms of causal role semantics (Adams & Aizawa, 2021). This is a broadly functionalist semantics for concepts where a functional role is first specified in structural, causal terms and a set of structured causes playing that role delivers the content of the concept; for a discussion of this topic, see Kiefer and Hohwy (2018).

7. For the idea of nested blankets, see Hohwy (2016), and for more comprehensive treatments, see Ciaunica, Levin, et al. (2023) and Kirchhoff and Kiverstein (2021).

8. I am here giving a philosophical account of what preference learning would look like for a self-evidencing agent. In much modeling work using active inference—such

as partially observable Markov decision process (POMDP) models—preferences are assumed to be known, as the focus is on the behavior of the system given some preferences. However, the empirical priors of a generative model are by definition learnable because they depend on empirical data, and one can then argue that prior preferences are conditioned upon latent states, which necessarily have empirical priors, which in turn means that prior preferences are learnable; for technical development of preference learning, which would be added to the update equations for POMDPs, see works led by Noor Sajid (e.g., Sajid, Ball, et al., 2021; Sajid, da Costa, et al., 2022; Sajid, Tigas, et al., 2021, 2022).

9. For Perrykkad's squares task studies, see Perrykkad, Robinson, et al. (2022), and Perrykkad et al. (2021).

10. For some of the neuroscience of networks involved in self-related processing, see Yeshurun et al. (2021), Koban et al. (2021), and Menon (2023). For a discussion of the way that people can "lose themselves" in depersonalization and meditation, see Deane et al. (2020), as well as Ciaunica (2024).

11. There might be some thorny cases challenging mineness, such as thought insertion in some types of psychosis, where the patient will have a thought, know that they intended to have it, know that they have it in their mind but believe that the thought is coming from someone else; the patient in one way senses the thought to be their own, but in another way senses it to be not their own. There is a rich body of literature on delusion formation in philosophy and neuropsychiatry, where predictive processing–type accounts began with Chris Frith (1992); see also Bortolotti and Miyazono (2015), Corlett (2019), Davies and Coltheart (2000), Hohwy (2004), Hohwy and Rosenberg (2005), Parrott (2021), Tappin and Gadsby (2019), and Williams, (2018).

12. A classic source for the distinction between minimal and narrative self is Shaun Gallagher (2000). For Dennett's take on the narrative self, see Dennett (1992); for discussions of narrativity and self, see Schechtman (2011, 2018). For antinarrativity, see Strawson (2004). For a more recent, flexible, and open-ended notion of self—the pattern theory of self—with connections to contemplative traditions, see Gallagher (2013) and Gallagher et al. (2023).

13. This model architecture is inspired by the sparse coupling of genotypes and phenotypes, proposed by Friston and colleagues; see Friston, Friedman, et al. (2023). Perhaps the notion of sparsely coupled blankets, for the self-model and world-model, could engage with debates about what is called the "deep self" (which often relate to moral responsibility); see Wolf (1993) and the discussion by Sripada (2016).

14. There is an interesting project here in discussing whether, and to what degree, this notion of a real self and the accompanying model architecture is in line with nondual and no-self discussions in contemplative studies and Indian philosophy; see, for example, Chadha (2023) and Josipovic (2010, 2021).

15. There is some debate about this distinction between having and being models in the predictive processing debate, but it is unresolved, see Seth and Tsakiris (2018, box 4), as well as Andrews (2021), Ramstead, Sakthivadivel, et al. (2023), and, for an instrumentalist position, van Es (2020a). It seems safe to say that the agent embodies a recognition model, where the causal dependencies of the internal states carry information about external states (the recognition model is the inversion of the generative model); which model the agent embodies, or in what sense it embodies a generative model, is a matter of empirical investigation, employing process theories and fitting data. Notice that in some robotics literature, "being a model" refers to the ability of a system, such as a biological organism or a robot, to respond to sensory contingencies without an explicit model (e.g., passive walkers); if such systems are actively self-evidencing, then it may be best to consider the entire system a model in the sense discussed here.

Chapter 8

1. I think that even in sleep, the agent is poised for action—namely, covert mental action associated with complexity reduction, and thus structure learning that will improve function for the next day. For this proposal and discussions of predictive processing and sleep and dreaming, see Bucci and Grasso (2017), Hobson and Friston (2012), Windt (2018), Windt et al., (2014).

2. I think this conceptual link between volition and counterfactuals relates to how robustly controlled a given action is; this theme is set out in the context of agents as dynamical systems in a compelling series of papers by Marius Usher and colleagues (Grinfeld et al., 2020; Usher, 2006, 2020).

3. The comments on deliberation here reflect some of the pitfalls when thinking about and doing research on impulsivity and self-control; for discussions of this topic, see work led by my past student Simon van Baal (e.g., van Baal, Walasek, Verdejo-Garcia, et al., 2025).

4. There are also various interesting discussions of addiction, as paradigms of involuntariness, in the predictive processing literature (e.g., Clark, 2018; Gerrans, 2022a; Levy, 2014; Miller et al., 2020; Schwartenbeck, FitzGerald, Mathys, Dolan, Wurst, et al., 2015).

5. For an overview of compatibilism and its different versions, see McKenna and Coates (2021).

6. This motif, where depriving the agent of a highly informative input places an extra inferential burden on the agent, which paradoxically adds capability, is consistent with the distinction developed by Friston and colleagues between "active" (or "conservative") things and "strange" things; see Friston, Da Costa, Sakthivadivel, et al. (2023) and our commentary on this paper (Hohwy & Sandved-Smith, 2024).

7. I think that a promising place to begin connecting literatures is Usher's work on dynamical systems, control, compatibilism, and responsibility (Grinfeld et al., 2020; Usher, 2006, 2020).

8. For the classic sources for this type of theory of emotions, see works by James and Lange (Lange, 1885; James, 1884). For types of constructionist theories, see Barrett (2017a), LeDoux (2020), Russell (2003), and Scherer (1999); for a wider set of approaches, see my overview (Hohwy, 2020a).

9. For my early ideas, see Hohwy (2011) and Hohwy (2013, chapter 12). There are several other accounts appealing to predictive processing (e.g., Barrett, 2017b; Hesp et al., 2021; Joffily & Coricelli, 2013; Van de Cruys, 2017; Wilkinson et al., 2019). A substantial proposal here is Lisa Feldman Barrett's theory of constructed emotion, labeled as active inference in Barrett (2017b). Barrett's account is not using active inference in the technical sense of planning and decision-making by selecting a policy against the objective function of expected free energy, which I am using in this chapter. Rather, Barrett is appealing to a predictive coding scheme that is geared to achieve the goal of maintaining allostasis in a forward-looking way; here, perceptual inference guides subsequent action and at the same time, through the embodied involvement of interoceptive signals, anticipates actions to deal with upcoming input. I think that there is an interesting project exploring the alignment between Barrett's theory and active inference in the technical sense (e.g., by writing the generative model in an emotion task and fitting data). For a discussion of Barrett's allostatic approach and of allostasis in active inference, see Corcoran and Hohwy (2018). Notice also that Barrett makes several important theoretical distinctions within the class of constructionist and appraisal theories.

10. Indeed, caffeine overuse and withdrawal are discussed in diagnostic manuals such as the *Diagnostic and Statistical Manual of Mental Disorders* and the *International Classification of Diseases.*

11. The theories of basic emotions include works such as Ekman (1992), Merker (2007), Panksepp (2004), and Solms (2021). See also, for example, work pointing to fast involvement of cortical areas, rather than just the midbrain (Li & Keil, 2023).

12. For this kind of idea applied to our engagement with horror fiction, see work by my colleague Mark Miller and collaborators (e.g., Miller et al., 2023).

13. Casting emotions in terms of the rate of uncertainty minimization is not uncommon in the predictive processing field. The first account in this vein is Joffily and Coricelli (2013), and several distinct developments can be found in a number of other studies (Allen et al., 2022; Clark et al., 2018; Hesp et al., 2021; Linson et al., 2020; Seth & Friston, 2016; Smith, Lane, et al., 2019; Smith, Parr, et al., 2019). Casper Hesp et al. (2021) in particular have used active inference to cast affect as "emotional valence [that] can be formalized as a state of self that is inferred on the basis of fluctuations in the estimated confidence (or precision) an agent has in her generative model of the

world that informs her decisions," providing a conceptual link between affect in terms of ambiguity and high- and low-arousal states (p. 401). This kind of work connects arousal to affect and mood, understood as fairly global states reflecting the model's overall prospects for future self-evidencing. This ties in well with the fine-graining that happens when specific emotions arise—namely, when active inference begins to specify particular policies guided by contextual cues.

14. A nice example of this situation, using active inference as the modeling tool, is the discussion of spider phobia in Smith et al. (2021).

15. There is a rich body of literature on biology and prosociality (e.g., Sober & Wilson, 1998). For work on evolution and the free energy principle, see Friston, Friedman, et al. (2023).

16. I do think that there are interesting avenues of research relating self-evidencing to debates on deliberative democracy (e.g., Dryzek et al., 2019; Landemore, 2012). A contemporary philosophical debate relating to participatory discourse and challenges to current beliefs focuses on conceptual engineering; for a predictive processing account, see Löhr and Michel (2023).

17. Longtermism is defended in MacAskill (2022). There is a burgeoning body of literature on the types of considerations discussed here, including from within the longtermism approach (e.g., Cremer & Kemp, 2021; Tarsney, 2023).

18. Jonathan Dancy is a main proponent of particularism (see Dancy, 1983).

19. In modern debates, the notion of reflective equilibrium was introduced by John Rawls (1971) and much discussed since; for a contemporary, systematic account, see Schmidt (2024).

Chapter 9

1. Sometimes the global state is called "creature consciousness," and sometimes this state is being referred to as being in the "awake state." Some of this terminology is not very useful; for example, someone who is dreaming seems to be in a global conscious state but is not awake. For discussions, see Mckilliam (2020), Bayne and Hohwy (2013), and Bayne, Hohwy, et al. (2016). There is discussion about exactly what background states are, and how they differ from each other; they are sometimes labeled differently, such as modes of consciousness (Bayne & Hohwy, 2016; Chalmers, 2000). It is somewhat unclear whether normal, everyday consciousness—someone just going about their business without any particular characteristics—deserves to have its own background states designated.

2. There is much discussion about what kinds of content can enter consciousness; for example, can we be conscious of causal relations, can we somehow experience the presence of unseen parts of objects (occlusions, the back of things), and can we

experience other people's mental states? There is also discussion and research on the possibility and nature of unconscious perception of content; and there is debate about the possibility of bare, contentless, conscious states, such as, perhaps, some types of meditation; see the extensive empirical and philosophical work of Thomas Metzinger (e.g., Metzinger, 2024).

3. The subjective character of consciousness also suggests that consciousness science differs from other established and fledgling scientific areas in at least one central respect. Its domain of research is subjective experience that, it seems, is truly accessible only to the one person having that experience. The experience, and its subjective phenomenology, must be inferred by the experimenter from verbal reports (e.g., "I see something red," "I am awake but feeling dizzy," "I feel like I had control over my movements"), or from other kinds of behavioral reports (e.g., button presses in response to seeing visual targets on a computer screen, or wincing in response to pain). The reliance on verbal reporting is sometimes taken to indicate that consciousness science is unscientific. Some researchers advocate relying only on actual behavior—a remnant of behaviorism specifically aimed at consciousness. I think that consciousness science has made significant progress in how to deal with this problem methodologically. With sufficient care, and for most subjective experiences, we can in fact trust verbal and behavioral reports. For an early argument in favour of trusting the subject's reports, see my paper with Chris Frith (Hohwy & Frith, 2004), which, incidentally, also invokes self-evidencing. Much of the debate in the science of consciousness about self in consciousness originates in Metzinger (2004).

4. For discussions of metacognition and consciousness with a computational perspective, see Fleming (2021) and Lau (2008); for a discussion of these aspects within the predictive processing framework, see Hohwy (2015).

5. For articulations of the neural correlates of consciousness, see Chalmers (2000), Hohwy (2009), Hohwy and Bayne (2015), and Klein et al. (2020); for early advocates of the approach, see Baars (1988) and Crick and Koch (1990). One should not get distracted by the term "correlate" which is not related to the dictum that "correlation is not causation"; it just speaks to cooccurrence that is not causal.

6. For an overview of these types of theories and a discussion of the theories of consciousness landscape, see Seth and Bayne (2022).

7. Seth and I have discussed predictive processing, theories and neural correlates (Hohwy & Seth, 2020; Seth & Hohwy, 2021).

8. For the canonical account of metaphysics in these terms, see Frank Jackson's classic book (Jackson, 1998).

9. The metaphysics here does not concern probability because of course it is highly probable that my physical twin is as conscious as I am. The metaphysical question has to do with whether the physical facts fix the truths about consciousness, or whether fixing those truths might require something above and beyond the physical

facts. The literature on the mind-body problem is vast, stretching back through the history of philosophy at least to René Descartes and Elizabeth of Bohemia. Contemporary discussions were kickstarted by Nagel's famous paper on what it is like to be a bat (Nagel, 1974).

10. For discussions of the problem of mental causation and a state-of-the-art discussion of philosophy of mind, see Bayne (2021) and Kim (1998).

11. Anil Seth (2016) argues that the hard problem of consciousness will melt away once we address these "real" problems of consciousness. There are various predictive processing approaches to consciousness with metaphysical aspirations. One approach is found in works by Mark Solms (e.g., Solms, 2019; Solms, 2021; Solms & Friston, 2018); here there is a direct claim that consciousness is identical to precision optimization. I think that it is correct that precision optimization is important for consciousness because it is important to all aspects of self-evidencing; but it cannot be a straightforward identity since it is easy to conceive something that optimizes beliefs about precisions but is not conscious. This identity thesis is in turn based on a proposal that affect is critical for consciousness, where affect is what is tied to precision optimization, such that conscious experience acquires affective salience when uncertainty from residual prediction error is felt. I agree that affect is important to consciousness, but, again, I do not think that affect in this computational sense entails phenomenality. Hence, I presented precision control in chapter 4 as a standard component of active inference long before considering consciousness, and we discussed affect in chapter 8. Another kind of approach is found in work by Andy Clark and his colleagues, who provide a solution to what is known as the "meta-problem of consciousness" (about why we think that there is a hard mind-body problem of consciousness) in terms of the way that our generative model posits qualia (conscious states) as simplifications of underlying complex patterns of responses to input. This is an interesting line of argument, but I do not believe that it is efficient; the problem that I see is that the line of argument overgeneralizes, predicting that we would see hard problems arise in all areas where generative models simplify and make abstractions, idealizations, or dimensionality reduction—but we do not in fact see such proliferation of hard problems; for this work, see Clark (2019, 2023) and Clark et al. (2019). Another approach is an intriguing argument from Alex Kiefer, who proposes a psychophysical identity theory for mind and brain, based on the way that the thermodynamic free energy of a systems encodes its variational free energy; I find Kiefer's metaphysical direction compelling, although his proposal does not yet speak directly to consciousness (Kiefer, 2020).

12. The best place to start on the motivations for panpsychism, in my view, is Nagel (1979); there is much literature that has been released since this discussion.

13. For a discussion of panpsychism in integrated information theory, see Tononi and Koch (2015). For a philosophical analysis of the theory, see work by my former student Niccolo Negro (2022).

14. For some discussion of implications along these broad-scope lines, see Friston et al. (2020).

15. See Friston (2018, p. 5).

16. The reference to miracles alludes to David Lewis's semantics for counterfactuals (Lewis, 1973).

17. There is some philosophical discussion and empirical research on the relations between free will and consciousness (e.g., Shepherd, 2012).

18. This model of rivalry appears in Hohwy et al. (2008); it was also discussed by Andy Clark in his influential clarion call for predictive processing (Clark, 2013b; see also Clark, 2016; Hohwy, 2013).

19. The interaction between conscious content and conscious states is an important area of development (e.g., Aru et al., 2012; Hohwy, 2009). I discuss several sources of evidence (see Hohwy, 2013, and elsewhere); there is compelling evidence to be found in work on other bistable stimuli, such as the bistable dancer studied by some of my colleagues (Johnston et al., 2017; Robinson et al., 2020). Impressive work has been done by several groups worldwide on the predictive coding elements of perceptual brain functions, see studies from the labs of Lucia Melloni, Peter Kok, Floris de Lange, Katharina Schmack, Phillip Sterzer, Phil Corlett, Lars Muckli, and Petra Vetter, among others. For a review of some of this literature, see Hodson et al. (2024). The research keeps evolving our understanding of how predictive coding might be implemented in the brain; for example, recent work by Andre Bastos suggests a notion of predictive routing (Bastos et al., 2020).

20. Testing this hypothesis was beyond us for some time, but Philipp Sterzer's group conducted an elegant experiment, led by Veith Weilnhammer, using computational models and functional magnetic resonance imaging (fMRI) to show that there is a signature of prediction error behaving in just this way for rivalry-like lissajous curves (Weilnhammer et al., 2017, 2018).

21. There are many excellent demonstrations of these various illusions (rivalry, Troxler fading, and others) available on the internet. There are many studies of rivalry exploring the effects of changing the stimulus properties. For the role of attention, see Zhang et al. (2011); for the effects of rewarding one percept, see Marx and Einhäuser (2015). A version of a rivalry-like bistable effect is continuous flash suppression where a dynamically moving stimulus in one eye can suppress a static stimulus in the other eye for extended periods of time (Tsuchiya & Koch, 2004). There is a large body of literature on multiple aspects of binocular rivalry, such as individual differences. One of the several excellent reviews is a good starting point; Randolph Blake is a leading authority (e.g., Blake, 2022).

22. The active inference model of Troxler fading and binocular rivalry is presented in Parr et al. (2019).

23. Some aspects of rivalry seem to persist in unconscious processing; see ingenious studies such as Xu et al. (2016) and Zou et al. (2016). This is part of the reason why active inference is not sufficient for consciousness in and of itself.

24. For work on bistability in animals and insects, see Toepfer et al. (2018) and Vetter et al. (2000). There is further development of the active inference models for bistability, such as in work led by Filip Novicky (e.g., Novicky et al., 2022).

25. Analogues of "To see is to look" can be framed for other sensory modalities: "To hear is to listen," "To feel is to touch," and so on; see Parr et al. (2022, p. vii). I have developed this kind of view in some detail in a paper that looks at several distinct properties of active inference, from conscious content as discussed here all the way to the sense of being (Hohwy, 2021a). See also our position paper, led by Christopher Whyte, that brings together all the modeling evidence for this kind of view (Whyte et al., 2024).

26. For the work by Marchi and myself, see Marchi and Hohwy (2020). Because the active sampling of predicted observations is done through precision control, and precision control is the substrate for attention (as discussed in chapter 4), this proposal is a good companion for theories of consciousness emphasizing intermediate-level attention. This kind of view is also developed by Kathryn Nave and her colleagues, although they frame it less in inferential and representionalist terms and more in terms of embodied and enactive cognition approaches (Nave et al., 2022).

27. For an example of this approach, consider that covert attention should be able to begin to generate conscious content about a visual target very quickly after the eyes move to it. This leads to the empirical prediction that entry into consciousness happens faster when much (low-ambiguity) active inference is involved compared to when less active inference is involved; or more simply, conscious content should be facilitated in more active rather than more passive conditions. Currently, my lab is involved in a large consortium project testing these ideas empirically.

28. For Andy Clark's articulation of these ideas, see Clark (2013a, pp. 470–471). For statements and framework along these lines, see earlier work (e.g., Clark, 2013b, 2016; Hohwy, 2013; Hohwy et al., 2008). Back in chapter 4, we encountered the notion of sensory attenuation, where executing an action requires attenuation of the current sensory input. Notice that this is consistent with the proposal here—namely, that conscious perception arises in the enaction of a policy, which is accompanied by attenuation of the current-but-soon-to-be-past experience, allowing the new experience to come in as the policy is enacted.

29. It would be possible to extend this account of online conscious perception to offline conscious experience, such as imagery, episodic memory, dreaming, and mind-wandering. For those offline conscious experiences, the current proposal would be that the precision control shifts to covertly amplify observations harvested from memory, from overtly or covertly sampling from the current sensorium. As discussed

in chapter 4, a central role is theorized for the hippocampus in effectuating these two precision-control roles; this account is worked out for memory (Barron et al., 2020), and I would propose extending this to the use of memory for counterfactual deliberation, as discussed earlier in this chapter. Here, one could also appeal to work done on the sense of presence of things in the sensorium (including other people's mental states; see Seth, 2014; Seth et al., 2012; Suzuki et al., 2023); for our proposal for the perceptual-like presence of mental states in other people's minds, see Palmer et al. (2015). There is also intriguing work on the role of metacognition in the sense of reality in online perception (Dijkstra & Fleming, 2023; Dijkstra et al., 2021). For further, philosophical discussion that forms bridges to topics about the sense of reality, see Gładziejewski (2022) and Teng (2022).

30. For discussions of these intriguing and often-troubling phenomena of the self, see Ciaunica, Seth, et al. (2022), Gerrans (2015b, 2022b); Josipovic (2021), and Letheby and Gerrans (2017).

31. For this proposal about unity of consciousness, see Hohwy (2013, chapter 10).

32. The global neuronal workspace theory is probably the most empirically well-established theory of consciousness; the first version was proposed by Bernie Baars (1997), and for the current version developed by Stan Dehaene, see Mashour et al. (2020) and Dehaene et al. (1998). Lately, Dehaene has suggested that ignition is indeed marked by perceptual inference (Melloni et al., 2023).

33. Research on the active inference account of the global workspace theory consciousness was led by Christopher Whyte (Whyte, 2019; Whyte & Smith, 2021; Whyte et al., 2022). For a review and further development, see (Whyte et al., 2024).

34. There is a sizable amount of literature on consciousness and predictive processing and/or active inference and the free energy principle (e.g., Clark, 2019, 2023; Ciaunica, Safron, et al., 2021; Deane, 2021; Dołęga & Dewhurst, 2020; Friston, 2018; Friston, Breakspear, et al., 2012; Friston et al., 2020; Hohwy, 2021a; Marchi & Hohwy, 2020; Nave et al., 2022; Ramstead, Albarracin, et al., 2023; Ramstead, Wiese, et al., 2023; Sandved-Smith et al., 2021; Seth, 2021; Seth & Tsakiris, 2018; Solms, 2019, 2021; Vilas et al., 2022; Whyte, 2019; Whyte & Smith, 2021; Whyte et al., 2022; Whyte et al., 2024; Wiese & Friston, 2020; Williford et al., 2018). For a review of Bayesian theories of consciousness, see Rorot (2021).

Chapter 10

1. Various good sources on wisdom include Glück and Sternberg (2019), Swartwood and Tiberius (2019), Tiberius (2008), Ryan (1999, 2020), and Zagzebski (1996). The consensus paper is Grossmann et al. (2020); of course, a consensus paper is not guaranteed to create consensus (despite the name), and there is significant and fruitful discussion in the area. For a different take, see Sternberg and Karami (2021). I think

that there is scope for interesting discussions about whether (and if so, how) self-evidencing speaks to different concepts of wisdom.

2. For more on Lars Sandved-Smith's metacognition model, see Sandved-Smith et al. (2021).

3. In fact, older people do appear to have lower learning rates, as Rosalyn Moran and colleagues (2014) have shown. And there is some indication that wisdom actually declines with age, expressed in Mara Mather's (2024, p. 13) autonomic compensation model, where "failing organs, the initial phases of Alzheimer's pathology, and other age-related diseases trigger noradrenergic hyperactivity," which is related to sympathetic activity and feeling more unsafe. To compensate, there is a cascade of neural activity, marked by age-related declines in nerve conduction, and leading the aging brain to a bias for "processing in favor of stimuli that tend to increase parasympathetic activity (e.g., stimuli that increase feelings of safety) and against stimuli that tend to increase sympathetic activity (e.g., threatening stimuli)." (Mather, 2024, p. 13). This could underwrite an appearance that older people are closed-minded and uninterested in new information due to brain malfunction—they are more agents affected by volatility than agents seeking volatility. Mather's conjecture is also a nice illustration of the interaction of interoception and exteroception in self-evidencing agents.

4. There is a strong conceptual link between the Aristotelian notion of the golden mean in behavior and cultivation of dispositions and the free energy principle; my past student Ricardo Perera (2022) sets this out in his PhD thesis.

5. I think that there is an interesting project in comparing these types of Bayesian, self-evidencing moves to epistemic discussions in the wisdom traditions, such as Zhuangzi's approach to humility and openmindedness (e.g., Singh, 2023).

6. For the active inference account of cognitive effort, see Parr, Holmes, et al. (2023).

7. For a multifactorial moral functioning framework that I believe is apt for considering moral aspiration for wisdom, see work led by my PhD student (and a senior Buddhist monk) Kevin Berryman, with Sara Lazar (Berryman et al., 2023).

8. This barely scratches the surface, and there is much debate about prosociality in relation to various instrumentalist imperatives and evolutional pressures (e.g., Sober & Wilson, 1998).

9. For places to start the long journey into discussions of meaning, see Cassedy (2022), Flanagan (2007), Landau (2017), Metz (2013), and Wolf et al. (2010).

10. For the mentioned meaning in life questionnaire, see Steger et al. (2006); other approaches include Costin and Vignoles (2020), and George and Park (2017). For the relation to well-being, there are a number of sources to consult (e.g., Huta & Waterman, 2014). For the philosophical view, see Wolf et al. (2010). There is significant

debate about the extent to which meaning in life is a subjective sense or experience, or mainly reflected in post hoc judgment (e.g., George & Park, 2016).

11. For an influential review of meaning in terms of these three components, see King and Hicks (2021).

12. For these questionnaires on comprehension, see Costin et al. (2020) and George et al. (2017).

13. For discussions of the themes around model maintenance, there are a number of sources (e.g., Chater & Loewenstein, 2016) and discussions of Occam's razor and Bayes, such as Jefferys and Berger (1992); this relates to the accuracy/complexity balance discussed in earlier chapters.

14. For these questionnaires about sense of purpose, see again Costin et al. (2020) and George et al. (2017).

15. For these questionnaires about existential mattering, see again (Costin et al., 2020; George et al., 2017). There is some debate about exactly what this notion means and whether it should be revised (e.g., Martela & Steger, 2023).

16. For examples of questionable meaning, see Wolf et al. (2010).

17. For discussions of these themes, see Dahl and Davidson (2019) and Van Gordon et al. (2022).

18. These mindfulness techniques are based on Buddhist contemplative practices and appear in secular forms in influential works (e.g., Kabat-Zinn, 2015). For our study on mindfulness and bodily illusions, see Guthrie et al. (2022). There are many illustrations of the rubber hand illusion on the internet; for a predictive processing take on several versions of this illusion, see Hohwy (2013).

19. For some of this exciting work, see papers by Ruben Laukkonen and Heleen Slagter, and by Lars Sandved-Smith and collaborators (Laukkonen & Slagter, 2021; Sandved-Smith et al., 2021). There is also work on the predictive processing aspects of meditation, such as whether meditators react less to prediction errors. The findings here are complex but broadly supportive of predictive processing; for research that I have been involved in, see work led by my colleague Neil Bailey (e.g., Bailey et al., 2023).

20. For discussions of neuroexistential angst and the perceived tension between meaning and science, see Caruso and Flanagan (2017) and Frank et al. (2024)

Chapter 11

1. For the notion of mortal computation, see Hinton (2022); for its application with relation to the free energy principle, see Ororbia and Friston (2023); for theorizing with respect to consciousness, see Kleiner (2024).

2. I am using the term "everyday experience" to avoid confusing different, technical uses of "lived experience." That term is used, for example, by Wilhelm Dilthey, and in Husserlian phenomenological traditions, such as in Francisco Varela's neurophenomenology. In these discourses, "lived experience" seems to be sometimes stemming from the German *Erlebnis,* in contradistinction to the mere "experience" of *Erfahrung.* These words can be translated easily into my native language, Danish, but not so easily into English, where I would say that they mean something like "ongoing experience" and "learned experience." Lived experience in these technical senses is sometimes seen to be in opposition to reductionist scientific treatment of experience, or as resulting only from special phenomenological training. But in a broader, everyday sense, lived experience need not be in tension with science, and neither is special training needed to appreciate or communicate lived experience.

3. The concern that cognitivism and representationalism are authoritarian was articulated by Richard Rorty (e.g., Rorty, 1999).

References

Adams, F., & Aizawa, K. (2021). Causal theories of mental content. In E. N. Zalta (Ed.), *The Stanford encyclopedia of philosophy*. https://plato.stanford.edu/archives/fall2021/entries/content-causal/.

Allen, M. (2020). Unravelling the neurobiology of interoceptive inference. *Trends in Cognitive Sciences, 24*, 265–266.

Allen, M., Levy, A., Parr, T., & Friston, K. J. (2022). In the body's eye: The computational anatomy of interoceptive inference. *PLoS Computational Biology, 18*, e1010490.

Anderson, M. L. (2017). Of Bayes and bullets. In T. K. Metzinger & W. Wiese (Eds.), *Philosophy and predictive processing*. MIND Group.

Andrews, M. (2021). The math is not the territory: Navigating the free energy principle. *Biology & Philosophy, 36*, 30.

Apps, M. A. J., & Tsakiris, M. (2014). The free-energy self: A predictive coding account of self-recognition. *Neuroscience & Biobehavioral Reviews, 41*, 85–97.

Aru, J., Bachmann, T., Singer, W., & Melloni, L. (2012). Distilling the neural correlates of consciousness. *Neuroscience & Biobehavioral Reviews, 36*, 737–746.

Ashby, W. R. (1947). Principles of the self-organizing dynamic system. *Journal of General Psychology, 37*, 125–128.

Austin, J. L. (1962). *Sense and sensibilia*. Clarendon Press.

Baars, B. J. (1988). *A cognitive theory of consciousness*. Cambridge University Press.

Baars, B. J. (1997). *In the theater of consciousness: The workspace of the mind*. Oxford University Press.

Bailey, N. W., Geddes, H., Zannettino, I., Humble, G., Payne, J., Baell, O., Emonson, M., Chung, S. W., Hill, A. T., Rogasch, N. C., Hohwy, J., & Fitzgerald, P. B. (2023). Meditators probably show increased behaviour-monitoring related neural activity. *Mindfulness, 14*, 33–49.

Baltieri, M., & Buckley, C. L. (2019). The dark room problem in predictive processing and active inference, a legacy of cognitivism? In *Artificial life conference proceedings* (pp. 40–47). MIT Press.

Barrett, L. F. (2017a). *How emotions are made: The secret life of the brain*. Pan Macmillan.

Barrett, L. F. (2017b). The theory of constructed emotion: An active inference account of interoception and categorization. *Social Cognitive and Affective Neuroscience, 12*, 1–23.

Barron, H. C., Auksztulewicz, R., & Friston, K. (2020). Prediction and memory: A predictive coding account. *Progress in Neurobiology, 192*, 101821.

Bastos, A., Lundqvist, M., Waite, A., Kopell, N., & Miller, E. (2020). Layer and rhythm specificity for predictive routing. *Proceedings of the National Academy of Sciences, 117*, 31459–31469.

Bastos, A., Usrey, W., Adams, R., Mangun, G., Fries, P., & Friston, K. (2012). Canonical microcircuits for predictive coding. *Neuron, 76*, 695–711.

Bayne, T. (2021). *Philosophy of mind: An introduction*. Routledge.

Bayne, T., & Hohwy, J. (2013). Consciousness: Theoretical approaches. In A. E. Cavanna, A. Nani, H. Blumenfeld, & S. Laureys (Eds.), *Neuroimaging of consciousness* (pp. 23–35). Springer.

Bayne, T., & Hohwy, J. (2016). Modes of consciousness. In W. Sinnott-Armstrong (Ed.), *Finding consciousness: The neuroscience, ethics and law of severe brain damage* (pp. 57–82). Oxford University Press.

Bayne, T., Hohwy, J., & Owen, A. M. (2016). Are there levels of consciousness? *Trends in Cognitive Sciences, 20*, 405–13.

Bechtel, W. (2007). *Mental mechanisms: Philosophical perspectives on cognitive neuroscience*. Routledge.

Beck, J., & Ramstead, M. (2025). Dynamic Markov blanket detection for macroscopic physics discovery. *arXiv preprint*. https://arxiv.org/abs/2502.21217

Behrens, T. E. J., Woolrich, M. W., Walton, M. E., & Rushworth, M. F. S. (2007). Learning the value of information in an uncertain world. *Nature Neuroscience, 10*, 1214–1221.

Beni, M. D. (2019). Cognitive Structural Realism: A Radical Solution to the Problem of Scientific Representation. Springer.

Beni, M. D. (2025). Scientific Explanation, Causality, and Agency: A Free Energy Account. Routledge.

Berryman, K., Lazar, S. W., & Hohwy, J. (2023). Do contemplative practices make us more moral? *Trends in Cognitive Sciences, 27*, 916–931.

Bickhard, M. (2009). The interactivist model. *Synthese, 166*, 547–591.

Bickhard, M. (2016). The anticipatory brain: Two approaches. In V. C. Müller (Ed.), *Fundamental issues of artificial intelligence* (pp. 261–283). Springer International Publishing.

Biderman, N., Bakkour, A., & Shohamy, D. (2020). What are memories for? The hippocampus bridges past experience with future decisions. *Trends in Cognitive Sciences, 24*, 542–556.

Biderman, N., Gershman, S. J., & Shohamy, D. (2023). The role of memory in counterfactual valuation. *Journal of Experimental Psychology: General, 156*, 1754–1767.

Bird, A. (2004). *Philosophy of science*. Routledge.

Bishop, C. M. (2007). *Pattern recognition and machine learning*. Springer.

Blake, R. (2022). The perceptual magic of binocular rivalry. *Current Directions in Psychological Science*. https://doi.org/09637214211057564.

Bogacz, R. (2017). A tutorial on the free-energy framework for modelling perception and learning, *Journal of Mathematical Psychology, 76*(Part B), 198–211.

Bongard, J., & Levin, M. (2021). Living things are not (20th century) machines: Updating mechanism metaphors in light of the modern science of machine behavior. *Frontiers in Ecology and Evolution, 9*, 650726.

Bortolotti, L., & Miyazono, K. (2015). Recent work on the nature and development of delusions. *Philosophy Compass, 10*, 636–645.

Bruineberg, J., Dołęga, K., Dewhurst, J., & Baltieri, M. (2022). The emperor's new Markov blankets. *Behavioral and Brain Sciences, 45*, e183.

Bruineberg, J., Rietveld, E., Parr, T., van Maanen, L., & Friston, K. J. (2018). Free-energy minimization in joint agent-environment systems: A niche construction perspective. *Journal of Theoretical Biology, 455*, 161–178.

Bucci, A., & Grasso, M. (2017). Sleep and dreaming in the predictive processing framework. In T. K. Metzinger & W. Wiese (Eds.), *Philosophy and predictive processing*. MIND Group.

Buckley, C. L., Kim, C. S., McGregor, S., & Seth, A. K. (2017). The free energy principle for action and perception: A mathematical review. *Journal of Mathematical Psychology, 81*, 55–79.

Caruso, G., & Flanagan, O. (2017). *Neuroexistentialism: Meaning, morals, and purpose in the age of neuroscience*. Oxford University Press.

Cassedy, S. (2022). *What do we mean when we talk about meaning?* Oxford University Press.

Chadha, M. (2023). *Selfless minds: A contemporary perspective on Vasubandhu's metaphysics*. Oxford University Press.

Chalmers, D. (2000). What is a neural correlate of consciousness? In T. Metzinger (Ed.), *Neural correlates of consciousness: Empirical and conceptual issues* (pp. 17–40). MIT Press.

Chater, N., & Loewenstein, G. (2016). The under-appreciated drive for sense-making, *Journal of Economic Behavior & Organization, 126*, 137–154.

Churchland, P. (2009). Inference to the best decision. In J. Bickle (Ed.), *Oxford handbook of philosophy and neuroscience* (pp. 419–430). Oxford University Press.

Ciaunica, A. (2024). Selfless minds, unlimited bodies?: Homeostatic bodily self-regulation in meditative experiences. *Journal of Consciousness Studies, 31*, 104–126.

Ciaunica, A., Constant, A., Preissl, H., & Fotopoulou, K. (2021). The first prior: From co-embodiment to co-homeostasis in early life. *Consciousness and Cognition, 91*, 103117.

Ciaunica, A., & Fotopoulou, A. (2017). The touched self: Psychological and philosophical perspectives on proximal intersubjectivity and the self. In C. Durt, C. Tewes & T. Fuchs (Eds.), *Embodiment, enaction, and culture: Investigating the constitution of the shared world* (pp. 173–192). MIT Press.

Ciaunica, A., Levin, M., Rosas, F., & Friston, K. (2023). Nested selves: Self-organisation and shared Markov blankets in prenatal development in humans. *Topics in Cognitive Science*. https://doi.org/10.1111/tops.12717.

Ciaunica, A., McEllin, L., Kiverstein, J., Gallese, V., Hohwy, J., & Woźniak, M. (2022). Zoomed out: Digital media use and depersonalization experiences during the COVID-19 lockdown. *Scientific Reports, 12*, 3888.

Ciaunica, A., Safron, A., & Delafield-Butt, J. (2021). Back to square one: The bodily roots of conscious experiences in early life. *Neuroscience of Consciousness, 2021*, niab037.

Ciaunica, A., Seth, A., Limanowski, J., Hesp, C., & Friston, K. J. (2022). I overthink—therefore I am not: An active inference account of altered sense of self and agency in depersonalisation disorder. *Consciousness and Cognition, 101*, 103320.

Ciaunica, A., Shmeleva, E. V., & Levin, M. (2023). The brain is not mental! Coupling neuronal and immune cellular processing in human organisms. *Frontiers in Integrative Neuroscience, 17*. https://doi.org/10.3389/fnint.2023.1057622.

Clark, A. (2013a). Expecting the world: Perception, prediction, and the origins of human knowledge. *Journal of Philosophy, 110*, 469–496.

Clark, A. (2013b). Whatever next? Predictive brains, situated agents, and the future of cognitive science. *Behavioral and Brain Sciences, 36*, 181–204.

Clark, A. (2016). *Surfing uncertainty: Prediction, action, and the embodied mind*. Oxford University Press.

Clark, A. (2017a). How to knit your own Markov blanket. In T. K. Metzinger & W. Wiese (Eds.), *Philosophy and predictive processing*. MIND Group.

Clark, A. (2017b). Predictions, precision, and agentive attention. *Consciousness and Cognition, 56*, 115–119.

Clark, A. (2018). A nice surprise? Predictive processing and the active pursuit of novelty. *Phenomenology and the Cognitive Sciences, 17*, 521–534.

Clark, A. (2019). Consciousness as generative entanglement. *Journal of Philosophy, 116*, 645–662.

Clark, A. (2022). Extending the predictive mind. *Australasian Journal of Philosophy, 102*, 119–130.

Clark, A. (2023). *The experience machine: How our minds predict and shape reality*. Pantheon Books.

Clark, A., & Chalmers, D. (1998). The extended mind. *Analysis, 58*, 7–19.

Clark, A., Friston, K., & Wilkinson, S. (2019). Bayesing qualia: Consciousness as inference, not raw datum. *Journal of Consciousness Studies, 26*, 19–33.

Clark, J. E., Watson, S., & Friston, K. J. (2018). What is mood? A computational perspective. *Psychological Medicine, 48*, 2277–2284.

Collerton, D., Barnes, J., Diederich, N. J., Dudley, R., ffytche, D., Friston, K., Goetz, C. G., Goldman, J. G., Jardri, R., Kulisevsky, J., Lewis, S. J. G., Nara, S., O'Callaghan, C., Onofrj, M., Pagonabarraga, J., Parr, T., Shine, J. M., Stebbins, G., Taylor, J.-P., Tsuda, I., & Weil, R. S. (2023). Understanding visual hallucinations: A new synthesis. *Neuroscience & Biobehavioral Reviews, 150*, 105208.

Constant, A., Ramstead, M. J. D., Veissière, S. P. L., Campbell, J. O., & Friston, K. J. (2018). A variational approach to niche construction. *Journal of the Royal Society Interface, 15*, 20170685.

Corcoran, A., & Hohwy, J. (2018). Allostasis, interoception, and the free energy principle: Feeling our way forward. In M. Tsakiris & H. de Preester (Eds.), *The interoceptive basis of the mind* (pp. 272–292). Oxford University Press.

Corcoran, A. W., Hohwy, J., & Friston, K. J. (2023). Accelerating scientific progress through Bayesian adversarial collaboration. *Neuron, 111*, 3505–3516.

Corcoran, A. W., Macefield, V. G., & Hohwy, J. (2021). Be still my heart: Cardiac regulation as a mode of uncertainty reduction. *Psychonomic Bulletin & Review, 28*, 1211–1223.

Corcoran, A. W., Perrykkad, K., Feuerriegel, D., & Robinson, J. E. (2023). Body as first teacher: The role of rhythmic visceral dynamics in early cognitive development. *Perspectives on Psychological Science*. https://doi.org/10.1177/17456916231185343.

Corcoran, A. W., Pezzulo, G., & Hohwy, J. (2020). From allostatic agents to counterfactual cognisers: Active inference, biological regulation, and the origins of cognition. *Biology & Philosophy, 35*, 32.

Corlett, P. R. (2019). Factor one, familiarity and frontal cortex: A challenge to the two-factor theory of delusions. *Cognitive Neuropsychiatry, 24*, 165–177.

Costin, V., & Vignoles, V. L. (2020). Meaning is about mattering: Evaluating coherence, purpose, and existential mattering as precursors of meaning in life judgments. *Journal of Personality and Social Psychology, 118*, 864–884.

Craver, C. (2007). *Explaining the brain: Mechanisms and the mosaic unity of neuroscience*. Oxford University Press.

Cremer, C. Z., & Kemp, L. (2021). *Democratising risk: In search of a methodology to study existential risk*. SSRN. https://ssrn.com/abstract=3995225.

Crick, F., & Koch, C. (1990). Towards a neurobiological theory of consciousness. *Seminars in the Neurosciences, 2*, 263–275.

Da Costa, L., Parr, T., Sajid, N., Veselic, S., Neacsu, V., & Friston, K. (2020). Active inference on discrete state-spaces: A synthesis. *Journal of Mathematical Psychology, 99*, 102447.

Dahl, C. J., & Davidson, R. J. (2019). Mindfulness and the contemplative life: Pathways to connection, insight, and purpose. *Current Opinion in Psychology, 28*, 60–64.

Dancy, J. (1983). Ethical particularism and morally relevant properties. *Mind, 92*, 530–547.

Davachi, L., & DuBrow, S. (2015). How the hippocampus preserves order: The role of prediction and context. *Trends in Cognitive Sciences, 19*, 92–99.

Davies, M., & Coltheart, M. (2000). Introduction: Pathologies of belief. *Mind and Language, 15*, 1–46.

Deane, G. (2021). Consciousness in active inference: Deep self-models, other minds, and the challenge of psychedelic-induced ego-dissolution. *Neuroscience of Consciousness, 2021*. https://doi.org/10.1093/nc/niab024.

Deane, G., Miller, M., & Wilkinson, S. (2020). Losing ourselves: Active inference, depersonalization, and meditation. *Frontiers in Psychology, 11*, 539726.

Dehaene, S., Kerszberg, M., & Changeux, J. P. (1998). A neuronal model of a global workspace in effortful cognitive tasks. *Proceedings of the National Academy of Science USA, 95*, 14529.

Dennett, D. (1992). The self as a center of narrative gravity. In F. Kessel, P. Cole, & D. Johnson (Eds.), *Self and consciousness: Multiple perspectives* (pp. 103–115). Erlbaum.

Dijkstra, N., & Fleming, S. M. (2023). Subjective signal strength distinguishes reality from imagination. *Nature Communications, 14*, 1627.

Dijkstra, N., Mazor, M., Kok, P., & Fleming, S. (2021). Mistaking imagination for reality: Congruent mental imagery leads to more liberal perceptual detection. *Cognition, 212*, 104719.

Di Paolo, E., Thompson, E., & Beer, R. (2022). Laying down a forking path: Tensions between enaction and the free energy principle. *Philosophy and the Mind Sciences, 3*. https://doi.org/10.33735/phimisci.2022.9187.

Dołęga, K., & Dewhurst, J. E. (2020). Fame in the predictive brain: A deflationary approach to explaining consciousness in the prediction error minimization framework. *Synthese, 198*, 7781–7806.

Douven, I. (2022). *The art of abduction*. MIT Press.

Drayson, Z. (2018). Direct perception and the predictive mind. *Philosophical Studies, 175*, 3145–3164.

Dryzek, J. S., Bächtiger, A., Chambers, S., Cohen, J., Druckman, J. N., Felicetti, A., Fishkin, J. S., Farrell, D. M., Fung, A., Gutmann, A., Landemore, H., Mansbridge, J., Marien, S., Neblo, M. A., Niemeyer, S., Setälä, M., Slothuus, R., Suiter, J., Thompson, D., & Warren, M. E. (2019). The crisis of democracy and the science of deliberation. *Science, 363*, 1144–1146.

Ekman, P. (1992). An argument for basic emotions. *Cognition & Emotion, 6*, 169–200.

Elkin, L., & Wiśniowska, K. (2022). Too rational: How predictive coding's success risks harming the mentally disordered and ill. *Journal of NeuroPhilosophy, 1*, 41–48.

Engel, M. M., Gadsby, S., Corcoran, A. W., Keizer, A., Dijkerman, H. C., & Hohwy, J. (2022). Waiting longer, feeling fatter: Effects of response delay on tactile distance estimation and confidence in females with anorexia nervosa. *Brain and Behavior, 12*, e2422.

Fabry, R. E. (2015). Enriching the notion of enculturation: Cognitive integration, predictive processing, and the case of reading acquisition. In T. K. Metzinger & J. M. Windt (Eds.), *Open mind*. MIND Group.

Fabry, R. E. (2017a). Predictive processing and cognitive development. In T. K. Metzinger & W. Wiese (Eds.), *Philosophy and predictive processing*. MIND Group.

Fabry, R. E. (2017b). Transcending the evidentiary boundary: Prediction error minimization, embodied interaction, and explanatory pluralism. *Philosophical Psychology, 30*, 395–414.

Facchin, M. (2021). Extended predictive minds: Do Markov blankets matter? *Review of Philosophy and Psychology, 14*, 909–938.

Feldman, H., & Friston, K. (2010). Attention, uncertainty and free-energy. *Frontiers in Human Neuroscience, 4*. https://doi.org/10.3389/fnhum.2010.00215.

Feyerabend, P. (1975). *Against method: Outline of an anarchistic theory of knowledge*. Verso Books.

Fields, C., Friston, K., Glazebrook, J. F., & Levin, M. (2022). A free energy principle for generic quantum systems. *Progress in Biophysics and Molecular Biology, 173*, 36–59.

Fink, S. B., & Zednik, C. (2017). Meeting in the dark room: Bayesian rational analysis and hierarchical predictive coding. In T. K. Metzinger & W. Wiese (Eds.), *Philosophy and predictive processing*. MIND Group.

Fisher, E. L., & Hohwy, J. (2024). The universal optimism of the self-evidencing mind. *Entropy, 26*, 518.

Fisher, E. L., Smith, R., Conn, K., Corcoran, A. W., Milton, L. K., Hohwy, J., & Foldi, C. J. (2024). Psilocybin increases optimistic engagement over time: Computational modelling of behavior in rats. *Translational Psychiatry, 14*, 394.

Fisher, E. L., Whyte, C., & Hohwy, J. (2024). An active inference model of the optimism bias. *Computational Psychiatry, 9*(1), 3–22.

Flanagan, O. (2007). *The really hard problem: meaning in a material world*. MIT Press.

Fleming, S. M. (2021). *Know thyself: The new science of self-awareness*. Basic Books.

Frank, A., Gleiser, M., & Thompson, E. (2024). *The blind spot: Why science cannot ignore human experience*. MIT Press.

Friston, K. (2010). The free-energy principle: A unified brain theory? *Nature Reviews: Neuroscience, 11*, 127–138.

Friston, K. (2011a). Embodied inference: Or "I think therefore I am, if I am what I think." In W. Wolfgang Tschacher & C. Bergomi (Eds.), *The implications of embodiment* (pp. 89–125). Imprint Academic.

Friston, K. (2011b). What is optimal about motor control? *Neuron, 72*, 488–498.

Friston, K. (2018). Am I self-conscious? (Or does self-organization entail self-consciousness?) *Frontiers in Psychology, 9*. https://doi.org/10.3389/fpsyg.2018.00579.

Friston, K. (2019a). Beyond the desert landscape. In *Andy Clark and his critics* (pp. 174–190). Oxford University Press.

Friston, K. (2019b). A free energy principle for a particular physics. *arXiv preprint*. https://arXiv:1906.10184.

Friston, K. (2022). The ultimate trick? Comment on "The Markov blanket trick: On the scope of the free energy principle and active inference" by Raja et al. *Physics of Life Reviews, 43*, 10–16.

Friston, K., Breakspear, M., & Deco, G. (2012). Perception and self-organised instability. *Frontiers in Computational Neuroscience, 6*. https://doi.org/10.3389/fncom.2012.00044.

Friston, K., Da Costa, L., Hafner, D., Hesp, C., & Parr, T. (2021). Sophisticated inference. *Neural Computation, 33*, 713–763.

Friston, K., Da Costa, L., Sajid, N., Heins, C., Ueltzhöffer, K., Pavliotis, G., & Parr, T. (2023). The free energy principle made simpler but not too simple. *Physics Reports, 1024*, 1–29.

Friston, K., Da Costa, L., Sakthivadivel, D. A. R., Heins, C., Pavliotis, G. A., Ramstead, M., & Parr, T. (2023). Path integrals, particular kinds, and strange things. *Physics of Life Reviews, 47*, 35–62.

Friston, K., FitzGerald, T., Rigoli, F., Schwartenbeck, P., & Pezzulo, G. (2017). Active inference: A process theory. *Neural Computation, 29*, 1–49.

Friston, K., Friedman, D. A., Constant, A., Knight, V. B., Fields, C., Parr, T., & Campbell, J. O. (2023). A variational synthesis of evolutionary and developmental dynamics. *Entropy, 25*, 964.

Friston, K., & Frith, C. (2015). A duet for one. *Consciousness and Cognition, 36*, 390–405.

Friston, K., Heins, C., Ueltzhöffer, K., Da Costa, L., & Parr, T. (2021). Stochastic chaos and Markov blankets. *Entropy, 23*, 1220.

Friston, K. J., Rosch, R., Parr, T., Price, C., & Bowman, H. (2017). Deep temporal models and active inference. *Neuroscience & Biobehavioral Reviews, 77*, 388–402.

Friston, K. J., Salvatori, T., Isomura, T., Tschantz, A., Kiefer, A., Verbelen, T., Koudahl, M., Paul, A., Parr, T., & Razi, A. (2025). Active inference and intentional behaviour. *Neural Computation*, 37(4): 666–700.

Friston, K., & Stephan, K. (2007). Free energy and the brain. *Synthese, 159*, 417–458.

Friston, K., Thornton, C., & Clark, A. (2012). Free-energy minimization and the dark room problem. *Frontiers in Psychology, 3*. https://10.3389/fpsyg.2012.00130.

Friston, K., Wiese, W., & Hobson, J. (2020). Sentience and the origins of consciousness: From Cartesian duality to Markovian monism. *Entropy, 22*, 516.

Frith, C. D. (1992). *The cognitive neuropsychology of schizophrenia*. Lawrence Erlbaum Associates.

Froese, T., Weber, N., Shpurov, I., & Ikegami, T. (2023). From autopoiesis to self-optimization: Toward an enactive model of biological regulation. *Biosystems, 230*, 104959.

Gadsby, S. (2017). Anorexia nervosa and oversized experiences. *Philosophical Psychology, 30*, 594–615.

Gadsby, S. (2021). Visual self-misperception in eating disorders. *Perception, 50*, 933–949.

Gallagher, S. (2000). Philosophical conceptions of the self: Implications for cognitive science. *Trends in Cognitive Sciences, 4*, 14–21.

Gallagher, S. (2013). A pattern theory of self. *Frontiers in Human Neuroscience, 7*. https://10.3389/fnhum.2013.00443.

Gallagher, S. (2023). *Embodied and enactive approaches to cognition*. Cambridge University Press.

Gallagher, S., Raffone, A., Berkovich-Ohana, A., Barendregt, H. P., Bauer, P. R., Brown, K. W., Giommi, F., Nyklíček, I., Ostafin, B. D., Slagter, H., Trautwein, F.-M., & Vago, D. R. 2023. The self-pattern and buddhist psychology. *Mindfulness, 15*, 795–803.

George, L. S., & Park, C. L. (2016). Meaning in life as comprehension, purpose, and mattering: Toward integration and new research questions. *Review of General Psychology, 20*, 205–220.

George, L. S., & Park, C. L. (2017). The multidimensional existential meaning scale: A tripartite approach to measuring meaning in life. *Journal of Positive Psychology, 12*, 613–627.

Gerrans, P. (2015a). All the self we need. In T. K. Metzinger & J. M. Windt (Eds.), *Open mind*. MIND Group.

Gerrans, P. (2015b). Metamisery and bodily inexistence. In T. K. Metzinger & J. M. Windt (Eds.), *Open mind*. MIND Group.

Gerrans, P. (2019). Depersonalization disorder, affective processing and predictive coding. *Review of Philosophy and Psychology, 10*, 401–418.

Gerrans, P. (2022a). Alienation and identification in addiction. *Philosophical Psychology, 37*, 684–706.

Gerrans, P. (2022b). A vessel without a pilot: Bodily and affective experience in the Cotard delusion of inexistence. *Mind & Language, 38*, 1059–1080.

Gerrans, P. (2024). *Anatomy of an avatar: The virtual self in pathologies of self awareness*. Oxford University Press.

Gigerenzer, G. (2000). *Adaptive thinking: Rationality in the real world*. Oxford University Press.

Gigerenzer, G. (2019). Axiomatic rationality and ecological rationality. *Synthese, 198*, 3547–3564.

Gigerenzer, G., & Selten, R. (Eds.). (2001). *Bounded rationality: The adaptive toolbox*. MIT Press.

Gładziejewski, P. (2016). Predictive coding and representationalism. *Synthese, 193*. 559–582.

Gładziejewski, P. (2022). The rational role of the perceptual sense of reality. *Mind & Language, 38*, 1021–1040.

Gładziejewski, P., & Miłkowski, M. (2017). Structural representations: Causally relevant and different from detectors. *Biology & Philosophy, 32*, 337–355.

Glück, J., & Sternberg, R. J. (Eds.). (2019). *The Cambridge handbook of wisdom*. Cambridge University Press.

Grinfeld, G., Lagnado, D., Gerstenberg, T., Woodward, J. F., & Usher, M. (2020). Causal responsibility and robust causation. *Frontiers in Psychology, 11*, 1069.

Grossmann, I., Weststrate, N. M., Ardelt, M., Brienza, J. P., Dong, M., Ferrari, M., Fournier, M. A., Hu, C. S., Nusbaum, H. C., & Vervaeke, J. (2020). The science of wisdom in a polarized world: Knowns and unknowns. *Psychological Inquiry, 31*, 103–133.

Gu, X., FitzGerald, T. H. B., & Friston, K. J. (2019). Modeling subjective belief states in computational psychiatry: Interoceptive inference as a candidate framework. *Psychopharmacology, 236*: 2405–2412.

Guthrie, T., Matthews, J. R., Chambers, R., Windt, J., & Hohwy, J. (2022). Changes in multisensory integration following brief state induction and longer-term training with body scan meditation. *Mindfulness, 13*, 1214–1231.

Harman, G. (1973). *Thought.* Princeton University Press.

Hempel, C. G. (1965). *Aspects of scientific explanation and other essays in the philosophy of science.* Free Press.

Hesp, C., Smith, R., Parr, T., Allen, M., Friston, K. J., & Ramstead, M. J. D. (2021). Deeply felt affect: The emergence of valence in deep active inference. *Neural Computation, 33*: 398–446.

Hills, T. T., Todd, P. M., Lazer, D., Redish, A. D., & Couzin, I. D. (2015). Exploration versus exploitation in space, mind, and society. *Trends in Cognitive Sciences, 19*, 46–54.

Hinton, G. (2022). The forward-forward algorithm: Some preliminary investigations. *arXiv preprint.* https://arXiv:2212.13345.

Hobson, J. A., & Friston, K. J. (2012). Waking and dreaming consciousness: Neurobiological and functional considerations. *Progress in Neurobiology, 98*, 82–98.

Hodson, R., Mehta, M., & Smith, R. (2024). The empirical status of predictive coding and active inference. *Neuroscience & Biobehavioral Reviews, 157*, 105473.

Hohwy, J. (2004). Top-down and bottom-up in delusion formation. *Philosophy, Psychiatry and Psychology, 11*, 65–70.

Hohwy, J. (2007). The sense of self in the phenomenology of agency and perception. *Psyche, 13*, 1–20.

Hohwy, J. (2009). The neural correlates of consciousness: New experimental approaches needed? *Consciousness and Cognition, 18*, 428–438.

Hohwy, J. (2011). Phenomenal variability and introspective reliability. *Mind & Language, 26*, 261–286.

Hohwy, J. (2012). Attention and conscious perception in the hypothesis testing brain. *Frontiers in Psychology, 3*. https://10.3389/fpsyg.2012.00096.

Hohwy, J. (2013). *The predictive mind.* Oxford University Press.

Hohwy, J. (2015). Prediction error minimization, mental and developmental disorder, and statistical theories of consciousness. In R. Gennaro (Ed.), *Disturbed consciousness: New essays on psychopathology and theories of consciousness* (pp. 293–324). MIT Press.

Hohwy, J. (2016). The self-evidencing brain. *Noûs, 50,* 259–285.

Hohwy, J. (2017a). How to entrain your evil demon. In T. K. Metzinger & W. Wiese (Eds.), *Philosophy and predictive processing.* MIND Group. https://predictive-mind.net/papers/how-to-entrain-your-evil-demon.

Hohwy, J. (2017b). Priors in perception: Top-down modulation, bayesian perceptual learning rate, and prediction error minimization. *Consciousness and Cognition, 47,* 75–85.

Hohwy, J. (2018). The predictive processing hypothesis. In A. Newen, L. Bruin, & S. Gallagher (Eds.), *The Oxford handbook of 4E cognition* (pp. 129–146). Oxford University Press.

Hohwy, J. (2019). Quick 'n' lean or slow and rich? Andy Clark on predictive processing and embodied cognition. In M. Colombo, E. Irvine, & M. Stapleton (Eds.), *Andy Clark and his critics* (pp. 191–205). Oxford University Press.

Hohwy, J. (2020a). New directions in predictive processing. *Mind & Language, 35,* 209–223.

Hohwy, J. (2021a). Conscious self-evidencing. *Review of Philosophy and Psychology, 13,* 809–882.

Hohwy, J. (2021b). Self-supervision, normativity and the free energy principle. *Synthese, 199,* 29–53.

Hohwy, J., & Bayne, T. (2015). The neural correlates of consciousness: Causes, confounds and constituents. In S. Miller (Ed.), *The constitution of phenomenal consciousness: Toward a science and theory* (pp. 155–176). John Benjamins.

Hohwy, J., & Frith, C. D. (2004). Can neuroscience explain consciousness? *Journal of Consciousness Studies, 11,* 180–198.

Hohwy, J., & Michael, J. (2017). Why would any body have a self? In F. Vignemont & A. Alsmith (Eds.), *The subject's matter: Self-consciousness and the body* (pp. 363–392). MIT Press.

Hohwy, J., & Palmer, C. (2014). Social cognition as causal inference: Implications for common knowledge and autism. In M. Gallotti & J. Michael (Eds.), *Perspectives on social ontology and social cognition* (pp. 167–189). Springer Netherlands.

Hohwy, J., & Rajan, V. (2011). Delusions as forensically disturbing perceptual inferences. *Neuroethics, 5,* 5–11.

Hohwy, J., Roepstorff, A., & Friston, K. (2008). Predictive coding explains binocular rivalry: An epistemological review. *Cognition, 108*, 687–701.

Hohwy, J., & Rosenberg, R. (2005). Unusual experiences, reality testing, and delusions of control. *Mind & Language, 20*, 141–162.

Hohwy, J., & Sandved-Smith, L. (2024). Less is more: Strangeness affords flexibility—a commentary on "path integrals, particular kinds, and strange things" by Friston, Da Costa, Sakthivadivel, Heins, Pavliotis, Ramstead, and Parr. *Physics of Life Reviews, 48*, 14–15.

Hohwy, J., & Seth, A. (2020). Predictive processing as a systematic basis for identifying the neural correlates of consciousness. *Philosophy and the Mind Sciences, 1*. https://10.33735/phimisci.2020.II.64.

Huebner, B., & Schulkin, J. (2022). *Biological cognition*. Cambridge University Press.

Hume, D. (1739–1740). *A treatise of human nature*. Clarendon Press.

Hurley, S. L. (1998). *Consciousness in action*. Harvard University Press.

Huta, V., & Waterman, A. S. (2014). Eudaimonia and its distinction from hedonia: Developing a classification and terminology for understanding conceptual and operational definitions. *Journal of Happiness Studies, 15*, 1425–1456.

Hutto, D. D. (2018). Getting into predictive processing's great guessing game: Bootstrap heaven or hell? *Synthese, 195*, 2445–2458.

Hutto, D., & Myin, E. (2013). *Radicalizing enactivism: Basic minds without content*. MIT Press.

Jackson, F. (1977). A causal theory of counterfactuals. *Australasian Journal of Philosophy, 55*, 3–21.

Jackson, F. (1998). *From metaphysics to ethics*. Oxford University Press.

James, W. (1884). What is an emotion? *Mind, 9*, 188–205.

Jefferys, W. H., & Berger, J. O. (1992). Ockham's razor and Bayesian analysis. *American Scientist, 80*, 64–72.

Joffily, M., & Coricelli, G. (2013). Emotional valence and the free-energy principle. *PLoS Computational Biology, 9*, e1003094.

Johnston, P., Robinson, J., Kokkinakis, A., Ridgeway, S., Simpson, M., Johnson, S., Kaufman, J., & Young, A. W. (2017). Temporal and spatial localization of prediction-error signals in the visual brain. *Biological Psychology, 125*, 45–57.

Jordan, R., & Keller, G. B. (2023). The locus coeruleus broadcasts prediction errors across the cortex to promote sensorimotor plasticity. *eLife, 12*, RP85111.

Josipovic, Z. (2010). Duality and nonduality in meditation research. *Consciousness and Cognition, 19*, 1119–1121.

Josipovic, Z. (2021). Implicit–explicit gradient of nondual awareness or consciousness as such. *Neuroscience of Consciousness, 2021*, niab031.

Kabat-Zinn, J. (2015). Mindfulness. *Mindfulness, 6*, 1481–1483.

Kahneman, D. (2011). *Thinking, slow and fast.* Farrar, Straus and Giroux.

Kahneman, D., & Frederick, S. (2002). *Heuristics and biases: The psychology of intuitive judgment.* Cambridge University Press.

Kahneman, D., Slovic, P., & Tversky, A. (1982). *Judgment under uncertainty: Heuristics and biases.* Macmillan Publishers Ltd.

Keogh, R., Wicken, M., & Pearson, J. (2023). Fewer intrusive memories in aphantasia: Using the trauma film paradigm as a laboratory model of PTSD. *PsyArXiv.* https://doi.org/10.31234/osf.io/7zqfe.

Kiefer, A. B. (2020). Psychophysical identity and free energy. *Journal of the Royal Society Interface, 17*, 20200370.

Kiefer, A., & Hohwy, J. (2018). Content and misrepresentation in hierarchical generative models. *Synthese, 195*, 2387–2415.

Kiefer, A., & Hohwy, J. (2022). Bayesian realism and structural representation. *Behavioral and Brain Sciences, 45*, e199.

Kikuchi, K., Galera-Laporta, L., Weatherwax, C., Lam, J. Y., Moon, E. C., Theodorakis, E. A., Garcia-Ojalvo, J., & Süel, G. M. (2022). Electrochemical potential enables dormant spores to integrate environmental signals. *Science, 378*, 43–49.

Kim, J. (1998). *Philosophy of mind.* Westview Press.

King, L. A., & Hicks, J. A. (2021). The science of meaning in life. *Annual Review of Psychology, 72*, 561–584.

Kirchhoff, M., (2025). The idealized mind. MIT Press.

Kirchhoff, M. D., & Froese, T. (2017). Where there is life there is mind: In support of a strong life-mind continuity thesis. *Entropy, 19*, 169.

Kirchhoff, M. D., & Kiverstein, J. (2021). How to determine the boundaries of the mind: A markov blanket proposal. *Synthese, 198*, 4791–4810.

Kirchhoff, M. D., Kiverstein, J., & Robertson, I. (2022). The literalist fallacy and the free energy principle: model-building, scientific realism, and instrumentalism. *British Journal for the Philosophy of Science.* https://doi.org/10.1086/720861.

Kirchhoff, M., Parr, T., Palacios, E., Friston, K., & Kiverstein, J. (2018). The Markov blankets of life: Autonomy, active inference and the free energy principle. *Journal of the Royal Society Interface, 15*, 20170792.

Kiverstein, J. (2020). Free energy and the self: An ecological–enactive interpretation. *Topoi, 39*, 559–574.

Kleckner, I. R., Zhang, J., Touroutoglou, A., Chanes, L., Xia, C., Simmons, W. K., Quigley, K. S., Dickerson, B. C., & Feldman Barrett, L. (2017). Evidence for a large-scale brain system supporting allostasis and interoception in humans. *Nature Human Behaviour, 1*, 0069.

Klein, C., Hohwy, J., & Bayne, T. (2020). Explanation in the science of consciousness: From the neural correlates of consciousness (NCCs) to the difference makers of consciousness (DMCs). *Philosophy and the Mind Sciences, 1*(II). https://doi.org/10.33735/phimisci.2020.II.60.

Kleiner, J. (2024). Consciousness qua mortal computation. *arXiv preprint*. https://arXiv:2403.03925.

Koban, L., Gianaros, P. J., Kober, H., & Wager, T. D. (2021). The self in context: Brain systems linking mental and physical health. *Nature Reviews Neuroscience, 22*, 309–322.

Kuhn, T. S. (1962). *The structure of scientific revolutions*. University of Chicago Press.

Lakatos, I. (1976). *Proofs and refutations: The logic of mathematical discovery*. Cambridge University Press.

Lammers, J., Crusius, J., & Gast, A. (2020). Correcting misperceptions of exponential coronavirus growth increases support for social distancing. *Proceedings of the National Academy of Sciences, 117*, 16264.

Landau, I. (2017). *Finding meaning in an imperfect world*. Oxford University Press.

Landemore, H. (2012). *Democratic reason: Politics, collective intelligence, and the rule of the many*. Princeton University Press.

Lange, C. (1885). *Om sindsbevægelser*. Jacob Lunds Forlag.

Lau, H. (2008). A higher order Bayesian decision theory of consciousness. *Progress in Brain Research, 168*, 35–48.

Laukkonen, R. E., & Slagter, H. A. (2021). From many to (n)one: Meditation and the plasticity of the predictive mind. *Neuroscience & Biobehavioral Reviews, 128*, 199–217.

Lavine, N., Reuben, M., & Clarke, P. B. (1997). A population of nicotinic receptors is associated with thalamocortical afferents in the adult rat: Laminal and areal analysis. *Journal of Comparative Neurology, 380*, 175–190.

Lawson, R. P., Mathys, C., & Rees, G. (2017). Adults with autism overestimate the volatility of the sensory environment. *Nature Neuroscience, 20*, 1293–1299.

LeDoux, J. (2020). *The deep history of ourselves: The four-billion-year story of how we got conscious brains*. Penguin.

Letheby, C., & Gerrans, P. (2017). Self unbound: Ego dissolution in psychedelic experience. *Neuroscience of Consciousness, 1*, nix016.

Levy, N. (2014). Addiction as a disorder of belief. *Biology & Philosophy, 29*, 337–355.

Lewis, D. (1973). *Counterfactuals*. Blackwwell.

Li, W., & Keil, A. (2023). Sensing fear: Fast and precise threat evaluation in human sensory cortex. *Trends in Cognitive Sciences, 27*, 341–352.

Limanowski, J., & Blankenburg, F. (2013). Minimal self-models and the free energy principle. *Frontiers in Human Neuroscience, 7*. https://10.3389/fnhum.2013.00547.

Limanowski, J., & Friston, K. (2018). "Seeing the dark": Grounding phenomenal transparency and opacity in precision estimation for active inference. *Frontiers in Psychology, 9*. https://10.3389/fpsyg.2018.00643.

Lin, Y.-T. (2015). Memory for prediction error minimization: From depersonalization to the delusion of non-existence. In T. K. Metzinger & J. M. Windt (Eds.), *Open mind*. MIND Group.

Linson, A., Parr, T., & Friston, K. J. (2020). Active inference, stressors, and psychological trauma: A neuroethological model of (mal)adaptive explore-exploit dynamics in ecological context. *Behavioural Brain Research, 380*, 112421.

Lipton, P. (2004). *Inference to the best explanation*. Routledge.

Löhr, G., & Michel, C. (2023). Conceptual engineering, predictive processing, and a new implementation problem. *Mind & Language, 39*, 201–219.

MacAskill, W. (2022). *What we owe the future*. Basic Books.

Marchi, F., & Hohwy, J. (2020). The intermediate scope of consciousness in the predictive mind. *Erkenntnis, 87*, 891–912.

Marreiros, A. C., Kiebel, S. J., Daunizeau, J., Harrison, L. M., & Friston, K. J. (2009). Population dynamics under the Laplace assumption. *Neuroimage, 44*, 701–714.

Martela, F., & Steger, M. F. (2023). The role of significance relative to the other dimensions of meaning in life—an examination utilizing the three dimensional meaning in life scale (3DM). *Journal of Positive Psychology, 18*, 606–626.

Marx, S., & Einhäuser, W. (2015). Reward modulates perception in binocular rivalry. *Journal of Vision, 15*, 11.

Mashour, G. A., Roelfsema, P., Changeux, J.-P., & Dehaene, S. (2020). Conscious processing and the global neuronal workspace hypothesis. *Neuron, 105*, 776–798.

Mather, M. (2024). The emotion paradox in the aging body and brain. *Annals of the New York Academy of Sciences, 1536*, 13–41.

Mathys, C. D., Lomakina, E. I., Daunizeau, J., Iglesias, S., Brodersen, K. H., Friston, K. J., & Stephan, K. E. (2014). Uncertainty in perception and the hierarchical Gaussian filter. *Frontiers in Human Neuroscience, 8*. https://doi.org/10.3389/fnhum.2014.00825.

Maturana, H. R., & Varela, F. (1972). *Autopoeisis and cognition*. D. Riedel Publishing Company.

Maturana, H. R., & Varela, F. J. (1987). *The tree of knowledge*. Shambhala Publications.

McDowell, J. (2008). Avoiding the myth of the given. In J. Lindgaard (Ed.), *John McDowell: Experience, norm, and nature*, (pp. 1–14). Blackwell Publishing.

McKenna, M., & Coates, J. (2021). Compatibilism. In E. N. Zalta (Eds.), *The Stanford encyclopedia of philosophy*. https://plato.stanford.edu/archives/sum2024/entries/compatibilism/.

Mckilliam, A. K. (2020). What is a global state of consciousness? *Philosophy and the Mind Sciences, 1*. https://doi.org/10.33735/phimisci.2020.II.58.

Melloni, L., Mudrik, L., Pitts, M., Bendtz, K., Ferrante, O., Gorska, U., Hirschhorn, R., Khalaf, A., Kozma, C., Lepauvre, A., Liu, L., Mazumder, D., Richter, D., Zhou, H., Blumenfeld, H., Boly, M., Chalmers, D. J., Devore, S., Fallon, F., de Lange, F. P., . . . & Tononi, G. (2023). An adversarial collaboration protocol for testing contrasting predictions of global neuronal workspace and integrated information theory. *PloS One, 18*, e0268577.

Menon, V. (2023). 20 years of the default mode network: A review and synthesis. *Neuron, 111*, 2469–2487.

Merker, B. (2007). Consciousness without a cerebral cortex: A challenge for neuroscience and medicine. *Behavioral and Brain Sciences, 30*, 63–81.

Metz, T. (2013). *Meaning in life*. Oxford University Press.

Metzinger, T. (2004). *Being no one: The self-model theory of subjectivity*. MIT Press.

Metzinger, T. K. (2013). The myth of cognitive agency subpersonal thinking as a cyclically recurring loss of mental autonomy. *Frontiers in Psychology, 4*. https://10.3389/fpsyg.2013.00931.

Metzinger, T. K. (2017). The problem of mental action. In T. K. Metzinger & W. Wiese (Eds.), *Philosophy and predictive processing*. MIND Group.

Metzinger, T. (2024). *The elephant and the blind.* MIT Press.

Miller, M., Kiverstein, J., & Rietveld, E. (2020). Embodying addiction: A predictive processing account. *Brain and Cognition, 138,* 105495.

Miller, M., White, B., & Scrivner, C. (2023). Surfing uncertainty with screams: Predictive processing, error dynamics and horror films. *Philosophical Transactions of the Royal Society B, 379,* 20220425.

Mirza, M. B., Adams, R. A., Friston, K., & Parr, T. 2019. Introducing a Bayesian model of selective attention based on active inference. *Scientific Reports, 9,* 13915.

Moran, R. J., Campo, P., Symmonds, M., Stephan, K. E., Dolan, R. J., & Friston, K. J. (2013). Free energy, precision and learning: The role of cholinergic neuromodulation. *Journal of Neuroscience, 33,* 8227–8236.

Moran, R. J., Symmonds, M., Dolan, R. J., & Friston, K. J. (2014). The brain ages optimally to model its environment: Evidence from sensory learning over the adult lifespan. *PLoS Computational Biology, 10,* e1003422.

Moutoussis, M., Fearon, P., El-Deredy, W., Dolan, R. J., & Friston, K. J. (2014). Bayesian inferences about the self (and others): A review. *Consciousness and Cognition, 25,* 67–76.

Mumford, D. (1992). On the computational architecture of the neocortex. II: The role of cortico-cortical loops. *Biological Cybernetics, 66,* 241–251.

Nagel, T. (1974). What is it like to be a bat? *Philosophical Review, 83,* 435–450.

Nagel, T. (1979). Panpsychism. In T. Nagel (Ed.), *Mortal questions* (pp. 181–195). Cambridge University Press.

Nau, M., Julian, J. B., & Doeller, C. F. (2018). How the brain's navigation system shapes our visual experience. *Trends in Cognitive Sciences, 22,* 810–825.

Nave, K. (2025). *A drive to survive: The free energy principle and the meaning of life.* MIT Press.

Nave, K., Deane, G., Miller, M., & Clark, A. (2022). Expecting some action: Predictive processing and the construction of conscious experience. *Review of Philosophy and Psychology, 13,* 1019–1037.

Negro, N. (2022). Can the integrated information theory explain consciousness from consciousness itself? *Review of Philosophy and Psychology, 14,* 1471–1489.

Newell, B. R., Lagnado, D. A., & Shanks, D. R. (2015). *Straight choices: The psychology of decision making.* Psychology Press.

Newen, A., Bruin, L., & Gallagher, S. (Eds.). (2018). *The Oxford handbook of 4E cognition.* Oxford University Press.

Novicky, F., Parr, T., Friston, K., Mirza, M. B., & Sajid, N. (2022). Bistable perception, precision and neuromodulation. *Cerebral Cortex, 34* (1). https://doi.org/10.1093/cercor/bhad401.

Oaksford, M., & Chater, N. (2007). *Bayesian rationality: The probabilistic approach to human reasoning*. Oxford University Press.

Oliver, G., Lanillos, P., & Cheng, G. (2019). Active inference body perception and action for humanoid robots. *arXiv preprint*. https://arXiv:1906.03022.

Orlandi, N. (2014). *The innocent eye: Why vision is not a cognitive process*. Oxford University Press.

Orlandi, N. (2018). Predictive perceptual systems. *Synthese, 195*, 2367–2386.

Ororbia, A., & Friston, K. (2023). Mortal computation: A foundation for biomimetic intelligence. *arXiv preprint*. https://arXiv:2311.09589.

Palacios, E. R., Razi, A., Parr, T., Kirchhoff, M., & Friston, K. (2020). On Markov blankets and hierarchical self-organisation. *Journal of Theoretical Biology, 486*, 110089.

Palmer, C. J., Seth, A. K., & Hohwy, J. (2015). The felt presence of other minds: Predictive processing, counterfactual predictions, and mentalising in autism. *Consciousness and Cognition, 36*, 376–389.

Panksepp, J. (2004). *Affective neuroscience: The foundations of human and animal emotions*. Oxford University Press.

Parr, T. (2022). Patterns and particles: Comment on "The Markov blanket trick: On the scope of the free energy principle and active inference" by Raja et al. *Physics of Life Reviews, 43*, 17–19.

Parr, T., Benrimoh, D., Vincent, P., & Friston, K. (2018). Precision and false perceptual inference. *Frontiers in Integrative Neuroscience, 12*. https://10.3389/fnint.2018.00039.

Parr, T., Corcoran, A., Friston, K., & Hohwy, J. (2019). Perceptual awareness and active inference. *Neuroscience of Consciousness, 1*. https://10.1093/nc/niz012.

Parr, T., & Friston, K. J. (2017a). The computational anatomy of visual neglect. *Cerebral Cortex, 28*, 777–790.

Parr, T., & Friston, K. J. (2017b). Uncertainty, epistemics and active inference. *Journal of the Royal Society Interface, 14*. https://10.1098/rsif.2017.0376.

Parr, T., & Friston, K. J. (2018). The anatomy of inference: Generative models and brain structure. *Frontiers in Computational Neuroscience, 12*. https://10.3389/fncom.2018.00090.

Parr, T., Holmes, E., Friston, K. J., & Pezzulo, G. (2023). Cognitive effort and active inference. *Neuropsychologia, 184*, 108562.

Parr, T., Pezzulo, G., & Friston, K. J. (2022). *Active inference: The free energy principle in mind, brain, and behavior*. MIT Press.

Parr, T., Rees, G., & Friston, K. J. (2018). Computational neuropsychology and Bayesian inference. *Frontiers in Human Neuroscience, 12*. https://10.3389/fnhum.2018.00061.

Parrott, M. (2021). Delusional predictions and explanations. *British Journal for the Philosophy of Science, 72*, 325–353.

Paulus, M. P., Feinstein, J. S., & Khalsa, S. S. (2019). An active inference approach to interoceptive psychopathology. *Annual Review of Clinical Psychology, 15*, 97–122.

Pearl, J. (1988). *Probabilistic reasoning in intelligent systems: Networks of plausible inference*. Morgan Kaufmann Publishers.

Pearl, J. (2000). *Causality*. Cambridge University Press.

Pearson, J. (2019). The human imagination: The cognitive neuroscience of visual mental imagery. *Nature Reviews Neuroscience, 20*, 624–634.

Pearson, J., Naselaris, T., Holmes, E. A., & Kosslyn, S. M. (2015). Mental imagery: Functional mechanisms and clinical applications. *Trends in Cognitive Sciences, 19*, 590–602.

Perera, R. (2022). *Bayesian hierarchical precision-weighted prediction and its penetrability consequences*. Monash University.

Perrykkad, K., & Hohwy, J. (2020a). Fidgeting as self-evidencing: A predictive processing account of non-goal-directed action. *New Ideas in Psychology, 56*, 100750.

Perrykkad, K., & Hohwy, J. (2020b). Modelling me, modelling you: The autistic self. *Review Journal of Autism and Developmental Disorders, 7*, 1–31.

Perrykkad, K., & Hohwy, J. (2022). How selves differ within and across cognitive domains: Self-prioritisation, self-concept, and psychiatric traits. *BMC Psychology, 10*, 165.

Perrykkad, K., Lawson, R. P., Jamadar, S., & Hohwy, J. (2021). The effect of uncertainty on prediction error in the action perception loop. *Cognition, 210*, 104598.

Perrykkad, K., Robinson, J. E., & Hohwy, J. (2022). Foraging for the self: Environment selection for agency inference. *Psychonomic Bulletin & Review, 30*, 608–620.

Perrykkad, K., Sherwell, C., Kirby, J., & Hohwy, J. (2023). Beliefs about action efficacy mediate the relationship between self-concept clarity and self-reported compassionate action. *Psychology of Consciousness: Theory, Research, and Practice*. http://10.1037/cns0000349.

Piccinini, G. (2020). *Neurocognitive mechanisms: Explaining biological cognition*. Oxford University Press.

Powers, W. T. (1973). *Behavior: The control of perception*. Aldine.

Quigley, K. S., Kanoski, S., Grill, W. M., Barrett, L. F., & Tsakiris, M. (2021). Functions of interoception: From energy regulation to experience of the self. *Trends in Neurosciences, 44*, 29–38.

Raginsky, M. (2023). Biological autonomy. *Biological Theory, 18*, 303–308.

Raja, V., Valluri, D., Baggs, E., Chemero, A., & Anderson, M. L. (2021). The Markov blanket trick: On the scope of the free energy principle and active inference. *Physics of Life Reviews, 39*, 49–72.

Ramstead, M. J., Albarracin, M., Kiefer, A., Klein, B., Fields, C., Friston, K., & Safron, A. (2023). The inner screen model of consciousness: Applying the free energy principle directly to the study of conscious experience. *arXiv preprint*. https://arXiv:2305.02205.

Ramstead, M. J. D., Constant, A., Badcock, P. B., & Friston, K. J. (2019). Variational ecology and the physics of sentient systems. *Physics of Life Reviews, 31*, 188–205.

Ramstead, M. J. D., Kirchhoff, M. D., Constant, A., & Friston, K. J. (2021). Multiscale integration: Beyond internalism and externalism. *Synthese, 198*, 41–70.

Ramstead, M. J. D., Kirchhoff, M. D., & Friston, K. J. (2019). A tale of two densities: Active inference is enactive inference. *Adaptive Behavior*. https://doi.org/10.1177/1059712319862774.

Ramstead, M. J. D., Sakthivadivel, D. A. R., Heins, C., Koudahl, M., Millidge, B., Da Costa, L., Klein, B., & Friston, K. J. (2023). On Bayesian mechanics: A physics of and by beliefs. *Interface Focus, 13*, 20220029.

Ramstead, M., Wiese, W., Miller, M., & Friston, K. J. (2023). Deep neurophenomenology: An active inference account of some features of conscious experience and of their disturbance in major depressive disorder. In T. Cheng, R. Sato, & J. Hohwy (Eds.), *Expected experiences* (pp. 9–46). Routledge.

Ransom, M., Fazelpour, S., & Mole, C. (2017). Attention in the predictive mind. *Consciousness and Cognition, 47*, 99–112.

Rawls, J. (1971). *A theory of justice*. Harvard University Press.

Redish, A. D. (2016). Vicarious trial and error. *Nature Reviews Neuroscience, 17*, 147–159.

Robinson, J. E., Breakspear, M., Young, A. W., & Johnston, P. J. (2020). Dose-dependent modulation of the visually evoked N1/N170 by perceptual surprise: A clear demonstration of prediction-error signalling. *European Journal of Neuroscience, 52*, 4442–4452.

Rorot, W. (2021). Bayesian theories of consciousness: A review in search for a minimal unifying model. *Neuroscience of Consciousness, 2021*, niab038.

Rorty, R. (1999). Pragmatism as anti-authoritarianism. *Revue Internationale de Philosophie, 53*, 7–20.

Russell, J. A. (2003). Core affect and the psychological construction of emotion. *Psychological Review, 110*, 145–172.

Ryan, S. (1999). What is wisdom? *Philosophical Studies, 93*, 119–139.

Ryan, S. (2020). Wisdom. In E. N. Zalta (Eds.), *The Stanford encyclopedia of philosophy*. https://plato.stanford.edu/archives/fall2023/entries/wisdom/.

Sajid, N., Ball, P. J., Parr, T., & Friston, K. J. (2021). Active inference: Demystified and compared. *Neural Computation, 33*, 674–712.

Sajid, N., da Costa, L., Parr, T., & Friston, K. (2022). Active inference, Bayesian optimal design, and expected utility. In C. M. Wu, E. Schulz, & I. C. Dezza (Eds.), *The drive for knowledge: The science of human information seeking* (pp. 124–146). Cambridge University Press.

Sajid, N., Tigas, P., & Friston, K. (2022). Active inference, preference learning and adaptive behaviour. *IOP Conference Series: Materials Science and Engineering, 1261*, 012020.

Sajid, N., Tigas, P., Zakharov, A., Fountas, Z., & Friston, K. (2021). Exploration and preference satisfaction trade-off in reward-free learning. *arXiv preprint*. https://arXiv:2106.04316.

Sales, A. C., Friston, K. J., Jones, M. W., Pickering, A. E., & Moran, R. J. (2019). Locus coeruleus tracking of prediction errors optimises cognitive flexibility: An active inference model. *PLOS Computational Biology, 15*, e1006267.

Sandved-Smith, L., Hesp, C., Mattout, J., Friston, K., Lutz, A., & Ramstead, M. J. D. (2021). Towards a computational phenomenology of mental action: Modelling meta-awareness and attentional control with deep parametric active inference. *Neuroscience of Consciousness, 2021*, niab018.

Sasan, D., Ward, P. G. D., Nash, M., Orchard, E. R., Farrell, M. J., Hohwy, J., & Jamadar, S. D. (2020). "Phantom kicks": Women's subjective experience of fetal kicks after the postpartum period. *Journal of Women's Health, 30*, 36–44.

Schechtman, M. (2011). The narrative self. In S. Gallagher (Ed.), *The Oxford handbook of the self* (pp. 394–416). Oxford University Press.

Schechtman, M. (2018). *The constitution of selves*. Cornell University Press.

Scherer, K. R. (1999). Appraisal theory. In T. Dalgliesh & M. J. Power (Eds.), *Handbook of cognition and emotion* (pp. 637–663). John Wiley & Sons.

Schlicht, T. (2018). Experiencing organisms: From mineness to subject of experience. *Philosophical Studies, 175*, 2447–2474.

Schlicht, T., & Venter, E. (2019). Getting the world right: Perceptual accuracy and the role of the perceiver in predictive processing models. *Journal of Consciousness Studies, 26*, 181–2006.

Schmidt, M. W. (2024). Defining the method of reflective equilibrium. *Synthese, 203*, 170.

Schulkin, J., & Sterling, P. (2019). Allostasis: A brain-centered, predictive mode of physiological regulation. *Trends in Neurosciences, 42*, 740–752.

Schwartenbeck, P., FitzGerald, T. H., Mathys, C., Dolan, R., Wurst, F., Kronbichler, M., & Friston, K. (2015). Optimal inference with suboptimal models: Addiction and active bayesian inference. *Medical Hypotheses, 84*, 109–117.

Schwartenbeck, P., FitzGerald, T. H. B., Mathys, C., Dolan, R., & Friston, K. (2015). The dopaminergic midbrain encodes the expected certainty about desired outcomes. *Cerebral Cortex, 25*, 3434–3445.

Schwartenbeck, P., & Friston, K. (2016). Computational phenotyping in psychiatry: A worked example. *eNeuro, 3*, ENEURO.0049-16.2016.

Sellars, W. (1956). *Empiricism and the philosophy of mind*. Harvard University Press.

Sengupta, B., Tozzi, A., Cooray, G. K., Douglas, P. K., & Friston, K. J. (2016). Towards a neuronal gauge theory. *PLoS Biology, 14*, e1002400.

Seth, A. K. (2014). A predictive processing theory of sensorimotor contingencies: Explaining the puzzle of perceptual presence and its absence in synesthesia. *Cognitive Neuroscience, 5*, 97–118.

Seth, A. K. (2015). Inference to the best prediction. In T. K. Metzinger & J. M. Windt (Eds.), *Open mind*. MIND Group.

Seth, A. (2016). The real problem. *Aeon*. https://aeon.co/essays/the-hard-problem-of-consciousness-is-a-distraction-from-the-real-one.

Seth, A. K. (2021). *Being you*. Faber & Faber.

Seth, A. K., & Bayne, T. (2022). Theories of consciousness. *Nature Reviews Neuroscience, 23*, 439–452.

Seth, A. K., & Friston, K. J. (2016). Active interoceptive inference and the emotional brain. *Philosophical Transactions of the Royal Society B: Biological Sciences, 371*, 20160007.

Seth, A. K., & Hohwy, J. (2021). Predictive processing as an empirical theory for consciousness science. *Cognitive Neuroscience, 12*, 89–90.

Seth, A. K., Millidge, B., Buckley, C. L., & Tschantz, A. (2020). Curious inferences: Reply to Sun and Firestone on "The dark room problem." *Trends in Cognitive Sciences, 24*, 681–683.

Seth, A. K., Suzuki, K., & Critchley, H. D. (2012). An interoceptive predictive coding model of conscious presence. *Frontiers in Psychology, 2*. https://doi.org/10.3389/fpsyg.2011.00395.

Seth, A. K., & Tsakiris, M. (2018). Being a beast machine: The somatic basis of selfhood. *Trends in Cognitive Sciences, 22*, 969–981.

Sharot, T. (2011). The optimism bias. *Current Biology, 21*, R941–R45.

Shepherd, J. (2012). Free will and consciousness: Experimental studies. *Consciousness and Cognition, 21*, 915–927.

Singh, D. (2023). Zhuangzi's epistemic perspectivism: Humility and open-mindedness as corrective virtues. *Asian Journal of Philosophy, 2*, 26.

Smith, R., Friston, K. J., & Whyte, C. J. (2022). A step-by-step tutorial on active inference and its application to empirical data. *Journal of Mathematical Psychology, 107*, 102632.

Smith, R., Lane, R. D., Parr, T., & Friston, K. J. (2019). Neurocomputational mechanisms underlying emotional awareness: Insights afforded by deep active inference and their potential clinical relevance. *Neuroscience & Biobehavioral Reviews, 107*, 473–491.

Smith, R., Moutoussis, M., & Bilek, E. (2021). Simulating the computational mechanisms of cognitive and behavioral psychotherapeutic interventions: Insights from active inference. *Scientific Reports, 11*, 10128.

Smith, R., Parr, T., & Friston, K. (2019). Simulating emotions: An active inference model of emotional state inference and emotion concept learning. *Frontiers in Psychology, 10*. https://10.3389/fpsyg.2019.02844.

Sober, E., & Wilson, D. S. (1998). *Unto others—the evolution and psychology of unselfish behavior*. MIT Press.

Solms, M. (2019). The hard problem of consciousness and the free energy principle. *Frontiers in Psychology, 9*. https://10.3389/fpsyg.2018.02714.

Solms, M. (2021). *The hidden spring: A journey to the source of consciousness*. Profile Books.

Solms, M., & Friston, K. (2018). How and why consciousness arises: Some considerations from physics and physiology. *Journal of Consciousness Studies, 25*, 202–238.

Sripada, C. (2016). Self-expression: A deep self theory of moral responsibility. *Philosophical Studies, 173*, 1203–1232.

Steger, M. F., Frazier, P., Oishi, S., & Kaler, M. (2006). The meaning in life questionnaire: Assessing the presence of and search for meaning in life. *Journal of Counseling Psychology, 53*, 80–93.

Sterling, P. (2012). Allostasis: A model of predictive regulation. *Physiology & Behavior, 106*, 5–15.

Sternberg, R. J., & Karami, S. (2021). What is wisdom? A unified 6p framework. *Review of General Psychology, 25*, 134–151.

Strawson, G. (2004). Against narrativity. *Ratio, 17*, 428–452.

Sun, Z., & Firestone, C. (2020). The dark room problem. *Trends in Cognitive Sciences, 24*, 346–348.

Suzuki, K., Mariola, A., Schwartzman, D. J., & Seth, A. K. (2023). Using extended reality to study the experience of presence. In C. Maymon, G. Grimshaw, & Y. C. Wu (Eds.), *Virtual reality in behavioral neuroscience: New insights and methods* (pp. 255–285). Springer.

Swartwood, J., & Tiberius, V. (2019). Philosophical foundations of wisdom. In J. Glück & R. J. Sternberg (Eds.), *The Cambridge handbook of wisdom* (pp. 10–39). Cambridge University Press.

Taine, H. (1870). *De L'intelligence*. Hachette.

Tappin, B. M., & Gadsby, S. (2019). Biased belief in the Bayesian brain: A deeper look at the evidence. *Consciousness and Cognition, 68*, 107–114.

Tarsney, C. (2023). The epistemic challenge to longtermism. *Synthese, 201*, 195.

Teng, L. (2022). A metacognitive account of phenomenal force. *Mind & Language, 38*, 1081–1101.

Tiberius, V. (2008). *The reflective life: Living wisely with our limits*. Oxford University Press.

Tiehen, J. (2022). Perception as controlled hallucination. *Analytic Philosophy, 64*, 355–372.

Toepfer, F., Wolf, R., & Heisenberg, M. (2018). Multi-stability with ambiguous visual stimuli in drosophila orientation behavior. *PLoS Biology, 16*, e2003113.

Tononi, G., & Koch, C. (2015). Consciousness: Here, there and everywhere? *Philosophical Transactions of the Royal Society B: Biological Sciences, 370*, 20140167.

Tschantz, A., Millidge, B., Seth, A., & Buckley, C. L. (2022). Hybrid predictive coding: Inferring, fast and slow. *PLoS Computational Biology, 19*, e1011280.

Tschantz, A., Seth, A. K., & Buckley, C. L. (2020). Learning action-oriented models through active inference. *PLOS Computational Biology, 16*, e1007805.

Tsuchiya, N., & Koch, C. (2004). Continuous flash suppression. *Journal of Vision, 4*, 61.

Usher, M. (2006). Control, choice, and the convergence/divergence dynamics: A compatibilistic probabilistic theory of free will. *Journal of Philosophy, 103*, 188–213.

Usher, M. (2020). Agency, teleological control and robust causation. *Philosophy and Phenomenological Research, 100*, 302–324.

van Baal, S. T., Walasek, L., & Hohwy, J. (2022). Modelling pandemic behaviour using an economic multiplayer game. *Scientific Reports, 12*, 13466.

van Baal, S. T., Walasek, L., Verdejo-Garcia, A., & Hohwy, J. (2025). Impulsivity and self-control as timeless concepts: A conceptual analysis of preferences in intertemporal choice. *Decision*. https://doi.org/10.1037/dec0000257.

Van de Cruys, S. (2017). Affective value in the predictive mind. In T. K. Metzinger & W. Wiese (Eds.), *Philosophy and predictive processing*. MIND Group.

Van de Cruys, S., Friston, K. J., & Clark, A. (2020). Controlled optimism: Reply to Sun and Firestone on "The dark room problem." *Trends in Cognitive Sciences, 24*, 680–681.

van Es, T. (2020a). Living models or life modelled? On the use of models in the free energy principle. *Adaptive Behavior*. https://doi.org/10.1177/1059712320918678.

van Es, T. (2020b). Minimizing prediction errors in predictive processing: From inconsistency to non-representationalism. *Phenomenology and the Cognitive Sciences, 19*, 997–1017.

Van Gordon, W., Sapthiang, S., & Shonin, E. (2022). Contemplative psychology: History, key assumptions, and future directions. *Perspectives on Psychological Science, 17*, 99–107.

van Rooij, I., Wright, C. D., Kwisthout, J., & Wareham, T. (2018). Rational analysis, intractability, and the prospects of "as if"-explanations. *Synthese, 195*, 491–510.

Varela, F. G., Maturana, H. R., & Uribe, R. (1974). Autopoiesis: The organization of living systems, its characterization and a model. *Biosystems, 5*, 187–196.

Varela, F. J. (1979). *Principles of biological autonomy*. Elsevier North-Holland.

Veissière, S. P. L., Constant, A., Ramstead, M. J. D., Friston, K. J., & Kirmayer, L. J. (2020). Thinking through other minds: A variational approach to cognition and culture. *Behavioral and Brain Sciences, 43*, e90.

Vetter, G., Haynes, J. D., & Pfaff, S. (2000). Evidence for multistability in the visual perception of pigeons. *Vision Research, 40*, 2177–2186.

Vilas, M. G., Auksztulewicz, R., & Melloni, L. (2022). Active inference as a computational framework for consciousness. *Review of Philosophy and Psychology, 13*, 859–878.

Vincent, P., Parr, T., Benrimoh, D., & Friston, K. J. (2019). With an eye on uncertainty: Modelling pupillary responses to environmental volatility. *PLOS Computational Biology, 15*, e1007126.

Wald, A. (1947). An essentially complete class of admissible decision functions. *Annals of Mathematical Statistics, 18*, 549–555.

Weber, L. A., Waade, P. T., Legrand, N., Møller, A. H., Stephan, K. E., & Mathys, C. (2023). The generalized hierarchical Gaussian filter. *arXiv preprint*. https://arXiv:2305.10937

Weilnhammer, V., Stuke, H., Hesselmann, G., Sterzer, P., & Schmack, K. (2017). A predictive coding account of bistable perception—A model-based FMRI study. *PLOS Computational Biology, 13,* e1005536.

Weilnhammer, V., Stuke, H., Sterzer, P., & Schmack, K. (2018). The neural correlates of hierarchical predictions for perceptual decisions. *Journal of Neuroscience, 38,* 5008–5021.

Whyte, C. J. (2019). Integrating the global neuronal workspace into the framework of predictive processing: Towards a working hypothesis. *Consciousness and Cognition, 73,* 102763.

Whyte, C., Corcoran, A., Robinson, J., Smith, R., Moran, R., Parr, T., Friston, K., Seth, A., & Hohwy, J. (2024). *On the minimal theory of consciousness implicit in active inference.* Preprint.

Whyte, C. J., Hohwy, J., & Smith, R. (2022). An active inference model of conscious access: How cognitive action selection reconciles the results of report and no-report paradigms. *Current Research in Neurobiology, 3,* 100036.

Whyte, C. J., & Smith, R. (2021). The predictive global neuronal workspace: A formal active inference model of visual consciousness. *Progress in Neurobiology, 199,* 101918.

Wiese, W. (2019). Explaining the enduring intuition of substantiality: The phenomenal self as an abstract "salience object." *Journal of Consciousness Studies, 26,* 64–87.

Wiese, W., & Friston, K. (2020). The neural correlates of consciousness under the free energy principle: From computational correlates to computational explanation. *Philosophy and the Mind Sciences, 2.* https://doi.org/10.33735/phimisci.2021.81.

Wilkinson, S., Deane, G., Nave, K., & Clark, A. (2019). Getting warmer: Predictive processing and the nature of emotion. In L. Candiotto (Ed.), *The value of emotions for knowledge* (pp. 101–119). Springer International Publishing.

Williams, D. (2018). Hierarchical Bayesian models of delusion. *Consciousness and Cognition, 61,* 129–147.

Williams, D. (2022). Is the brain an organ for free energy minimisation? *Philosophical Studies,* 179, 1693–1714.

Williford, K., Bennequin, D., Friston, K., & Rudrauf, D. (2018). The projective consciousness model and phenomenal selfhood. *Frontiers in Psychology, 9.* https://10.3389/fpsyg.2018.02571.

Wilson, T. D., & Gilbert, D. T. (2005). Affective forecasting: Knowing what to want. *Current Directions in Psychological Science, 14,* 131–134.

Windt, J. M. (2018). Predictive brains, dreaming selves, sleeping bodies: How the analysis of dream movement can inform a theory of self- and world-simulation in dreams. *Synthese,* 195, 2577–2625.

Windt, J. M., Harkness, D. L., & Lenggenhager, B. (2014). Tickle me, I think I might be dreaming! Sensory attenuation, self-other distinction, and predictive processing in lucid dreams. *Frontiers in Human Neuroscience, 8*. https://10.3389/fnhum.2014.00717.

Wittgenstein, L. (1953). *Philosophical investigations*. Basil Blackwell.

Wolf, S. (1993). *Freedom within reason*. Oxford University Press.

Wolf, S., Macedo, S., Koethe, J., Adams, R. M., Arpaly, N., & Haidt, J. (2010). *Meaning in life and why it matters*. Princeton University Press.

Woźniak, M. (2018). "I" and "me": The self in the context of consciousness. *Frontiers in Psychology, 9*. https://doi.org/10.3389/fpsyg.2018.01656.

Woźniak, M., & Hohwy, J. (2020). Stranger to my face: Top-down and bottom-up effects underlying prioritization of images of one's face. *PloS One, 15*, e0235627.

Wozniak, M., McEllin, L., Hohwy, J., & Ciaunica, A. (2023). Depersonalization affects self-prioritization of bodily, but not abstract self-related information. *Journal of Experimental Psychology: Human Perception and Performance, 49*, 1447.

Woźniak, M., Schmidt, T. T., Wu, Y.-h., Blankenburg, F., & Hohwy, J. (2021). *Differences in working memory coding of biological motion attributed to oneself and others*. bioRxiv. https://doi.org/10.1101/2021.08.16.456121.

Xu, H., Han, C., Chen, M., Li, P., Zhu, S., Fang, Y., Hu, J., Ma, H., & Lu, H. D. (2016). Rivalry-like neural activity in primary visual cortex in anesthetized monkeys. *Journal of Neuroscience, 36*, 3231–3242.

Yeshurun, Y., Nguyen, M., & Hasson, U. (2021). The default mode network: Where the idiosyncratic self meets the shared social world. *Nature Reviews Neuroscience, 22*, 181–192.

Zagzebski, L. T. (1996). *Virtues of the mind: An inquiry into the nature of virtue and the ethical foundations of knowledge*. Cambridge University Press.

Zarkali, A., Adams, R. A., Psarras, S., Leyland, L.-A., Rees, G., & Weil, R. S. (2019). Increased weighting on prior knowledge in Lewy body-associated visual hallucinations. *Brain Communications, 1*, fcz007.

Zhang, P., Jamison, K., Engel, S., He, B., & He, S. (2011). Binocular rivalry requires visual attention. *Neuron, 71*, 362–369.

Zou, J., He, S., & Zhang, P. (2016). Binocular rivalry from invisible patterns. *Proceedings of the National Academy of Sciences, 113*, 8408–8413.

Index

Publisher contact:
The MIT Press
Massachusetts Institute of Technology
77 Massachusetts Avenue, Cambridge, MA 02139
mitpress.mit.edu

EU Authorised Representative:
Easy Access System Europe, Mustamäe tee 50,
10621 Tallinn, Estonia
gpsr.requests@easproject.com

Printed by Integrated Books International,
United States of America